Supply Chain Management Models: Forward, Reverse, Uncertain, and Intelligent

Foundations with Case Studies

Supply Chain Management Models: Forward, Reverse, Uncertain, and Intelligent

Foundations with Case Studies

Hamed Fazlollahtabar
Department of Industrial Engineering,
College of Engineering, Damghan University,
Damghan, Iran

CRC Press
Taylor & Francis Group
Boca Raton London New York

CRC Press is an imprint of the
Taylor & Francis Group, an **informa** business

CRC Press
Taylor & Francis Group
6000 Broken Sound Parkway NW, Suite 300
Boca Raton, FL 33487-2742

First issued in paperback 2019

ISBN-13: 978-1-138-57017-7 (hbk)
ISBN-13: 978-0-367-89246-3 (pbk)

Library of Congress Cataloging-in-Publication Data

Names: Fazlollahtabar, Hamed, author.
Title: Supply chain management models : forward, reverse, uncertain, and intelligent foundations with case studies / Hamed Fazlollahtabar.
Description: Boca Raton, FL : CRC Press, 2017. | Includes index.
Identifiers: LCCN 2017033596 | ISBN 9781138570177 (hardback)
Subjects: LCSH: Business logistics. | Industrial procurement.
Classification: LCC HD38.5 .F39 2017 | DDC 658.7--dc23
LC record available at https://lccn.loc.gov/2017033596

Visit the Taylor & Francis Web site at
http://www.taylorandfrancis.com

and the CRC Press Web site at
http://www.crcpress.com

Contents

Section II   Reverse Supply Chain Models

Section III   Uncertain and Intelligent Supply Chain Models

Preface

Supply chain management (SCM) is a systematic flow of materials, information, and finances as they progress in a process from supplier to manufacturer, wholesaler, retailer and consumer. It includes coordination and integration of elements so that the objectives of the chain are satisfied. The concept of SCM is based on two focal points. The first point is the aggregated efforts by several units made on a product to be delivered to end user. These units are referred to, collectively, as the supply chain. The second point is that the enterprises pay more attention to their inner activities rather than their supply chains. The outcome is a malfunctioned supply chain leading to inefficiency of activities within a supply chain. Supply chain management, then, is the active management of supply chain activities to maximize customer value and obtain a sustainable competitive advantage. It indicates a comprehensive attempt by corresponding companies to program, develop and activate supply chains as effectively as possible. The enterprises conducting the supply chains are "linked" together through physical flows and information flows. Physical flows involve the transformation, movement, and storage of goods and materials. These activities are a visible part of the supply chain. Information flows allow the different supply chain practitioners to coordinate their long-term plans, and to control the gradual flow of goods and material forward and reverse in the supply chain.

No model can include all aspects of supply chain processes. To compromise the dilemma between model complexity and reality, a model builder should define the scope of the supply chain model in such a way that it is reflective of key real-world dimensions, yet not too complicated to solve. Different models of supply chains are classified by the problems they consider in three levels: strategic, tactical and operational. Problems in competitive strategic analysis include location-allocation decisions, demand planning, distribution channel planning, strategic alliances, new product development, outsourcing, supplier selection, information technology (IT) selection, pricing and network restructuring. Although most supply chain issues are strategic by nature, there are also some tactical problems. These include inventory control, production/distribution coordination, order/freight consolidation, material handling, equipment selection and layout design. The operational problems include vehicle routing/scheduling, workforce scheduling, record keeping and packaging. It should be noted that the stated problems and models are not always clear because some supply chain problems involve hierarchical, multi-echelon planning that overlaps different decision levels. Sufficient knowledge about the required components of supply chains leads to better management for achieving specific supply chain goals. Lack of specific goals, in turn, means difficulty in developing appropriate performance measures that can be targeted or benchmarked by a supply chain practitioner. While performance measures preview the desired outcome of the supply chain model, it is very important for a model builder to identify key components of a supply chain.

Goal setting is the first step of supply chain modeling. To set the supply chain goals, a model builder first needs to figure out what will be the major driving forces (drivers) behind the supply chain linkages. These drivers include customer service initiatives, monetary value, information/knowledge transactions and risk elements.

Customer service initiatives. Though difficult to quantify, the ultimate goal of a supply chain is customer satisfaction. Put simply, customer satisfaction is the degree to which customers are satisfied with the product and/or service received.

Monetary value. The monetary value is generally defined as a ratio of revenue to total cost. A supply chain can enhance its monetary value through increasing sales revenue, market share and labor productivity, while reducing expenditures, defects and duplication. Because such value directly reflects the cost efficiency and profitability of supply chain activities, this is the most widely used objective function of a supply chain model.

Information/knowledge transactions. Information serves as the connection between the various phases of a supply chain, allowing supply chain partners to coordinate their actions and increase inventory visibility. Therefore, successful supply chain integration depends on the supply chain partners' ability to synchronize and share "real-time" information. Such information encompasses data, technology, know-how, designs, specifications, samples, client lists, prices, customer profiles, sales forecasts and order history.

Risk elements. The important leverage gained from the supply chain integration is the mitigation of risk. In the supply chain framework, a single supply chain member does not have to stretch beyond its core competency, since it can pool the resources shared with other supply chain partners. On the other hand, a supply chain can pose greater risk of failure due to its inherent complexity and volatility. Researchers noted that a supply chain would be a veritable hive of risks unless information is synchronized, time is compressed and tensions among supply chain members are recognized. They also observed that supply chain risks (emanating from sources external to the firm) would be always greater than risks which arose internally, as less was known about them. Thus, a model builder needs to profile the potential risks involved in supply chain activities.

Supply chain constraints represent restrictions (or limitations) placed on a range of decision alternatives that the firm can choose. Thus, they determine the feasibility of some decision alternatives. These constraints include:

Capacity. The supply chain member's financial, production, supply and technical (EDI or bar coding) capability determines its desired outcome in terms of the level of inventory, production, workforce, capital investment, outsourcing and IT adoption. This capacity also includes the available space for inventory stocking and manufacturing.

Service compliance. Because the ultimate goal of a supply chain is to meet or exceed customer service requirements, this may be one of the most important constraints to satisfy. Typical examples are delivery time windows, manufacturing due dates, maximum holding time for backorders and the number of driving hours for truck drivers.

The extent of demand. The vertical integration of a supply chain is intended to balance the capacity of supply at the preceding stage against the extent of consumption (i.e., demand) of the downstream supply chain members at the succeeding stage. Thus, this constraint can be added to the supply chain model.

Because decision variables generally set the limits on the range of decision outcomes, they are functionally related to supply chain performances. Thus, the performance measures (or objectives) of a supply chain are generally expressed as functions of one or more decision variables. Though not exhaustive, the following illustrates these decision variables:

Location. This type of variable involves determining where plants, warehouses (or distribution centers ([DCs]), consolidation points and sources of supply should be located.

Allocation. This type of variable determines which warehouses (or DCs), plants and consolidation points should serve which customers, market segments and suppliers.

Network structuring. This type of variable involves centralization or decentralization of a distribution network and determines which combination of suppliers, plants, warehouses, and consolidation points should be utilized or phased-out. This type of variable may also involve the exact timing of expansion or elimination of manufacturing or distribution facilities.

Number of facilities and equipment. This type of variable determines how many plants, warehouses and consolidation points are needed to meet the needs of customers and market segments. This type of variable may also determine how many lift trucks are required for material handling.

Number of stages (echelons). This variable determines the number of stages that will comprise a supply chain. This variable may involve either increasing or decreasing the level of horizontal supply chain integration by combining or separating stages.

Service sequence. This variable determines delivery or pickup routes and schedules of vehicles serving customers or suppliers.

Volume. This variable includes the optimal purchasing volume, production, and shipping volume at each node (e.g., a supplier, a manufacturer, a distributor) of a supply chain.

Inventory level. This variable determines the optimal amount of every raw material, part, work-in-process, finished product and stock-keeping unit (SKU) to be stored at each supply chain stage.

Size of workforce. This variable determines the number of truck drivers or order pickers needed for the system.

The extent of outsourcing. This type of variable determines which suppliers, IT service providers, and third-party logistics providers should be used for long-term outsourcing contacts and how many (e.g., single versus multiple sourcing) of those should be utilized.

Thus, in this book different models are configured so that the inclusion of various variables and constraints and mainly the drivers are considered. Models for supply chain management are proposed in three categories—forward, reverse and intelligent. In forward supply chain management models such as strategic marketing, clustering, performance evaluation, quality management, system dynamics, life cycle and customer satisfaction, inventory management, pricing, utility, vehicle routing, food time-windows and two-stage are considered in multi-layer and multi-product in multi-echelon supply network. The organization of chapters in the forward SCM section is so that the macro and strategic topics are inserted first, and then the micro and tactical topics are followed. At the end of the first section some specific models in operational level are also considered. Numerical examples are also given for chapters that obviously provide a different concept; that is, for chapters introducing similar concepts in incremental manner, numerical example is presented for the more comprehensive model.

For reverse supply chain and logistics networks, models including data mining, vehicle routing, disposal, wastes and disassembly are investigated. In this section, some general models that are practically important are studied. The chapters are set so that the general collection of returning products, carrying them to disassembly centers, disposal or resending to the supply chain as a closed loop supply chain are considered. Also, customer satisfaction, revenue and cost management, and green supply chains are studied.

Also, for IT- based (electronic) supply chains and the application of artificial intelligence in modeling the supply chain problems such as intelligent information system, agent-based and web-based models, real-time decision support are designed and modeled. For uncertain systems, fuzzy, genetic algorithm, and immunity-based models are studied.

Different numerical studies are illustrated for chapters of the book—some of them are real application studies; some are step-by-step implementations; in some chapters hypothetical examples are given; and finally some are computationally studies.

Acknowledgments

I would like to express my gratitude to the many people who saw me through this book: to all those who provided support, talked things over, read, wrote, offered comments, allowed me to quote their remarks and assisted in the editing, proofreading and design.

I would like to thank Damghan University for enabling us to publish this book. Above all I want to thank my family, who supported and encouraged me in spite of all the time it took me away from them. It was a long and difficult journey for them.

I would like to thank Prof. Nezam Mahdavi-Amiri for helping me in the process of selection and editing. Thanks to my graduate students who helped in the process of modeling and verification.

Thanks to Prof. Abdolreza Sheikholeslami—without him this book would never find its way to the academician.

I would like to thank my graduate students who helped in numerical study parts: Lida Mohammadi, Somaye Fouladi, Reza Khoshnoudi, Saeb Sadeghi, Zahra Hosseini, Hoda Mahmoudi, Hamed Hajmohammadi, and S. Hosna Shafieian.

Last and not least: I beg forgiveness of all those who have been with me over the course of the years and whose names I have failed to mention.

Hamed Fazlollahtabar
Department of Industrial Engineering, College of Engineering,
Damghan University, Damghan, Iran

Author

Hamed Fazlollahtabar had graduated in BSc and MSc of Industrial Engineering at Mazandaran University of Science and Technology, Babol, Iran at 2008 and 2010, respectively. He received his PhD of Industrial and Systems Engineering at Iran University of Science and Technology, Tehran, Iran at 2015. He has just completed the postdoctoral research fellowship at Sharif University of Technology, Tehran, Iran, in Reliability Engineering for Complex Systems. He is currently a member of the Department of Industrial Engineering, Damghan University. He is on the editorial board of several journals and technical committees of conferences. His research interests are robot path planning, reliability engineering, supply chain planning, and business intelligence and analytics. He has published more than 230 research papers in international books, journals, and conference proceedings. Also, he published five books, three of which are internationally distributed to academicians.

ORCID: 0000-0003-3053-4399

Section I

Forward Supply Chain Models

1

Multi-Layer Multi-Product Supply Chain: Strategic Marketing Model

SUMMARY In this chapter, we propose a multi-layer supply chain that consists of material supplier, manufacturer, distributor, retailer and end customer having multiple products. The aim is to integrate marketing actions in an intelligent manner to increase the profit in a programming horizon using strategic planning. Quantitative approaches, namely QSPM and AHP, are used for quantification and prioritization.

1.1 Introduction

A supply chain is the material flow, information, funds and services from raw materials suppliers through factories and warehouses to the final customers (Liang and Cheng, 2009; Taleizadeh et al., 2011). Multi-layer and multi-product chain involves many tasks such as purchases, cash flow, material transport, planning and production control, inventory control and logistics and distribution and delivery. The process of applying new approaches to production and operations management indicates there is an increasing trend in the use of supply chain management approach among the various industrial companies and service which aims to reduce costs and increase their market share and competitiveness. The expression of each issue must first have an understanding of the factors under consideration. The understanding of the definition of each factor influencing the factors affecting system performance is important (Bello et al., 2004; Fandela and Stammenb, 2004; Meloa et al., 2005). The multi-layered supply chain, depending on the industry and its products, can include layers of suppliers, manufacturers, distributors, retailers and customers. Economic enterprises, especially manufacturing enterprises, to accomplish their mission require extensive interaction with other firms supplying materials and parts required for production and to support their operations. The provision of simple and direct purchase from the market needs collaboration and cooperation with the firm (Jiuh-Biing, 2005). Relationships with various suppliers, to the extent that some levels of participation and close cooperation between the two sides will promote demands that the traditional systems of internal and external purchasing systems support specific types of cooperation, can be used. Usually a large part of economic institutions through the purchase of goods and services required is supplied from domestic sources. In this case, it is necessary to identify internal sources, ordering and purchasing process directly or through contract to be carried out. Procurement from foreign sources of economic institutions is inevitable. In addition to identifying the process and vendor selection, checking out other steps to obtain necessary permits clearance and domestic transportation is done in accordance with official regulations. All operations relating to their unique purchasing system are included. As mentioned, in many cases a firm needs manufacturing and purchasing functions beyond the direct needs, such as providing some of the items needed, or sometimes there is no actual provision. In this case, we identified the need to cooperate with potential suppliers. Some programs that provide production and inventory data will be directly affected by consumer applications. The following relationships providing unique management system designed and built able to cover all functions and workflows are the suppliers and recipients.

Businesses, particularly manufacturing firms, need to interact extensively with other agencies to carry out their mission to provide materials and parts required to produce and support their operations. Operations provide a simple and direct purchase from the market to meet demand for the firm's collaboration with a supplier that is selected. Therefore, all operations related to the purchase of the system are uniquely included. Some operations are related to financing a partnership amongst suppliers. But, for some suppliers it is challenging to have a sustainable cooperation due to different ordering patterns and fluctuations between supply and demand (Das, 2011).

1.2 Problem Definition

A supply chain is the flow of materials, information, money and services from raw materials suppliers to producer and warehouses, to the end customers and includes the organizations and processes by which goods and services are produced and delivered to consumers. Trends in production and operations management using modern approaches suggest that there is an increasing path available from different companies in the industrial and service supply chain management that aims to reduce costs and increase their competitiveness and market share. Today, the supply chain is faced with serious challenges. In this chapter, we propose a multi-layer supply chain that consists of material supplier, manufacturer, distributor, retailer and end customer having multiple products. The process begins with the raw material provision toward the distributor and finally the consumer. Because marketing plays an important role in better presentation of products and requirement planning for production, we design a marketing decision making mechanism. Better implementation of marketing programs require an elaborated multi-stage and a range of short-term and long-term action plans. Thus, strategic planning has been employed to propose an innovative marketing model for a multi-layer supply chain. The benefits of the proposed mechanism are reduction of costs and decrease of products' time-to-market.

1.3 Strategic Marketing Model

Industrial marketing will facilitate the process of exchange between manufacturers and enterprise customers. Nature of industrial marketing can provide value for customers by offering goods and services perceived by organizations. The marketing industry, in many cases, distinguishes between the retailers offering similar products and customer receiving services. We accept the product on the market, or do anything that needs to provide an idea for a product that is responsive to the needs of the customer. Marketing ideas and a mechanism to deliver these ideas lead to comments to create understanding, change attitudes and beliefs of people and organizations. One of the problems with our society is that we have an idea but do not believe in it. A framework for idea fulfillment in marketing through a supply chain is given in Figure 1.1.

Studies show companies that have given the constant changes and developments over the market will grow. New ideas lead to economic growth and future success of firms if they are considered as operational strategies. New ideas and resources for further development is a first step in accurately determining the following:

What is the product being sold?

What are the existing and potential customers?

What is the market and how will it change?

The importance of planning for marketing has been increased in recent years. Marketing must specify the identity of each customer, and the characteristics of each should be separately examined. The closer the relationship between enterprise and customers, the more reliable their interactions. Establishing an intimate relationship requires detailed information from the customer. Internet marketing and email marketing are new facilities provided to achieve this goal.

Email marketing is the process of traditional marketing through Internet technology. This reciprocal relationship is established between you and your customers. The electronic marketing can be defined as an advanced interactive media in order to attract a person or organization.

Email marketing is a function that is related not only to selling products and services. It is the administrative process for handling communication between an organization and its customers. Given this market structure, we consider a new way of marketing with effective parameters we designed. The marginal effects of this process on the system are competitors, environmental constraints (such as rules of some manufacturing, healthcare, government role, etc.), rapidly changing market demands and needs and rapid changes. An integrated internal and environmental process of marketing in supply chain is depicted in Figure 1.2.

As shown in Figure 1.3, sometimes suppliers of goods and services affect the environment or the environment takes effect. On the other hand, sometimes the manufacturer will exchange with their environments. The input and output of the system are factors effective on the layers of the supply chain.

Distribution channels are set to meet the needs of producers and the marketing mix design. Factors considered here include company resources, customer behavior, competitor strategies and products. A channel with external exchange and also internal ones is designed. Input–output systems can be distributed outside information, money, services, and programs. Figure 1.4 shows the distribution layer. Inputs and outputs of a distribution channel can be evaluated. The main purpose of this system is to optimize the interaction of input–output. Some of the input–output should be minimized or maximized and some others in a particular situation remains stable. The input–output data are often received from market. Figure 1.4 indicates an internal communication system.

The last layer is retail sector in the distribution system and the boundary between producers and consumers. Data entry of retailers can be received from the manufacturer or distributers. It has great importance in improving the quality of retail services. The service

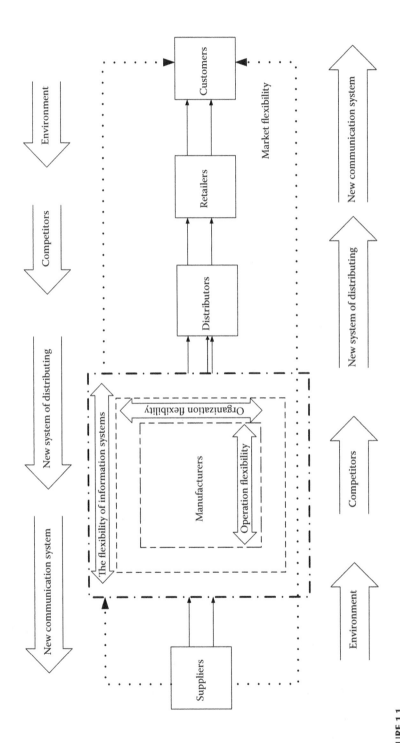

FIGURE 1.1
A framework.

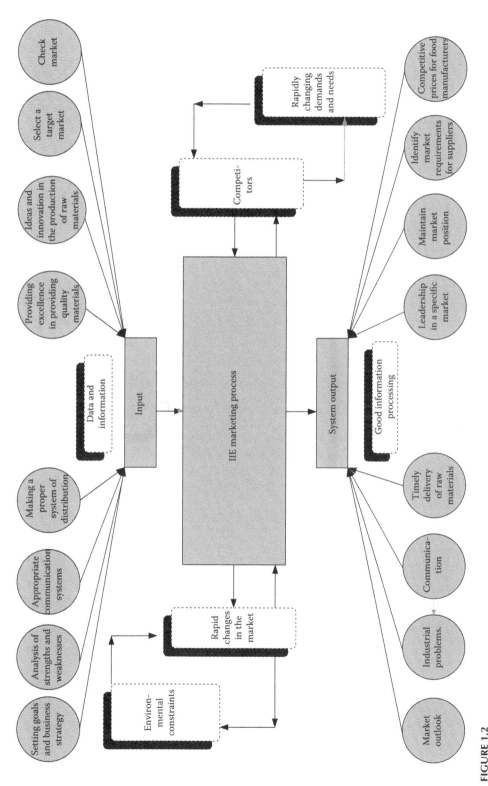

FIGURE 1.2
Integrated internal and environmental (IIE) market structure in supply chain.

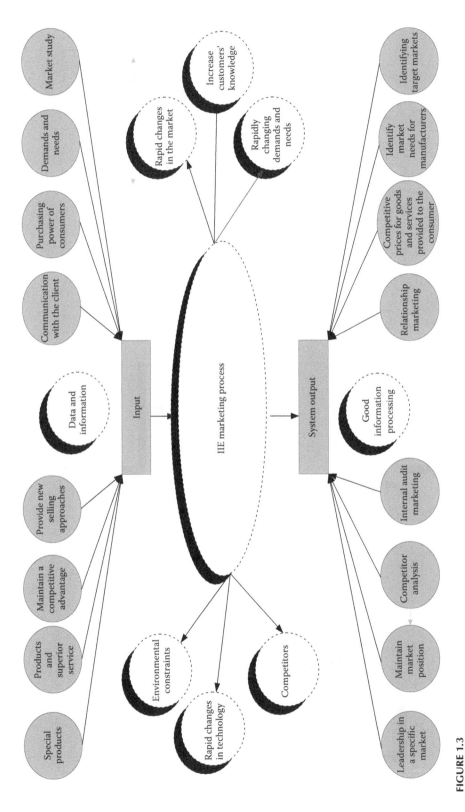

FIGURE 1.3
IIE market structure in supplier layer.

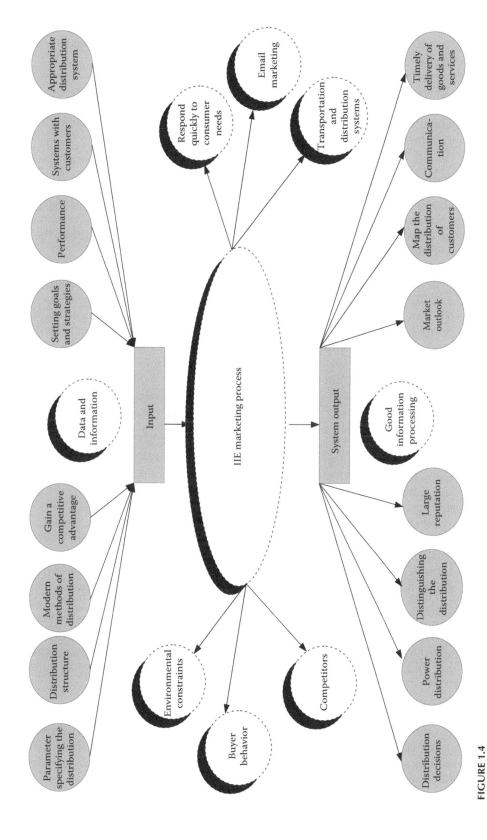

FIGURE 1.4
IIE market structure in the distribution layer.

processor produced the goods in the supply layer employing skilled manpower. It is one of the most important assets of a retail layer. The output layer consists of parts that have been carefully processed. This layer is shown in Figure 1.5.

In the present era of knowledge and information technologies, customers are expecting higher levels of services. Therefore, organizations and management are trying to fortify themselves to satisfy the needs. Necessary design and planning services are tailored to the needs and demands of customers, including issues that we should pay more attention to. The customer layer is detailed in Figure 1.6.

1.3.1 Quantitative Strategic Planning Matrix (QSPM)

Quantitative Strategic Planning Matrix (QSPM) is a high-level strategic management approach for evaluating possible strategies. A QSPM provides an *analytical method* for comparing feasible alternative actions (David, 1986). The QSPM method falls within the so-called Stage 3 of the strategy formulation analytical framework.

When company executives think about what to do, and which way to go, they usually have a *prioritized list* of strategies. If they like one strategy over another one, they move it up on the list. This process is very much intuitive and subjective. The QSPM method introduces some numbers into this approach, making the technique a little more "expert." The QSPM approach attempts to objectively select the best strategy using input from other management techniques and some easy computations. In other words, the QSPM method uses inputs from Stage 1 analyses, matches them with results from Stage 2 analyses, and then decides objectively among alternative strategies.

Stage 1 strategic management tools: The first step in the overall strategic management analysis is to identify *key strategic factors*. This can be done using, for example, the EFE matrix and IFE matrix.

Stage 2 strategic management tools: After we identify and analyze key strategic factors as inputs for QSPM, we can formulate the type of the strategy we would like to pursue. This can be done using the Stage 2 strategic management tools, for example the SWOT analysis (or TOWS), SPACE matrix analysis, BCG matrix model, or the IE matrix model.

Stage 3 strategic management tools: The Stage 1 strategic management methods provided us with key strategic factors. Based on that analysis, we formulated possible strategies in Stage 2. Now, the task is to compare in QSPM alternative strategies and decide which one is the most suitable for our goals.

The Stage 2 strategic tools provide the needed information for setting up the Quantitative Strategic Planning Matrix. The QSPM method allows us to evaluate alternative strategies objectively.

Conceptually, the QSPM in Stage 3 determines the relative attractiveness of various strategies based on the extent to which key external and internal critical success factors are capitalized upon or improved. The relative attractiveness of each strategy is computed by determining the cumulative impact of each external and internal critical success factor.

1.3.2 Analytic Hierarchy Process (AHP)

To weigh the parameters, we take a multi-criteria decision-making (MCDM) approach. MCDM, dealing primarily with problems of evaluation or selection, is a rapidly developing area in operations research and management science. The analytical hierarchy process (AHP), developed by Saaty (1980), is a technique of considering data or information for a decision in a systematic manner. AHP is mainly concerned with the way

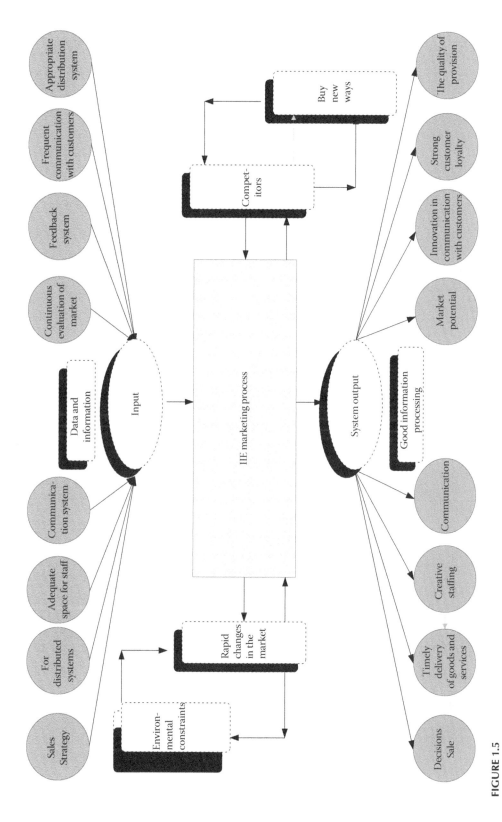

FIGURE 1.5
Structure of retail marketing IIE layer.

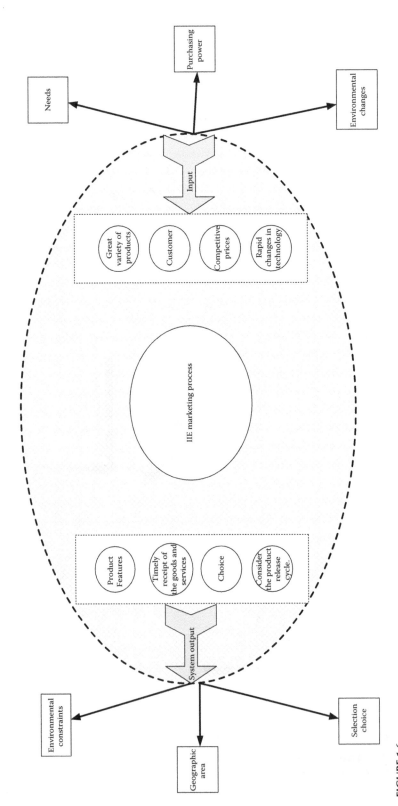

FIGURE 1.6
Structure of customer layer marketing IIE.

to solve decision problems with uncertainties in multiple criteria characterization. It is based on three principles: (1) constructing the hierarchy, (2) priority setting, and (3) logical consistency. We apply AHP to weigh the parameters. In AHP, multiple paired comparisons are based on a standardized comparison scale of nine levels (see Table 1.1; Saaty, 1980).

We are now ready to give an algorithm for computing parameter weights using the AHP. The following notations are used:

Notations and Definitions

n number of criteria

i number of parameters

p index for parameters, $p = 1$ or 2

d index for criteria, $1 \leq d \leq D$

R_{pd} the weight of pth item with respect to dth criterion

w_d the weight of dth criterion

Algorithm: PWAHP (Compute Parameter Weights Using the AHP)

Step 1: Define the decision problem and the goal.
Step 2: Structure the hierarchy from the top through the intermediate to the lowest level.
Step 3: Construct the parameter-criteria matrix using Steps 4–8 using the AHP.

(Steps 4–6 are performed for all levels in the hierarchy.)

Step 4: Construct pair-wise comparison matrices for each of the lower levels for each element in the level immediately above by using a relative scale measurement. The decision maker has the option of expressing his or her intensity of preference on a nine-point scale. If two criteria are of equal importance, a value of 1 is set for the corresponding component in the comparison matrix, while a 9 indicates an absolute importance of one criterion over the other (Table 1.1 shows the measurement scale defined by Saaty, 1980).

Step 5: Compute the largest eigenvalue by the relative weights of the criteria and the sum taken over all weighted eigenvector entries corresponding to those in the next lower level of the hierarchy.

TABLE 1.1

Scale of Relative Importance

Intensity of Importance	Definition of Importance
1	Equal
2	Weak
3	Moderate
4	Moderate plus
5	Strong
6	Strong plus
7	Very strong or demonstrated
8	Very, very strong
9	Extreme

Analyze pair-wise comparison data using the eigenvalue technique. Using these pair-wise comparisons, estimate the parameters. The eigenvector of the largest eigenvalue of matrix A constitutes the estimation of relative importance of the attributes.

Step 6: Construct the consistency check and perform consequence weights analysis as follows:

$$A = (a_{ij}) = \begin{bmatrix} 1 & w_1/w_2 & \cdots & w_1/w_n \\ w_2/w_1 & 1 & \cdots & w_2/w_n \\ \vdots & \vdots & \ddots & \\ w_n/w_1 & w_n/w_2 & & 1 \end{bmatrix}.$$

Note that if the matrix A is consistent (that is, $a_{ik} = a_{ij} \ldots a_{jk}$, for all $i, j, k = 1, 2, \ldots, n$), then we have (the weights are already known),

$$a_{ij} = \frac{w_i}{w_j}, \quad i, j = 1, 2, \ldots, n, \tag{1.1}$$

If the pair-wise comparisons do not include any inconsistencies, then $\lambda_{\max} = n$. The more consistent the comparisons are, the closer the value of computed $\lambda_{\max}$ is to n. Set the consistency index (CI), which measures the inconsistencies of pair-wise comparisons, to be:

$$CI = \frac{(\lambda_{\max} - n)}{(n-1)}, \tag{1.2}$$

and let the consistency ratio (CR) be:

$$CR = 100 \left(\frac{CI}{RI} \right), \tag{1.3}$$

where n is the number of columns in A and RI is the random index, being the average of the CI obtained from a large number of randomly generated matrices.

Note that RI depends on the order of the matrix, and a CR value of 10% or less is considered acceptable.

Step 7: Form the parameter-criteria matrix as specified in Table 1.2.

Step 8: As a result, configure the pair-wise comparison for criteria-criteria matrix as in Table 1.3.

The w_d are gained by a normalization process. The w_d are the weights for criteria.

Step 9: Compute the overall weights for the parameters, using Tables 1.2 and 1.3:

TABLE 1.2

The Parameter-Criteria Matrix

	C_1	C_2	$\cdots$	C_d
Parameter 1	R_{11}	R_{12}	$\cdots$	R_{1d}
Parameter 2	R_{21}	R_{22}	$\cdots$	R_{2d}

TABLE 1.3

The Criteria-Criteria Pair-Wise Comparison Matrix

	C_1	C_2	...	C_d	w_d
Criteria 1	1	a_{12}	...	a_{1d}	w_1
Criteria 2	$1/a_{12}$	1	...	a_{2d}	w_2
$\vdots$	$\vdots$	$\vdots$	$\vdots$	$\vdots$	$\vdots$
Criteria d	$1/a_{1d}$	$1/a_{2d}$	...	1	w_d

$$\psi = \text{Total weight for parameter } 1 = R_{11} \times w_1 + R_{12} \times w_2 + \cdots + R_{1d} \times w_d,$$
$$\psi' = \text{Total weight for parameter } 2 = R_{21} \times w_1 + R_{22} \times w_2 + \cdots + R_{2d} \times w_d, \tag{1.4}$$

where considering $\Sigma_i w_i = 1$ and normalizing the columns of the R matrix so that $R_{11} + R_{21} = 1, ..., R_{1d} + R_{2d} = 1$, we have $\psi' + \psi = 1$.

1.4 Discussions

According to the research conducted it is worth saying that the proposed structure of the organization with extensive supply chain performance is effective. Organizations need to strengthen their supply chain with respect to appropriate technical, financial and human resource elements for development and improvement purposes. By improving each of these capabilities, the ability to enhance is increased. In the other words, a strong supply chain together with all these capabilities should insert the concepts in all layers associated with an intelligent management. We can enhance the performance layer intervals for each industry based on their corresponding experts opinions. Then, continuous improvement is audited and possible deviation can be determined and handled for maximum compliance.

References

Bello DC, Lohtia R, Sangtani V. An institutional analysis of supply chain innovations in global marketing channels. *Industrial Marketing Management* 2004;33:57–64.

Das K. Integrating effective flexibility measures into a strategic supply chain planning model technology systems. *European Journal of Operational Research* 2011;211(1):170–183.

David F. The strategic planning matrix—A quantitative approach. *Long Range Planning* 1986; 19(5):102.

Fandela G, Stammenb M. A general model for extended strategic supply chain management with emphasis on product life cycles including development and recycling. *International Journal of Production Economics* 2004;89:293–308.

Jiuh-Biing S. A multi-layer demand-responsive logistics control methodology for alleviating the bullwhip effect of supply chains. *European Journal of Operational Research* 2005;161(3):797–811.

Liang TF, Cheng HW. Application of fuzzy sets to manufacturing/distribution planning decisions with multi-product and multi-time period in supply chains. *Expert Systems with Applications* 2009;36:3367–3377.

Melo MT, Nickel S, Saldanha da Gama F. Dynamic multi-commodity capacitated facility location: A mathematical modeling framework for strategic supply chain planning. *Computers & Operations Research* 2005;33:181–208.

Saaty TL. *The Analytic Hierarchy Process.* McGraw-Hill, New York, 1980.

Taleizadeh AA, Niaki STA, Barzinpour F. Multiple-buyer multiple-vendor multi-product multi-constraint supply chain problem with stochastic demand and variable lead-time. *Applied Mathematics and Computation* 2011;217:9234–9253.

2

Multi-Layer and Multi-Product Supply Chain: Performance Evaluation Model

SUMMARY This chapter presents clustering of elements in a multi-layer and multi-product bi-direction supply chain for purification of interactions using data mining. The objective is to improve the performance of the supply chain and prevent the bottlenecks using some operational specifications related to each layer of the proposed supply chain. A developed version of k-mean clustering technique is illustrated.

2.1 Introduction

Supply chains are generally viewed as a network of materials and information flows both in and between facilities, including manufacturing and assembly plants and distribution centers (Thomas and Griffin, 1996; Sabri and Beamon, 2000). Supply chain has cross-boundary and multi-aspect features that always include multiple suppliers and multiple distributors. Most likely, customer information of a business organization tends to accumulate as time passes and consequently a huge amount of customer data might have accumulated in databases (Shaw et al., 2009). A larger amount of untreated customer information stored in database is wasteful unless useful knowledge has been extracted. Knowledge provides power in many manufacturing contexts enabling and facilitating the preservation of valuable heritage, new learning, solving intricate problems, creating core competencies and initiating new situations for both individuals and organizations now and in the future (Choudhary et al., 2007). Data mining thus has become an indispensable tool in understanding needs, preferences and behaviors of customers. It is also used in pricing, promotion and product development. Conventionally, data mining techniques have been used in banking, insurance and retail business. This is largely because of the fact that the implementation of these techniques showed quick returns. Data mining is being used for customer profiling where characteristics of good customers are identified with the goals of predicting new customers and helping marketing departments target new prospects. The effectiveness of sales promotions/product positioning can be analyzed using market-basket analysis to determine which products are purchased together or by an individual over time, which products to stock in a particular store, and where to place products in each store (Groth, 2000; Kopanakis and Theodoulidis, 2003). In addition, data mining is used in a variety of other industries such as the financial, healthcare, and telecommunications industries, among others. There are a lot of opportunities and applications of data mining even beyond the obvious. One of the potential areas is supply chain management (SCM).

On the other hand, issues related to supply chain management have been successfully dealt with the exploitation of data mining techniques. Customer and supplier categorization, market basket analysis, and inventory scheduling are typical problems where data mining is applied, to provide efficient solutions (Symeonidis, 2006). Wang and Wang (2005) suggested that cluster analysis could be used to cluster all suppliers and to establish a supplier evaluation index, to effectively manage suppliers. Bottani and Rizzi (2008) pointed out that suppliers with similar characteristics could be clustered by using cluster analysis to reduce supplier combinations. Sung and Ramayya (2007) stated that cluster analysis could effectively differentiate supplier types. Basic time series analysis will be used in this research as one of the "traditional" methods against which the performance of other advanced techniques will be compared. The latter include neural networks (NN), recurrent neural networks (RNN), and support vector machines (SVM). NN and RNN are frequently used to predict time series. In particular, RNN are included in the analysis because the manufacturer's demand is considered a chaotic time series. RNN perform back-propagation of error through time that permits learning patterns through an arbitrary depth in the time series. This means that even though we provide a time window of data as the input dimension to the RNN, it can match pattern through time that extends further than the provided current time window because it has recurrent connections. SVM, a more recent learning algorithm that has been developed from statistical learning theory, has a very strong mathematical foundation, and has been previously applied to time series analysis. According to Wu et al. (2000), because SCM is fundamentally concerned with coherence among multiple globally distributed decision makers, a multi-agent modeling framework based on explicit communication between constituent agents (such as manufacturers, suppliers, retailers and customers) seems very appealing. Many researchers have focused on performance measurement (PM) framework designs to exploit these new frameworks in action rather than on how to sustain or increase supply chain (SC) performance in the long term, or how to improve and validate the present PM results. In this context, we introduce a data mining–based framework that enables agents to successfully improve its performance when participating in global supply chains.

There is a growing consensus in the literature regarding the advantages information sharing provides for the supply chain partners. Researchers suggest that closer information-based linkages become a prevalent way of effectively managing supply chains that seek improved performance through effective use of resources and capabilities. Information sharing significantly contributes in reducing supply chain costs, improving partner relationships, increasing material flow, enabling faster delivery, improving order fulfillment rate thus contributing to customer satisfaction, enhancing channel coordination, and facilitating the achievement of an competitive advantage. Many researchers agree that information sharing is a key driver of an effective and efficient supply chain by speeding up the information flow, shortening the response time to customer needs, providing enhanced coordination and collaboration and sharing the risks as well as the benefits. For supply chain performance optimization, identifying important measures at multiple levels is more important than just maximizing or minimizing the identified indicators. One approach towards evaluating important indicators is the fuzzy logic technique, which is a problem-solving tool for handling vague and imprecise information, to get a definite decision. Although specific applications of the fuzzy logic tool for decision-making have been presented in the hierarchical measurement system, there have been few studies of using this tool in performance management, in practice, in comparison to other practical areas. A reliable performance measurement is beneficial in evaluating the SCM effectiveness and efficiency. Supposing we have a good performance measurement; we can profoundly understand the current performance status of the SC network easily so as to effectively recognize our strengths, weaknesses, threats, and opportunities. However, it has been a challenge to establish an "appropriate" collaborative network for the SC network. Kittelson et al. (2003) pointed out that PM, among collaborative SC network, is crucial for management. There have been many attempts to apply and explore artificial intelligence (AI) and data mining techniques to make up for the typical techniques in optimizing the PM in SCM with a better development roadmap. Bevilacqua et al. (2006) employed a fuzzy-quality function deployment (QFD) approach to supplier selection, and Jain et al. (2007) proposed supplier selection using a fuzzy association rules mining approach. On the process view Lau et al. (2009) developed a process mining system for supporting knowledge discovery in a supply chain network using fuzzy association rules to fine-tune the configuration of process parameters. Huang et al. (2008) developed a fuzzy neural network optimized by particle swarm optimization to solve the problem of demand uncertainty in SC. Moreover, the number of publications in operational optimization in terms of scheduling, routing, and inventory using genetic algorithms (GA) has increased according to its performance. In addition, Almejalli et al. (2007) has applied fuzzy neural network and GA for real time identification of road traffic control measures.

2.2 Problem Definition

To maintain competitiveness, manufacturers are seeking to deliver high-quality products at affordable prices to customers. In order to improve the performance in a multi-layer and multi-product supply chain, we consider a five-layer supply chain as shown in Figure 2.1 where manufacturers produce different products. The problem of supply chain performance can be shown as follows:

The counters for different elements of the model are given below.

Suppliers Layer = $\{s_1, s_2, s_3, ..., s_i\}$

Manufacturers Layer = $\{m_1, m_2, m_3, ..., m_j\}$

Distributors Layer = $\{d_1, d_2, d_3, ..., d_k\}$

Retailers Layer = $\{r_1, r_2, r_3, ..., r_l\}$

Customers Layer = $\{c_1, c_2, c_3, ..., c_m\}$

Each of these layers makes decisions based on the information they have about the prior and next layers. It can be shown that these series of decisions do indeed lead to the best overall performance from start to finish. This is achieved by selecting the best available option at each stage. To clarify, let us consider a manufacturer that should select an appropriate supplier through the suppliers' layer in multi-layer supply chain as shown in Figure 2.1. It is vital for both operational and informational performance of supply chains. We will use data mining techniques in order to improve supply chain performance. Indeed we apply clustering in each layer separately to lead to better decision making for each layer and finally improvement in the whole supply chain. The clustering is based on some operational specifications. With regard to the use of data mining in recent years in various industries, considering production is on the rise also, all the platforms for data mining, including high-speed data processing, data warehousing, data analysis and data mining software, are available. So, data mining can be used to improve the performance of

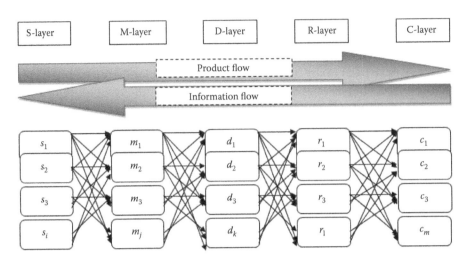

FIGURE 2.1
Forward and backward flow in multi-layer supply chain.

the supply chain in manufacturing. The purpose of our study is providing a framework for improving performance of multi-product multi-layer supply chain using data mining techniques and knowledge discovery. Through this framework, we can improve supply chain performance and meet customer needs. Also, the concept of knowledge discovery used in our supply chain management framework lead to customer satisfaction.

2.3 Data Mining Model

In this section, we explain our proposed framework and provide an algorithm for application. The proposed framework is given in Figure 2.2. First, for clustering each layer, we need some performance indicators. So, by reviewing the literature in the performance indicators field for supply chain, we select some of important indicators in each layer and start clustering by the proposed developed k-means algorithm. Finally, we use a questionnaire

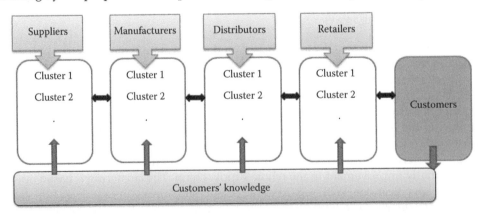

FIGURE 2.2
Proposed framework.

to collect the feedback of customer knowledge about products and services in order to transfer their knowledge to the related layer in supply chain and use their feedback for performance improvement.

The common *k*-means clustering algorithm is given below:

```
1: Select k points as the initial centroids
2: repeat
3:    Form k clusters by assigning all points to the closest centroid.
4:    Recompute the centroid of each cluster.
5: until The centroids don't change
```

But, due to drawbacks of the algorithm for our problem, a developed version of the *k*-means algorithm is provided by adding some steps as follows:

Solutions to Initial Centroids Problem:

- Multiple runs
    - Helps, but probability is not on your side
- Sample and use hierarchical clustering to determine initial centroids
- Select more than *k* initial centroids and then select among these initial centroids
    - Select most widely separated
- Post processing
- Bisecting *k*-means
    - Not as susceptible to initialization issues

In order to select the appropriate initial centroids (*k*), we use hierarchical clustering to determine initial centroids for Step 1 in the *k*-means algorithm. In backward flow, in order to extract customer knowledge, we use a questionnaire and analysis of these filled-out questionnaires in an Excel package, then show results in pie charts.

We illustrate the framework using a multi-product and multi-layer supply chain. We perform clustering for the supplier layer. First, we should determine the number of clusters by performing a hierarchical clustering on our data.

We introduce selected performance indicators for each layer as follows:

Supplier layer:

- Warehousing cost
- Inventory cost
- Total logistic cost
- Rejection percent
- On time delivery percent
- Delivery lead time

Manufacturer layer:

- Delivery lead time
- On time delivery percent
- Rejection percent

- Total logistic cost
- Inventory cost
- Warehousing cost
- Failure rate
- Production cost
- Delivery time

Distributor layer:

- Warehousing cost
- Inventory cost
- Total logistic cost
- Rejection percent
- On time delivery percent
- Delivery lead time
- Distribution cost

Retailer layer:

- On time delivery percent
- Delivery lead time
- Warehousing cost
- Inventory cost
- Rejection percent

By collecting the required data, one can easily run the proposed data mining model to cluster the performance indicators in each layer and then, by asking customer satisfaction metrics, the backward information flow can be conducted to strengthen the overall performance of the supply chain.

2.4 Discussions

We presented clustering of elements in a multi-layer and multi-product bi-direction supply chain for purification of interactions using data mining. The goal was to improve the performance of the supply chain and prevent the bottlenecks using some operational specifications related to each layer of the proposed supply chain. The advantages of the proposed methodology were:

- Reviewing a large number of key performance indicators in the supply chain literature
- Comprehensive performance evaluation of a multi-layer multi-product supply chain
- Providing real-time assessment of the supply chain with regard to dynamism of data

- Leveling and prioritizing the elements in each layer
- Considering customers' feedback to include customer satisfaction at the same time

Also, we developed the *k*-mean algorithm by combining a hierarchical procedure to break through the drawbacks.

References

Almejalli KA, Dahal KP, Hossain MA. Real time identification of road traffic control measures. In: Fink A and Rothlauf F (eds.). *Advances in Computational Intelligence in Transport, Logistics, and Supply Chain Management*. Studies in computational intelligence (series). Heidelberg: Springer, 2007, pp. 63–80.

Bevilacqua M, Ciarapica FE, Giacchetta G. A fuzzy-QFD approach to supplier selection. *Journal of Purchasing and Supply Management* 2006;12(1):14–27.

Bottani E, Rizzi A. An adapted multi-criteria approach to suppliers and products selection—An application oriented to lead-time reduction. *International Journal of Production Economics* 2008;111:763–781.

Choudhary AK, Harding JA, Lin HK. Engineering moderator to universal knowledge moderator for moderating collaborative projects. In: *Proceedings of GLOGIFT 07* November 15–17, 2007, UP Technical University, Noida, 2007:529–537.

Groth R. *Data Mining: Building Competitive Advantage*. Prentice Hall PTR, Upper Saddle River, NJ, USA, 2000.

Huang S, Chiu N, Chen L. Integration of the grey relational analysis with genetic algorithm for software effort estimation. *European Journal of Operational Research* 2008;188(3):898–909.

Jain V, Wadhwa S, Deshmukh SG. Supplier selection using fuzzy association rules mining. *International Journal of Production Research* 2007;45(6):1323–1353.

Kittelson & Associates, KFH Group, Parsons Brinckerhoff Quade & Douglass and Hunter–Zaworski, K. *Transit Capacity and Quality of Service Manual (TCQSM)*. Transit Cooperative Research Program Report 100, 2nd edition, Part 3. Federal Transit Administration, 2003.

Kopanakis L, Theodoulidis B. Visual data mining modeling techniques for the visualization of mining outcomes. *Journal of Visual Languages & Computing* 2003;14(6):543–589.

Lau HCW, Ho GTS, Zhao Y, Chung NSH. 2009. Development of a process mining system for supporting knowledge discovery in a supply chain network. *International Journal of Production Economics* 2009;122(1):176–187.

Sabri E, Beamon E. 2000. A multi-objective approach to simultaneous strategic and operational planning in supply chain design. *Omega* 2000;28(5):581–598.

Shaw MJ, Subramanian C, Tan GW. Knowledge management and data mining for marketing. *Decision Support Systems* 2009;31:127.

Sung HH, Ramayya K. A hybrid approach to supplier selection for the maintenance of a competitive supply chain. *Expert Systems with Applications* 2007;34:1303–1311.

Symeonidis LA, Nikolaidou V, Mitkas AP. Exploiting data mining techniques for improving the efficiency of a supply chain management agent. *IEEE/WIC/ACM International Conference on Web Intelligence and Intelligent Agent Technology*, 2006;23:26.

Thomas D, Griffin P. Coordinated supply chain management. *European Journal of Operational Research* 1996;94(1):1–15.

Wang ST, Wang ZJ. Study of the application of PSO algorithms for nonlinear problems. *Journal of Huazhong University of Science and Technology* 2005;33:4–7.

Wu J, Ulieru M, Cobzaru M, Norrie D. Supply chain management systems: State of the art and vision. In: *International Conference on Management of Innovation and Technology (ICMIT)*, IEEE Computer Society, Hong Kong, China, 2000.

3

Multi-Layer Supply Chain: Mathematical Evaluation Model

SUMMARY This chapter proposes a systematic approach that helps to analyze and select the right key performance indicators (KPIs) to improve supply chain (SC) performance. A mathematical model is formulated to maximize the overall performance of the supply chain.

3.1 Introduction

Using measurements to support manufacturing operations dates back to the late 19th and early 20th centuries with Fredrick W. Taylor applying scientific methods to running a business. His ideas for time and motion studies of operations were successfully used to scientifically manage production lines and warehouse operations. Today, performance measurement has become a part of all business processes, which are striving to be more efficient and cost effective. Over the last decade, companies have spent a lot of time and money to improve their supply chains. Their efforts have been made easier by the enterprise resource planning/supply chain management (ERP/SCM) software vendors, which have developed sophisticated software solutions, both for supply chain operations and supply chain planning. Whereas, all these software solutions enable companies to drastically improve their supply chain performance, yet they do not provide adequately the tools needed to measure the improvements (or performance levels). Thus, companies need to develop their own set of performance metrics or key performance indicators (KPIs), so as to know how close or how far they are from meeting set objectives. In the context of a dynamic supply chain, continuously improving performance has become a critical issue for most suppliers, manufacturers, and the related retailers to gain and sustain competitiveness. In practice, supply chain based companies (e.g., Dell, Wal-Mart, Samsung, Toyota, Lenovo, Gome, etc.) have used different performance management tools to support their supply chain strategies. Monitoring and improvement of performance of a supply chain has become an increasingly complex task. A complex performance management system includes many management processes, such as identifying measures, defining targets, planning, communication, monitoring, reporting and feedback. These processes have been embedded in most information system solutions, such as i2, SAP, Oracle EPM, etc. These system solutions measure and monitor KPIs, which are crucial for optimizing supply chain performance.

Performance measurement is critical for companies to improve supply chains' effectiveness and efficiency (Beamon, 1999). Decision-makers in supply chains usually focus on developing measurement metrics for evaluating performance (Beamon, 1999). In practice, once the supply chain performance measures are developed adequately, managers have to identify the critical KPIs that need to be improved. However, it is difficult to figure out the intricate relationships among different KPIs and the order of priorities for accomplishment of individual KPIs. As a matter of fact, determination of priorities within a given set of KPIs has become a bottleneck for many companies in their endeavors for improving their supply chain management (SCM). As these problems have received relatively less attention in previous research, significant gaps remain between practical needs and their effective solutions. To address these issues, this chapter proposes a systematic approach that helps to analyze and select the right KPIs to improve supply chain performance.

Improving supply chain performance is a continuous process that requires both an analytical performance measurement system and a mechanism to initiate steps for realizing KPI goals; herein we call the mechanism to achieve KPI goals as "KPI accomplishment," which connects planning and execution, and builds steps for realization of performance goals into routine daily work. To measure supply chain performance, there are a set of variables that capture the impact of actual working of supply chains on revenues and costs of the whole system. These variables as drivers of supply chain performance are always derived from supply chain management practices. After identifying KPIs, managers have to achieve improvement in them, through continuous planning, monitoring and execution. According to the results of selected KPIs' accomplishment, managers may create current reports on KPIs, to compare multiple plans of supply chain management. In this performance management cycle, there are many challenges, both in performance measurement and its improvement.

Once critical KPIs have been identified and selected effectively, another challenge is the coordination of the parallel steps required for accomplishment of improvement in identified KPIs. Generally speaking, there are two methodological streams to cope with this problem in previous literature. One stream involves finding out the bottlenecks in the supply chain by implementing the KPIs. For instance, the theory of constraints (TOC) is a set of concepts and tools that can be used to implement the widely used continuous improvement management philosophy. TOC improves performance in a system by focusing attention of management on the system's constraints. Thus, by preventing distractions from the primary purpose and concentrating limited resources on efficacious management of the constraint, decision-makers are able to gain significant leverage, sufficient to attain the desired performance levels. In the TOC theory, the method is to find a suitable approach to identify and solve bottlenecks in production, delivery, and service processes. However, the TOC method does not deal with selection of crucial bottlenecks and it doesn't provide the optimal solution of performance improvement for each KPI. Sometimes, the KPIs are coupled or correlated, and it is hard to find the precise bottleneck; improving one KPI might undermine performance of another one.

The second stream focuses on performance optimization; the optimization philosophy assumes that there is an optimal performance point, with maximizing or minimizing the identified indicators. Although the performance optimization approach, in theory, is widely accepted by researchers, it is difficult to ensure that an optimized KPI accomplishment strategy is implemented by different members of the supply chain. First, it is difficult to apply in practice, in terms of both data acquisition and computing. It is also difficult for decision-makers to understand in real SCM situations. Second, it does not take into account the relationships among indicators. Though classified into different categories, different measures in a measurement system are often correlated. The correlations among different measures arise from the inherent internal relations of different SCM processes, and the interdependent influences of different KPIs' accomplishment tasks. Therefore, a feasible methodology of identifying and analyzing the relationships among KPIs related to different SCM processes is important and necessary for improving SCM performance. For supply chain performance optimization, identifying important measures at multiple levels is more important than just maximizing or minimizing the identified indicators. One approach towards evaluating important indicators is the fuzzy logic technique, which is a problem-solving tool for handling vague and imprecise information to get a definite decision. Although specific applications of the fuzzy logic tool for decision-making have been presented in the hierarchical measurement system (Chan and Qi, 2003), there have been few studies of using this tool in performance management, in practice, in comparison to other practical areas, e.g., project management.

In practice, organizations are prone to making rushed decisions when faced with continuously changing goals and tight deadlines. Managers are short of time to compare all the options when situations demand immediate solutions. Therefore, it is important to describe the mutually dependent relationships among KPIs and to optimize their accomplishment based on their complex interdependence. However, most previous research does not provide specific operational procedures for analyzing KPI accomplishment. Considering pros and cons of different methods, this chapter provides a framework of supply chain performance measurement and improvement, based on a systematic approach to analyzing KPI accomplishment.

3.2 Problem Definition

In this chapter, after determining key performance indices, we are seeking to find the effective indices on the performance of the supply chain. To configure the problem, consider a three-layer supply chain in which a mathematical model is used to analyze the impacts of performance indices. The objective of the mathematical model is to maximize the profit of the whole supply chain. The effective performance indices lead to customer satisfaction, and therefore investigating which set of indices is effective in an appropriate layer of the SCM. Here, a three-layer supply chain is designed, having a supplier, producer and customer. Raw materials are provided by suppliers and then transferred to the producer to perform the processing required for a final product. Then the produced products are sent to customers to complete the chain. To maximize customer satisfaction, effective performance indicators in each layer are obtained and the profit of the whole chain is optimized.

3.3 Mathematical Model for Performance Evaluation

In the proposed three-layer supply chain the following performance indices are considered:

Supplier: Cooperation of suppliers, delivery of defect-free products by suppliers, assistance of supplier in solving technical issues, capability of supplier quality, cycle time of purchase order, time the order is received, good record of cooperation, investment of supplier, delivery cost.

Producer: Time cycle, total time of cash flow, diversity of products and services, deviations from budget, cost-saving innovations, accuracy of prediction methods, new product development cycle time, ordering methods, main produced schedule, rate of return on investment, levels of inventory turnover, lead time, minimizing the time between order and delivery, rate of return, guarantee, good performance of the product, transportation costs.

Customer: Customer perception of product value, degree of flexibility satisfying customer needs, supply rate, customer satisfaction, minimizing response time to customer, flexibility orders.

Thus a mathematical model is developed for performance evaluation.

Indices

$i = 1, 2, ..., I$ Key performance indicators
$j = 1, 2, ..., J$ Supply chain layers

Parameters

Initiation cost for each index i in layer j c_{ij}
The significance of each of the indicators w_{ij}
The funds available to each layer B_j
Risk launch R_{ij}
Economic profit percent is allowed θ
Random variable corresponding to each index y_j

Decision Variables

If select key performance indicator i in layer j $X_{ij} = 1$
Otherwise $X_{ij} = 0$
The amount of any proceeds E_{ij}

3.3.1 Objective Functions

$$\text{Max } Z_1 = W_{ij}X_{ij}$$

Weights for indicators,

$$\text{Min } Z_2 = R_{ij}X_{ij}$$

Risk of performance indicator in each layer,

$$\text{Max } Z_3 = E_{ij}X_{ij} - C_{ij}X_{ij}$$

Profit of the supply chain.

3.3.2 Constraints

$$\sum_i C_{ij} X_{ij} \leq B_j \quad \forall j$$

These constraints reflect the investment for indicators in each layer, limited to the available budget in each layer.

$$\sum_j \sum_i W_{ij} = 1$$

This constraint shows that the total weights sum up to 1.

$$\lambda(y) = |t - y_j|^2$$

This constraint reflects the loss function for any indicator.

$$\Lambda = \int_\infty \lambda(y) f(y) dy \quad \Leftrightarrow R_{ij}$$

This constraint implies the risk function using the probability density function associated with each indicator.

$$E_{ij} X_{ij} - C_{ij} X_{ij} \leq (\theta + 1) B_j$$

The above equation shows that the profit is confined with a coefficient $(1 + \theta)$ of the available budget.

$$E_{ij} \geq 0$$

The above relation certifies that the earning for each indicator in each layer is more than or equal to zero.

$$X_{ij} \in \{0, 1\}$$

The above relation represents the sign of the binary decision variable.

The supply chain of this research is a three-layer supply chain that includes a supplier, producer and customer. Suppliers offer basic material to producers for making products. The producer prepares final products to be dispatched to customers. In the following data collection process has been expressed. After identifying key performance indicators of the supply chain by the review of the literature, variables were examined and 32 factors were chosen as supply chain factors affecting the supply chain layers, shown in Table 3.1.

TABLE 3.1

Key Performance Indicators Identified by the Review of the Literature

Supplier		Producer		Customer	
1	Cooperation of Suppliers	1	Time Cycle	1	Customer Perception of Product Value
2	Delivery of Defect-Free Products by Suppliers	2	Total Time of Cash Flow	2	Degree of Flexibility Satisfying Customer Needs
3	Assistance Supplier Solving Technical Issues	3	Diversity of Products and Services	3	Supply Rate
4	Ability of Supplier Quality	4	Deviations from Budget	4	Customer Satisfaction
5	Cycle Time Purchase Order	5	Cost-Saving Innovations	5	Minimizing Response Time to Customer
6	Time the Order Is Received	6	The Accuracy of Prediction Methods	6	Flexibility Orders
7	Good Record of Cooperation	7	New Product Development Cycle Time		
8	Investment Supplier	8	Ordering Methods		
9	Delivery Cost	9	Main Produced Schedule		
		10	Rate of Return on Investment		
		11	Levels of Inventory Turnover		
		12	Lead Time		
		13	Minimizing the Time Between Order and Delivery		
		14	Rate of Return		
		15	Guarantee		
		16	Good Performance of the Product		
		17	Transportation Costs		

3.4 An Example

The supply chain of this chapter is a three-layer supply chain that includes suppliers, producers and customers. The material flow initiates from suppliers to producers, and the final products are sent to customers. In the following data collection process has been expressed. After identifying key performance indicators of the supply chain by the review of the literature, variables were examined and the 32 factors were chosen as supply chain factors affecting the supply chain layers that are given in Table 3.1. The costs of each indicator in each layer are given in Table 3.2.

Due to importance of each performance indicator a corresponding weight is allocated to them accordingly. To do that, a general ranking method of Technique for Order Preference by Similarity to Ideal Solution (TOPSIS) is used. The pairwise comparison matrices are filled with respect to four criteria of quality, price, product differences and safety. The obtained weights are shown in Table 3.3.

TABLE 3.2

The Cost of Setting Up Key Performance Indicators

	Supplier	Producer	Customer
1	9	12	16
2	20	6	7
3	16	9	7
4	7	4	15
5	13	13	13
6	19	11	10
7	10	8	
8	12	11	
9	12	14	
10		5	
11		10	
12		9	
13		19	
14		18	
15		15	
16		7	
17		11	

TABLE 3.3

Weight of Key Performance Indicators

W_{ij}	Supplier	Producer	Customer
1	0.019	0.073	0.030
2	0.023	0.034	0.047
3	0.004	0.027	0.008
4	0.010	0.009	0.057
5	0.069	0.040	0.021
6	0.018	0.005	0.047
7	0.015	0.073	
8	0.012	0.027	
9	0.032	0.053	
10		0.069	
11		0.031	
12		0.023	
13		0.020	
14		0.010	
15		0.063	
16		0.007	
17		0.024	

Risk of implementing each performance indicator is followed by uniform probability distribution. An example for a typical uniform distribution follows here:

1.

$$f(y) = \frac{1}{b-a}$$

$$\Rightarrow f(y_1) = \frac{1}{80-30} \rightarrow f(y_1) = 0.02$$

Then the loss function is formed as
2.

$$\lambda(y) = |t - y_j|^2 \quad (t = 0.2)$$

$$\Rightarrow \quad \lambda(y_1) = |0.2 - y_1|^2$$

And finally the risk is computed by
3.

$$\Lambda_1 = \int_0^{0.2} 0.02\left(0.04 - 0.4y_1 + y_1^2\right) dy$$

$$\Rightarrow R_{11} = 0.02\left[0.04y_1 - 0.2y_1^2 + \frac{y_1^3}{3}\right]\Big|_0^{0.2} = 0.000053333$$

And the rest of the performance indicators' risk computations is given in Table 3.4. The budget available for each layer is shown in Table 3.5.

Finally, the percentage of profit allowed to be considered is $\theta = 0.25$. As a result, after implementing the problem in LINGO optimization software, the effective KPIs are obtained, and the corresponding earning is also in hand. The results are shown in Tables 3.6 and 3.7.

3.5 Discussions

In this chapter, a collection of key performance indices (KPIs) is extracted from the related literature. Then, a mathematical model was developed for a three-layer supply chain including supplier, producer and customer. The objectives were to maximize the profit, maximize the effective KPI importance weights and minimize the risk of investment. The target was to improve the whole supply chain. The decision variables were the allocation of KPI in each layer and the economic earning reasonable for each indication. The results imply that the model can be employed as a helpful decision aid for managerial decision-making in real world industries.

TABLE 3.4

Initiation Risk of KPI *Rij*

	a	b	$f(y) = 1/(b-a)$	$\Lambda = \int_0^{0.2} f(y) * (0.04 - 0.4y_1 + y_1^2)dy$
Supplier:				
1	30	80	0.02	5.3333E − 05
2	15	63	0.02083	5.5556E − 05
3	30	90	0.01667	4.4444E − 05
4	12	82	0.01429	3.8095E − 05
5	6	64	0.01724	4.5977E − 05
6	10	75	0.01538	4.1026E − 05
7	33	51	0.05556	1.4815E − 04
8	5	73	0.01471	3.9216E − 05
9	48	91	0.02326	6.2016E − 05
Producer:				
1	12	71	0.01695	4.5198E − 05
2	20	98	0.01282	3.4188E − 05
3	18	49	0.03226	8.6022E − 05
4	21	77	0.01786	4.7619E − 05
5	35	85	0.02	5.3333E − 05
6	10	60	0.02	5.3333E − 05
7	45	96	0.01961	5.2288E − 05
8	33	78	0.02222	5.9259E − 05
9	30	80	0.02	5.3333E − 05
10	17	82	0.01538	4.1026E − 05
11	15	68	0.01887	5.0314E − 05
12	17	91	0.01351	3.6036E − 05
13	34	88	0.01852	4.9383E − 05
14	41	93	0.01923	5.1282E − 05
15	23	79	0.01786	4.7619E − 05
16	9	77	0.01471	3.9216E − 05
17	11	89	0.01282	3.4188E − 05
Customer:				
1	16	86	0.01429	3.8095E − 05
2	25	91	0.01515	4.0404E − 05
3	14	73	0.01695	4.5198E − 05
4	33	89	0.01786	4.7619E − 05
5	12	90	0.01282	3.4188E − 05
6	10	90	0.0125	3.3333E − 05

TABLE 3.5

The Budget Allocated to Each Layer in the Supply Chain

B1	100
B2	300
B3	100

TABLE 3.6

The Obtained Effective KPI

X_{ij}	$j = 1$	$j = 2$	$j = 3$
$i = 1$	0	0	0
$i = 2$	0	1	1
$i = 3$	0	0	0
$i = 4$	1	0	0
$i = 5$	1	0	0
$i = 6$	0	0	1
$i = 7$	0	1	
$i = 8$	0	0	
$i = 9$	0	0	
$i = 10$		1	
$i = 11$		0	
$i = 12$		0	
$i = 13$		0	
$i = 14$		0	
$i = 15$		0	
$i = 16$		1	
$i = 17$		0	

TABLE 3.7

The Obtained Earnings

E_{ij}	$j = 1$	$j = 2$	$j = 3$
$i = 1$	0	0	0
$i = 2$	0	383	132
$i = 3$	0	0	0
$i = 4$	132	0	0
$i = 5$	138	0	0
$i = 6$	0	0	135
$i = 7$	0	382	
$i = 8$	0	0	
$i = 9$	0	0	
$i = 10$		380	
$i = 11$		0	
$i = 12$		0	
$i = 13$		0	
$i = 14$		0	
$i = 15$		0	
$i = 16$		382	
$i = 17$		0	

Bibliography

Beamon MB. Measuring supply chain performance. *International Journal of Operations & Productions Management* 1999;19(3):275–292.

Bradford KD, Stringfellow A, Weitz B. Managing conflict to improve the effectiveness of retail Networks. *Journal of Retailing* 2004;80(3):181–195.

Burgess K, Singh PJ, Koroglu R. Supply chain management: A structured literature review and implications for future research. *International Journal of Operations & Production Management* 2006;26(7):703–729.

Chae BK. Developing key performance indicators for supply chain: an industry perspective. *Supply Chain Management: An International Journal* 2009;14(6):422–428.

Chan FTS, Qi HJ. An innovative performance measurement method for supply chain management. *Supply Chain Management: An International Journal* 2003;8(3):209–223.

Clemencic A. Management of the supply chain-Case of Danfoss District Heating Business Area. *Master's Degree Thesis*, 2006.

Cooper MC, Lambert DM, Pagh JD. Supply chain management: More than a new name for logistics. *International Journal of Logistics Management* 1997;8(1):1–13.

Gulledge T, Chavusholu T. Automating the construction of supply chain key performance indicators. *Industrial Management & Data Systems* 2008;108(6):750–774.

Ip WH, Chan SL, Lam CY. Modeling supply chain performance and stability. *Industrial Management & Data Systems* 2011;111(8):1332–1354.

Jian C, Xiangdong L, Zhihui X, Jin L. Improving supply chain performance management: A systematic approach to analyzing iterative KPI accomplishment. *Decision Support Systems* 2009;46:212–521.

Laudon K, Lauden J. 2001. *Management Information System*, Prentice-Hall, 4,2002- Shields M6.E-Busines and ERP, Whilley, 2001.

Mabert VA, Vnkataremanam MA. Special research focus on supply chain linkage: Challenges for design & management in 21 century. *Decision Sciences* 2002.

Maskell BH. *Performance Measurement for World Class Manufacturing*. Productivity Press, Portland, Oregon, 1991.

Morgan C. Supply network performance measurement: future challenges? *The International Journal of Logistics Management* 2007;18(2):255–273.

Najmi M, Rigas J, Fan IS. A framework to review performance measurement systems. *Business Process Management Journal* 2005;11(2):109–122 (in Persian).

4

Supply Chain Inventory Planning: System Dynamics Model

SUMMARY This chapter models a system dynamic process for inventory planning in a supply chain. The dynamical tools of casual loop and state flow are employed for analyzing the effectiveness of variables. The model is implemented in Vensim simulation environment.

4.1 Introduction

Fierce global competition over sophisticated customers demanding increasing customization and constantly faster response in addition to advancements in information and communication technology have resulted in making supply chain networks critical contributors in the production and distribution of goods in contemporary markets. The growing interest in supply chain networks has in turn pointed out the importance of relying on efficient management practices specially designed to manage the complexity, enormity and breadth of scope of the supply chain structures. Supply chain management (SCM) has evolved to one of the most prevailing 21st century manufacturing paradigms focusing on the design, organization and operation of supply chains.

In a supply chain there are three types of flow—that of materials, information and finance. The objective of this chapter is to model a manufacturing supply chain, which is assumed to have a moderate complexity of four echelons, to measure its performance under different operational conditions and finally identify and understand its dynamic behavior. A system dynamics (SD) approach is used to build the model and measure the system's performance. SD is a methodology that is capable of studying and modeling complex systems, as in our study, for supply chain networks. The operations performed within a supply chain are a function of a great number of key variables which often seem to have a strong interrelationship.

Systems dynamics aims to provide a holistic view of the system and to identify how these interrelationships affect the system as a whole. The ability of understanding the whole system as well as analyzing the interactions between various components of the integrated system and eventually supplying feedback without breaking it into its components make SD an ideal methodology for modeling supply networks. A structure is provided to transform the system from a mental model to a computer-based level and rates model. Further experimentation is then carried out on the model involving a number of designed scenarios. Conclusions are drawn on the behavior displayed.

The area of supply chain management is being increasingly investigated by both academia and industry. The successful implementation of supply chain improvement programs has pointed out the benefits of efficient SCM. All supply chains are different and a lot of companies struggle to understand the dynamics of their supply chain.

The traditional supply chain problems studied in the literature are more related to location/allocation decisions, demand planning, distribution channel planning, strategic alliances, new product development, outsourcing, supplier selection, pricing, and network structuring at the strategic level. The tactical level problems cover inventory control, production distribution coordination, order/freight consolidation, material handling, equipment selection and layout design. The problems addressed at the operational level include vehicle routing/scheduling, workforce scheduling, record keeping, and packaging.

Various alternative methods have been proposed for modeling supply chains. According to Beamon (1998), they can be grouped into four categories: deterministic models where all the parameters are known, stochastic models where at least one parameter is unknown but follows a probabilistic distribution, economic game-theoretic models and models based on simulation, which evaluate the performance of various supply chain strategies. The majority of these models are steady-state models based on average performance or steady-state conditions. However, static models are insufficient when dealing with the dynamic characteristics of the supply chain system, which are due to demand fluctuations, lead-time delays, sales forecasting, etc. In particular, they are not able to describe, analyze and find remedies for a major problem in supply chains.

System Dynamics is a powerful methodology for obtaining insights into problems of dynamic complexity and policy resistance. Forrester (1961) introduced SD in the 1960s as a modeling and simulation methodology in dynamic management problems. The system under study in this chapter is dynamic and full of feedback; therefore SD becomes an appropriate modeling and analysis tool.

The first published work in system dynamics modeling related to supply chain management is found in *Industrial Dynamics: A Major Breakthrough for Decision Makers* (Forrester, 1958). Forrester (1961) expanded on his basic model through further and more detailed analysis, and establishes a link between the use of the model and management education. Figure 4.1 shows the classic supply chain model that was used by Forrester in his simulation experiments.

There is a downstream flow of material from the factory via the factory warehouse, the distributor and the retailer to the customer. Orders (information flow) flow upstream and there is a delay associated with each echelon in the chain, representing, for instance, the production lead time or delays for administrative tasks such as order processing. Researchers since have coined the expression of the 'Forrester Supply Chain' or Forrester Model, which essentially is a simple four-level supply chain (consisting of factory, a warehouse, a distributor and a retailer).

Using the Forrester Model as an example, Forrester (1961) describes the modeling process used in modelling continuous processes, while clearly emphasizing the importance of information feedback to the SD method (Towill, 1996). Pointing out that the first step in an SD study is the problem identification and the formulation of questions to be answered, he illustrates the stages of model conceptualization, model parameterization, and model testing through various experiments. Forrester (1958) disapproves of the approach taken by operations research (OR) in the 1950s, where OR methods are applied to isolated company problems. He suggests that the success of industrial companies depends on the interaction between the flows of information, materials, orders, money, manpower, and capital equipment, and states that the understanding and control of these flows is the main task of management.

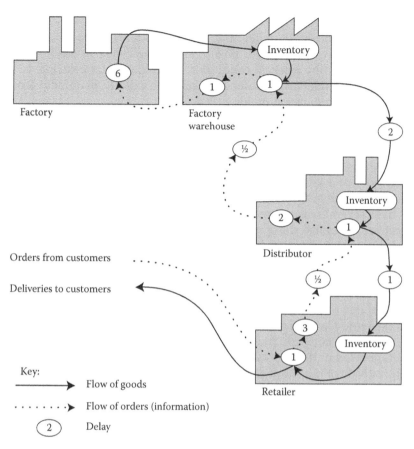

FIGURE 4.1
The Forrester supply chain.

The Forrester Model received much criticism over the years, which served as a basis for applying and extending Forrester's research further. Despite its simplicity, the Forrester Model yielded important insights into supply chain dynamics. Demand amplification, a fundamental problem in supply chains, has only recently been recognized to the full extent of the problem (Towill, 1996). Forrester accidentally established the ground rules for effective supply chain design when he showed that medium period demand amplification was an SD phenomenon which could be tackled by reducing and eliminating delays and the proper design of feedback loops (Towill, 1996).

4.2 Problem Definition

SD is a method to cope with complex system problems, with the combination of quantitative and qualitative methods, based on feedback control theory, using computer simulation technology as its measures (Özbayrak, 2007). According to an SD viewpoint, a system dynamic model analyzes a system's objects changing trends by simulating them dynamically in order to study and future action plan and to study the future action plan and assist the corresponding decision-making. The model is characterized by the object studied as a dynamic system (Towill, 1996). The dynamic system has a certain internal structure and is affected by external conditions. Its fundamental principle is: use system modeling, ending the model to computer and verify the validity, in order to provide a basis to work out strategy and decision-making (Feng, 2012).

The most important problem discussed in this chapter is the cost associated with fluctuations in warehouse inventory planning in the supply chain. Costs due to inventory stocking in warehouse, costs due to discount on sales to reduce inventory and even costs of perished products, shortage costs of product, lost customers costs and decreased market share when inventory is lower than market need, make forecasting of these fluctuations a very significant issue in supply chain. Undesired behavior and fluctuation in warehouse inventories have several reasons and different variables from the supply chain are influenced. Therefore, to resolve this problem, identification of important effective variables on the supply chain, defining variables' behavior, and formulating the way they influence inventory fluctuation and simulating inventory behavior in distribution centers' warehouses are necessary.

The overall supply chain model considers only four echelons—suppliers, manufacturers, distributors and retailers—and its dynamics are studied from the operational perspective. None of the companies that form the network has a partnership with any of the companies within the network; in addition, the companies forming the echelons of the supply chain modeled in this work are assumed to have no interactions whatsoever with any company outside of the supply chain considered, and hence no dedicated supply system is available for any of them. The central company cooperates with warehouses and distribution centers.

4.3 System Dynamic Model

The demand at the retailers' follows a random uniform distribution that represents customer demand. Consumers' demands are considered during a year which is altered by any seasonal or other statistically defined patterns of consumer behavior. The production capacity of the manufacturer is known per week. Production time follows a normal distribution and there are no assembly operations. The suppliers order and receive raw materials from an external source. The model only includes information and material flows. Cash flow is not considered in this work due to the added complexity. The system will be simulated for specified period of time. The process is modelled using Vensim software.

In this section, we will identify the important variables in the system. To identify effective variables, experts' opinions and study of related literature were used. Table 4.1 shows influencing variables on the proposed supply chain.

4.3.1 Dynamic Hypothesis

The original source of variables' behavior is influenced more than anything by sales behavior in distribution centers; this behavior depends on several factors, such as seasonal changes, discounts and promotions, price changes, etc. The effect of these variables is bidirectional and as we discuss later in this chapter, cause and effect variables are interchanged alternatively.

TABLE 4.1

Effective Variables on Supply Chain of Kaleh Company

The Important Variables in the System
Sale rate
Product inventory in distribution center
Product inventory in central warehouse
Safety stock in distribution center
Safety stock in central warehouse
Distribution center warehouses empty space
Central warehouses empty space
Product transfer
Production volume
Production rate
Transportation time from central warehouse to distribution center's warehouse
Order volume for production
Production time
Order volume for purchasing raw material

4.3.2 Casual-Loop Diagram (CLD)

In order to design casual-loop diagram, defined variables in dynamic theory and influencing factors on their behavior are evaluated and variables relations type and their feedback loops are designed step by step to reach the final casual-loop diagram. In continue, a general configuration of supply chain's cause and effect model is depicted and we discuss the most important loops within the model.

4.3.3 Goal Seeking Loop of Sale–Inventory

In this loop, the effect of sales on inventory and inventory on sales can be observed. Here, the effect of demand is shown as an incremental factor (+) on sales. Of course, this behavior is possible if there are enough inventories in distribution centers, otherwise the effect of demand on sales is not incremental and some adjustments are needed. To specify the time and amount of this adjustment, inventories of distribution centers' warehouses and demand are assessed and the amount of shortage is determined. In cases where there is a shortage, available inventories are sold and the remaining demands become lost sales and cannot be recovered. As shown in Figure 4.2, the relation of sales and inventory is decreasing and an increase in sale decreases inventory in distribution center. An inventory decrease in distribution center will cause an increase in shortage due to decreasing relation of inventory and shortage and consequently decreases sales. Hence, this loop behave to balance the variables and is goal seeking loop of sale–inventory.

4.3.4 Goal Seeking Loop of Transmission

In this loop, the effects of various factors on transfer as the only incremental factor on inventory of distribution center can be seen. As shown, the following factors are influencing on determining the amount of a transfer.

Warehouse fullness at distribution center indicates the amount of space needed for stock is sufficient or not.

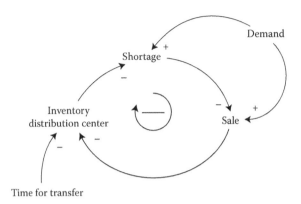

FIGURE 4.2
Goal seeking loop of sale-inventory.

Transfer time which is the time required to transfer inventory from central warehouse to distribution center. Here we only consider the average transfer time of different distribution centers. The existence of this factor causes a lag in relation of transfer and inventory variables in distribution center. In this loop, warehouse fullness is higher when the inventory levels are higher in distribution center or warehouse capacity is smaller. Also, with higher fullness percentage, the transfer amount decreases even if higher inventory needed, and it indicates a decreasing relation between fullness percentage and transfer variables (see Figure 4.3).

4.3.5 Goal Seeking Loop of Transmission Material

Within this loop, the transfer order of materials to supplier is determined as a key variable based on demand not responded, inventories in supplier's warehouse and the production

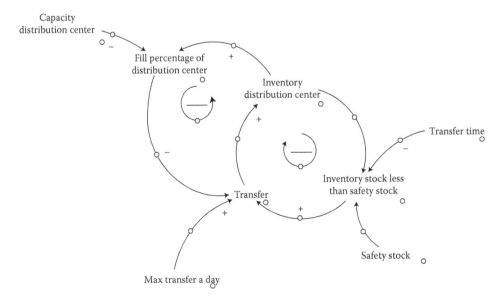

FIGURE 4.3
Goal seeking loop of transmission.

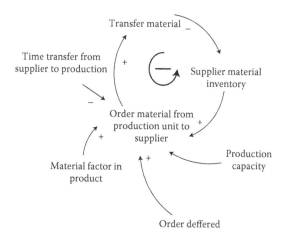

FIGURE 4.4
Goal seeking loop of transmission material.

capacity. After ordering materials to supplier, they are transferred to the production units and it causes a decrease in inventory of supplier's warehouse. Moreover, if there is no order for transferring materials, they are transferred to the production units with amount up to production capacity until the inventory at supplier warehouse becomes zero. This issue exists to sustain production and leads inventory to increase and prevent production shut down (see Figure 4.4).

4.3.6 Growth Loop of Sale-Transfer

In this loop sales feedback to transfer is created via forecasting sales and determining safety stock. By increasing forecasted sales, safety stock also increases. By increasing safety stock, the amount of shortage decreases which in turn causes a higher transfer in following periods. Therefore the relations of mentioned variables are of incremental type and this issue leads to creation of a growth loop from sales to transfer and from transfer to sales. This growth loop in cases where there is enough inventory is central warehouse, adequate warehouse space, and existence of enough demand, leads to incrementally higher transfer and sales up to completely responding to market needs. It should be noted that this increasing growth could transform to incremental decline as well (see Figure 4.5).

4.3.7 Growth Loop of Sales-Production

In this loop, in addition to variables related to sales, intermediary variables between sales and production and production variables are entered as well which leads to creation of a growth loop with some internal loops that finally all of them create a growth sales loop causing increased forecasted sales and safety stock. The relations of these variables, as shown in Figure 4.6, is incremental and by increasing lagged orders, the amount of transferred materials to product ion units and in turn production rate and product inventory in warehouse increase.

4.3.8 State-Flow Diagram

State-flow diagrams are about physical structure of feedback loops and involve casual diagrams that draw entering data for a decision policy. In these diagrams, the focus is on data

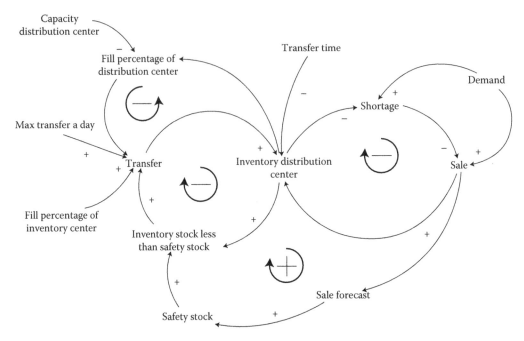

FIGURE 4.5
Growth loop of sale-transfer.

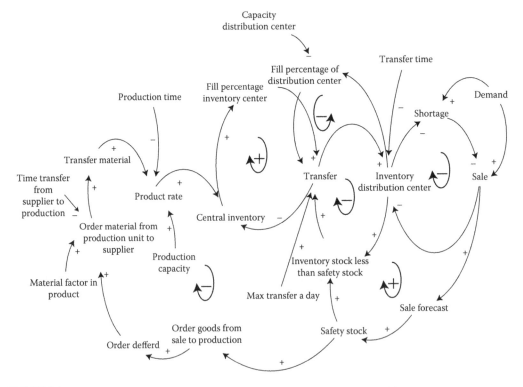

FIGURE 4.6
Growth loop of sales-production.

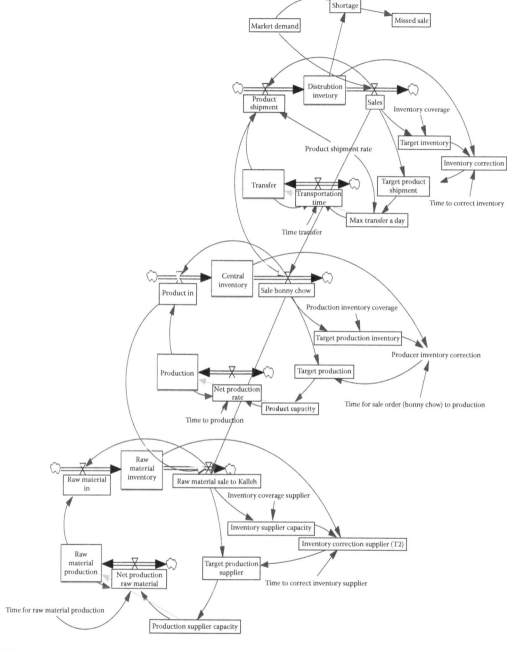

FIGURE 4.7
State-flow diagram.

that the modeler uses for the decision-making process. In state-flow diagrams, the system state variables are the number of customers, material inventory, debts, etc. System flow variables are such as production rate, material transfer rate, sales rate, etc. In these systems, decisions are determined based on state variables and executed based on flow variables (see Figure 4.7).

4.4 Discussions

In this chapter, the aim was to identify influencing variables on supply chain and their relations and effectiveness on the whole supply chain. The model was designed based on an SD method using Vensim software. The most important characteristic of this study is the possibility of implementation of the model in the reality which could prevent wrong decision in complex systems. In the designed model of this study, only one supplier was considered, which can be extended to several horizontal and vertical suppliers to the chain so that the accessibility time to the raw material for production can be reduced. Moreover, in distribution centers we only considered one distributor, which can be modeled as multi-distributor system so that to cluster products to distributors to prevent shortage.

References

Beamon BM. Supply chain design and analysis: Models and methods. *International Journal of Production Economics* 1998;55:281–94.

Feng Y. System dynamics modeling for supply chain information sharing. *Physics Procedia* 2012;25(1):1463–1469.

Forrester JW. Industrial dynamics: A major breakthrough for decision makers. *Harvard Business Review* 1958;36(4):37–66.

Forrester JW. *Industrial Dynamics*. Cambridge, MA: MIT Press, 1961.

Özbayrak M. Systems dynamics modelling of a manufacturing supplychain system. *Simulation Modelling Practice and Theory* 2007;15:1338–1355.

Towill DR. Industrial dynamics modelling of supply chains. *Logistics Information Management* 1996;9(4): 43–56.

5

Supplier Evaluation: Six Sigma Model

SUMMARY In this chapter, a qualitative mechanism based on quality control charts to evaluate the service and flexibility of suppliers in the layers of providing raw material, manufacturing process, and distribution is considered. Also, 0/1 integer programming is applied to identify the optimal supplier in any layers.

5.1 Introduction

The retailing and wholesaling industries were the first to hear the changing needs of their customers. Their customers wanted shorter delivery lead-times. Integrating the logistics and physical distribution functions solved this need. Manufacturers and service providers quickly followed this path with the integration of suppliers to reduce cost and improve quality and delivery time. They took this relationship with their suppliers one step further by building a partnership, which shares costs, risks and profits, while focusing on serving the customer. This new way of doing business has become what is known as supply chain management (SCM). This chain is traditionally characterized by a forward flow of materials and a backward flow of information. For years, researchers and practitioners have primarily investigated the various processes of the supply chain individually. Recently, however, there has been increasing attention placed on the performance, design and analysis of the supply chain as a whole. From a practical standpoint, the supply chain concept arose from a number of changes in the manufacturing environment, including the rising costs of manufacturing, the shrinking resources of manufacturing bases, shortened product life cycles, the leveling of the playing field within manufacturing, and the globalization of market economies. The current interest has sought to extend the traditional supply chain to include reverse logistics, to include product recovery for the purposes of recycling, re-manufacturing, and re-use. Within manufacturing research, the supply chain concept grew largely out of two-stage multi-echelon inventory models, and it is important to note that considerable progress has been made in the design and analysis of two-echelon systems. Most of the research in this area is based on the classic work of Clark and Scarf (1960) and Clark and Scarf (1962). The interested reader is referred to Federgruen (1993) and Bhatnagar et al. (1993) for comprehensive reviews of models of this type. More recent discussions of two-echelon models may be found in Diks et al. (1996) and van Houtum et al. (1996).

In the case of outsourcing engagement, achieving business objectives requires effective use of integrated supply management network. These include information integration, knowledge integration and design integration. It is obvious that in these cases, there is risk of knowledge leak, as the companies cannot have complete domination over knowledge inflows and outflows. These supply management risks must be carefully identified and evaluated. Some of these issues have been considered in the literature (Christopher and Peck, 2004; Gaudenzi and Borghesi, 2006; Narasimhan et al., 2008).

Much of the literature on supply chain risk has dealt with various types of risk and sources of such risk (Norrman and Lindroth, 2004; Speckman and Davis, 2004). They have not taken a multidimensional view of risk that encompasses supply management processes, objectives and risk source. A large part of the literature has focused on identifying sources of uncertainty and the risks that emanate from them. To do this effectively, it is necessary to develop a consistent methodology for risk identification. Several authors have addressed this issue (see, e.g., Chopra and Sodhi, 2004; Wu and Knott, 2006). Risk identification is succeeded by quantification of risks that can be used in deriving risk mitigation strategies (Cachon, 2004; Sodhi, 2005). According to Wu et al. (2007) locating items of a supply chain is considered as a key location if the interruption of its activities results in a major disruption in the flow of goods and services. Examples can include a sole sourced supplier, a major distribution center and production plants. The authors propose utilizing some of the approaches developed in extant research such as Disruption Analysis Network methodology (Wu et al., 2007), which assists in identifying how the effects of disruptions propagate throughout a supply chain, and supply chain mapping analysis (Gardner and Cooper, 2003) for identifying key locations. After identifying key locations, the authors suggest applying approaches proposed by Mitroff and Alpaslan (2003), Chopra and Sodhi (2004) and Svensson (2004) for potential threat identification.

Here, we propose a qualitative mechanism based on quality control charts to evaluate the service and flexibility of suppliers in the layers of providing raw material, manufacturing process, and distribution. Also, 0/1 integer programming is applied to identify the optimal supplier in any layers.

5.2 Problem Definition

We consider a buyer who wants to evaluate suppliers and identifies the optimal collection of them. As stated in the introduction, a supply network is a group of activities including providing raw material, manufacturing process, and distribution. For simplification, we separate the three segments of the supply network in layers. The first layer is a group of raw material providers, second layer is a group of manufacturers, and third layer is a group of distributors. The proposed configuration is presented in Figure 5.1.

In each layer, we investigate the required data of service and flexibility and record the number of nonconformities. Then, the C-control chart is configured to determine the control area and also the outliers. In this stage each supplier that is out of the control limit is omitted from the evaluation. After that, for more precise investigations, we apply another criterion entitled process capability. For each of the suppliers the process capability is measured and the values are recorded. The same procedure is exerted for other layers. Now, we have the six sigma-based suppliers in each layer. But the problem is how to choose a set of suppliers that have both capable service and flexibility, simultaneously. Here, we apply an integrated 0/1 integer programming and calculus normalization process to gain the collection of optimal suppliers. The mechanism of identifying the set of optimal suppliers is presented in Figure 5.2.

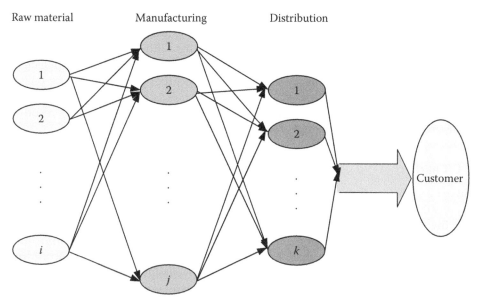

FIGURE 5.1
The proposed configuration of supply network segments.

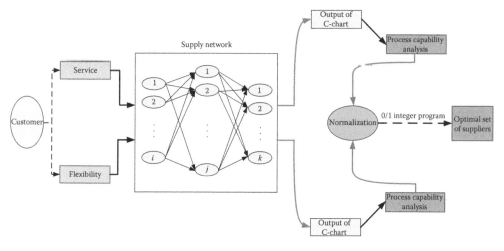

FIGURE 5.2
The mechanism of identifying the set of optimal suppliers.

5.3 Process Control Model

5.3.1 Quality Control

Statistical process control (SPC) is an effective method of monitoring a process through the use of control charts. Control charts enable the use of objective criteria for distinguishing background variation from events of significance based on statistical techniques. Much of its power lies in the ability to monitor both process center and its variation about that center. By collecting data from samples at various points within the process, variations in the process that may affect the quality of the end product or service can be detected and corrected, thus reducing waste as well as the likelihood that problems will be passed on to the customer. With its emphasis on early detection and prevention of problems, SPC has a distinct advantage over quality methods, such as inspection, that apply resources to detecting and correcting problems in the end product or service.

5.3.2 C-Chart

In industrial statistics, the c-chart is a type of control chart used to monitor "count"-type data, typically total number of nonconformities per unit. It is also occasionally used to monitor the total number of events occurring in a given unit of time. The c-chart differs from the p-chart in that it accounts for the possibility of more than one nonconformity per inspection unit. The p-chart models "pass"/"fail"-type inspection only. Nonconformities may also be tracked by type or location, which can prove helpful in tracking down assignable causes. Examples of processes suitable for monitoring with a c-chart include:

- Monitoring the number of voids per inspection unit in injection molding or casting processes
- Monitoring the number of discrete components that must be re-soldered per printed circuit board
- Monitoring the number of product returns per day

The Poisson distribution is the basis for the chart and requires the following assumptions:

- The number of opportunities or potential locations for nonconformities is very large
- The probability of nonconformity at any location is small and constant
- The inspection procedure is same for each sample and is carried out consistently from sample to sample

The control limits for this chart type are as follows:

$$UCL = \bar{c} + 3\sqrt{\bar{c}}$$

$$CL = \bar{c}$$

$$LCL = \bar{c} - 3\sqrt{\bar{c}}$$

where $\bar{c}$ is the estimate of the long-term process mean established during control-chart setup.

5.3.3 Process Capability Index

Process capability indices (PCIs) are summary statistics that measure the actual or the potential performance of the process characteristics relative to the target and specification limits. A process capability index (PCI) is a number that summarizes the behavior of a product or process characteristic relative to specifications.

Generally, this comparison is made by forming the ratio of the width between the process specification limits to the width of the natural tolerance limits. These indices help us to decide how well the process meets the specification limits (Montgomery, 2005). Several PCIs such as C_p, C_{pk}, and C_{pm} are used to estimate the capability of a process (Kotz and Johnson, 2002). C_p is defined as the ratio of specification width over the process spread. The specification width represents customer and/or product requirements. The process variations are represented by the specification width. If the process variation is very large, the C_p value is small and it represents a low process capability.

Abbasi (2009) proposed an artificial neural network that used skewness, kurtosis, and upper specification limit as input variables to estimate PCI for right skewed distributions without appeal to probability density function of the process. He also presented a simulation methodology using the actual data from a manufacturing process for different non-normal distributions to validate the proposed methodology in a case study. Pearn et al. (2009) considered the supplier selection problem for two one-sided processes, and presented an exact analytical approach based on the hypothesis test of PCIs to solve the problem. Critical values of the tests were calculated to determine the best decision.

They also investigated the sample size required for a designated selection power and confidence level. The one supplier that has a significantly higher capability value was selected as the best decision. Castagliola and Castellanos (2008) proposed a new approach based on the use of Johnson's system of distributions in order to transform the bivariate non-normal distribution into an approximate bivariate normal distribution for the estimation of bivariate PCIs in the case of a bivariate normal distribution. They also extended the proposed approach to the estimation of bivariate PCIs in the case of non-normal bivariate distributions.

Process capability compares the output of an *in-control* process to the specification limits using *capability indices*. The comparison is made by forming the ratio of the spread between the process specifications (the specification "width") to the spread of the process values, as measured by 6 process standard deviation units (the process "width").

A process capability index uses both the process variability and the process specifications to determine whether the process is "capable." We are often required to compare the output of a stable process with the process specifications and make a statement about how well the process meets specification. To do this we compare the natural variability of a stable process with the process specification limits. A capable process is one where almost all the measurements fall inside the specification limits. There are several statistics that can be used to measure the capability of a process: C_p. Most capability indices estimates are valid only if the sample size used is large enough. Large enough is generally thought to be about 50 independent data values. The C_p statistics assume that the population of data values is normally distributed. Assuming a two-sided specification, if μ and σ are the mean and standard deviation, respectively, of the normal data and USL, and LSL are the upper and lower specification limits, respectively, then the population capability index is defined as follows:

$$C_p = \frac{|USL - LSL|}{6\sigma}$$

Note that $||$ indicates the absolute value certifying the non-negativity of the C_p, and

- $C_p < 1$ means the process variation exceeds specification, and a significant number of defects are being made.

- $C_p = 1$ means that the process is just meeting specifications. A minimum of 0.3% defects will be made and more if the process is not centered.
- $C_p > 1$ means that the process variation is less than the specification, however, defects might be made if the process is not centered on the target value.

For the mathematical program, the following notations are given:

Indices

i	Number of raw material suppliers;	$i = 1, 2, ..., I.$
j	Number of manufacturing suppliers;	$j = 1, 2, ..., J.$
k	Number of distributors;	$k = 1, 2, ..., K.$
l	Number of layers;	$l = 1, 2, 3.$

Parameters

C_{ps}^M Process capability of supplier Mth for service ($M = 1, ..., i$ (in layer 1); 1, ..., j (in layer 2); 1, ..., k (in layer 3).

C_{pf}^M Process capability of supplier Mth for flexibility ($M = 1, ..., i$ (in layer 1); 1, ..., j (in layer 2); 1, ..., k (in layer 3).

To gain a better standard in supplier selection, the following limits are considered:

$$C_{ps}^M \geq 1.33,$$
$$C_{pf}^M \geq 1.33.$$

Decision Variable

$$X_{lM} = \begin{cases} 1 & \text{if } M\text{th supplier in } l\text{th layer is chosen} \\ 0 & \text{o.w} \end{cases}$$

5.3.4 Normalization Process

Service and flexibility can not be integrated in the present form, because they are two different kinds of variables with different units. Hence, we apply a normalization process to turn them into a unique unit as follows:

$$C_{ps}^{NM} = \frac{C_{ps}^M}{\sqrt{\sum_M \left(C_{ps}^M\right)^2}}, \tag{5.1}$$

$$C_{pf}^{NM} = \frac{C_{pf}^M}{\sqrt{\sum_M \left(C_{pf}^M\right)^2}}, \tag{5.2}$$

where C_{ps}^{NM} and C_{pf}^{NM} are normalized values for the supplier Mth's process capability of service and flexibility, respectively. Therefore, the normalized process capability of each supplier is gained using the following formulae:

$$C_p^{TM} = C_{ps}^{NM} + C_{pf}^{NM},$$ (5.3)

where C_p^{TM} is total process capability for each supplier Mth.
Here, we present the mathematical model.

$$\text{Max} \sum_M \sum_{l=1}^{3} C_p^{TM} \times X_{lM}.$$ (5.4)

s.t.

$$\sum_{M=1}^{i,j,k} X_{lM} = 1, \quad \forall l,$$ (5.5)

$$X_{lM} \in \{0,1\}.$$ (5.6)

The aim of the objective function is to maximize the process capability of the suppliers which will be selected. Note that only one supplier in each layer should be selected. The output of the model will be a set of suppliers in different levels. The related numerical example for this chapter is illustrated in Chapter 6 due to the more comprehensive model developed in that chapter.

5.4 Discussions

The proposed model of this chapter evaluates the suppliers in a supplier network considering the quality of their services and flexibility. The suppliers are divided into three layers. In each layer the various suppliers are analyzed considering the process capability, which is six sigma criterion. Regarding to the different nature of service and flexibility, a mathematical normalization has been applied to gain the total process capability for each supplier. To identify a set of optimal supplier considering process capability aspect, 0/1 integer programming has been applied. The efficiency and validation of the proposed mechanism is tested via an illustrative example. The applications of such a mechanism are in the supply networks where several criteria are significant in determining decision variables and for designing robust and under control supply networks. Managers and strategic planners of supply networks can also make use of the proposed mechanism in order to justify the optimal decisions. As future study, it is aimed to consider other control charts and to include other real criteria of supply network in the model.

References

Abbasi A. A neural network applied to estimate process capability of nonnormal processes. *Expert Systems with Applications* 2009;36:3093–3100.

Bhatnagar R, Pankaj C, Goyal SK. Models for multi-plant coordination. *European Journal of Operational Research* 1993;67:141–160.

Cachon G. The allocation of inventory risk in a supply chain: Push, pull, and advance-purchase discount contracts. *Management Science* 2004;50(2):222–238.

Castagliola P, Castellanos JVG. Process capability indices dedicated to bivariate non-normal distributions. *Journal of Quality in Maintenance Engineering* 2008;14(1):87–101.

Chopra S, Sodhi MS. Managing risk to avoid supply chain breakdown. *MIT Sloan Management Review* 2004;46(1):53–61.

Christopher M, Peck H. Building the resilient supply chain. *The International Journal of Logistics Management* 2004;15(2):1–14.

Clark AJ, Scarf H. Optimal policies for a multi-echelon inventory problem. *Management Science* 1960;6(4): 475–490.

Clark AJ, Scarf H. Approximate solutions to a simple multi-echelon inventory problem. In: Arros KJ, Karlin S, Scarf H. (eds.), *Studies in Applied Probability and Management Science*, Stanford, CA: Stanford University Press, 1962:88–110.

Diks EB, de Kok AG, Lagodimos AG. Multi-echelon systems: A service measure perspective. *European Journal of Operational Research* 1996;95:241–263.

Federgruen A. Centralized planning models for multi-echelon inventory systems under uncertainty. In: Graves S, Rinnooy Kan A, Zipkin P (eds.), *Logistics of Production and Inventory*, Amsterdam: North Holland 1993:133–173.

Gardner JT, Cooper MC. Strategic supply chain mapping approaches. *Journal of Business Logistics* 2003;24(2):37–64.

Gaudenzi B, Borghesi A. Managing risks in the supply chain using the AHP method. *International Journal of Logistics Management* 2006;17(1):114–136.

Kotz S, Johnson N. Process capability indices—A review 1992–2000. *Journal of Quality Technology* 2002;34:2–19.

Mitroff II, Alpaslan MC. Preparing for evil. *Harvard Business Review* 2003;81(4):109–115.

Montgomery DC. *Introduction to Statistical Quality Control*. New York, NY: John Wiley and Sons, 2005.

Narasimhan R, Narayanan S, Srinivasan R. Role of integrative supply management practices in strategic outsourcing. *Working Paper, Broad Graduate School of Management* 2008.

Norrman A, Lindroth R. Categorization of supply chain risk and risk management. In: *Supply Chain Risk*. Brindley C. (ed.), Hampshire, UK: Ashgate Publishing, 2004:14–27.

Pearn WL, Hung HN, Cheng YC. Supplier selection for one-sided processes with unequal sample sizes. *European Journal of Operational Research* 2009;195:381–393.

Sodhi M. Managing demand risk in tactical supply chain planning for a global consumer electronics company. *Production and Operations Management* 2005;14(1):69–79.

Speckman RE, Davis EW. Risky business: Expanding the discussion on risk and the extended enterprise. *International Journal of Physical Distribution and Logistics Management* 2004;34(5):414–433.

Svensson G. Key areas, causes and contingency planning of corporate vulnerability in supply chains. *International Journal of Physical Distribution and Logistics Management* 2004;34(9):728–748.

van Houtum GJ, Inderfurth K, Zijm WHM. Materials coordination in stochastic multi-echelon systems. *European Journal of Operational Research* 1996;95:1–23.

Wu B, Knott A. Entrepreneurial risk and market entry. *Management Science* 2006;52(9):1315.

Wu T, Blackhurst J, O'Grady P. Methodology for supply chain disruption analysis. *International Journal of Production Research* 2007;45(7):1665–1682.

6

Supplier Selection and Order Allocation: Process Performance Index

SUMMARY   In this chapter, a multi-objective mathematical programming approach is proposed to select the most appropriate supply network elements. The process performance index (PPI) is used as an assessment tool for the supply network elements (i.e., supplier, distributor, retailer, and customer), and the AHP is used to integrate the objectives of the proposed mathematical program into a single one.

6.1 Introduction

Supplier selection has a critical effect on the competitiveness of the entire supply chain network. Lewis (1943, p. 249) suggested "it is probable that of all the responsibilities which may be said to belong to the purchasing officers, there is none more important than the selection of a proper source. Indeed, it is in some respects the most important single factor in purchasing." England and Leenders (1975) made the same suggestion by stating "supplier selection is purchaser's most important responsibility."

There is a plethora of research on the supplier selection process. Weber et al. (1991) grouped the quantitative methods for supplier selection into three categories: linear weighting models, mathematical programming models, and statistical models. In linear weighting models, a weight is assigned to each criterion and a total score for each supplier is determined by summing up its performance on the criteria multiplied by these weights. Analytic hierarchy process (AHP) (Wang and Yang, 2009; Chamodrakas et al., 2010; Labib, 2011; Wang et al., 2010) and interpretive structural modeling (Mandal and Deshmukh, 1994; Kannan et al., 2009) are among the most widely used linear weighting models in supplier selection. In mathematical programming models, several suppliers are selected in order to maximize an objective function subject to supplier/buyer constraints. The objective function could be a single criterion or multiple criteria. Mathematical programming models used in supplier section include linear programming, mixed integer programming and goal programming (Ghodsypour and O'Brien, 1998; Cakravastia et al., 2002; Oliveria and Lourenço, 2002; Dahel, 2003; Yan et al., 2003; Razmi and Rafiei, 2010). Statistical approaches include methods such as cluster analysis and stochastic economic order quantity model (Mummalaneni et al., 1996; Verma and Pullman, 1998; Tracey and Tan, 2001). The unconventional supplier selection methods include cost-based models such as activity based cost approach (Roodhooft and Konings, 1997), total cost of ownership (Degraeve et al., 2000), and transaction cost theory (Qu and Brocklehurst, 2003); neural networks (Wray et al., 1994; Albino et al., 1998; Choy et al., 2002; Lee and Yang, 2009); and fuzzy sets (Amid et al., 2006; Bevilacqua et al., 2006; Chen et al., 2006; Florez-Lopez, 2007; Büyüközkan and Çifçi, 2011).

The use of hybrid methods for supplier selection is not new. Wang et al. (2005) proposed a methodology derived from AHP and pre-emptive goal programming. Haq and Kannan (2006) developed an integrated supplier selection and multi-echelon distribution inventory framework by combining fuzzy AHP and genetic algorithm. Ramanathan (2006) proposed a data envelopment analysis (DEA) model to generate local weights of alternatives from pair-wise comparison judgment matrices used in the AHP. Sevkli et al. (2007) applied the DEA methodology developed by Ramanathan (2006) into an integrated DEA-AHP framework to select suppliers in a well-known Turkish company operating in the appliance industry. Shin-Chan Ting and Cho (2008) used AHP, in consideration of both quantitative and qualitative criteria, to identify a set of candidate suppliers. A multi-objective linear programming model, with multiple objectives and a set of system constraints, was then formulated and solved to allocate the optimum order quantities to the candidate suppliers. Chan and Kumar (2007) and Chan et al. (2008) developed a fuzzy based AHP framework to efficiently tackle both quantitative and qualitative decision factors involved in the selection of global suppliers. They showed that fuzzy AHP is an efficient tool for handling the fuzziness of the data involved in the global supplier selection process.

Sevkli et al. (2008) proposed a hybrid method of AHP and fuzzy linear programming for supplier selection. The weights of the various criteria, taken as local weights from a given judgment matrix, were calculated using AHP. The criteria weights were then considered as the weights of the fuzzy linear programming model. Tsai and Hung (2009) proposed a fuzzy goal programming approach that integrated activity-based costing and performance evaluation in a value chain for optimal supplier selection and flow allocation. Sen et al. (2010) proposed a methodology that utilized a fuzzy AHP method to determine the weights of the pre-selected decision criteria, a max-min approach to maximize and minimize the supplier performances against these weighted criteria, and a non-parametric statistical test to identify an effective supplier set. Liao and Kao (2010) integrated the Taguchi loss function, AHP, and multi-choice goal programming to solve the supplier selection problem. The advantage of their proposed method was that it allowed decision makers (DMs) to set multiple aspiration levels for the decision criteria. Kuo et al. (2010) developed an integrated fuzzy AHP and fuzzy DEA for assisting organizations in supplier selection decisions. Fuzzy AHP was first applied to find the indicators' weights through an expert questionnaire survey. Then, these weights were integrated with fuzzy DEA. They used α-cut set and extension principle of fuzzy set theory to simplify the fuzzy DEA as a pair of traditional DEA models. Finally, fuzzy ranking using maximizing and minimizing set method was utilized to rank the evaluation samples.

Amid et al. (2011) developed a weighted max-min fuzzy model to handle the vagueness of input data and different weights of criteria in supplier selection problems. They used AHP to determine the weights of criteria and the proposed model to find the appropriate order to each supplier. Büyüközkan and Çifçi (2011) developed a novel approach based on fuzzy analytic network process (ANP) within multi-person decision-making schema under incomplete preference relations. Nobar et al. (2011) developed a new conceptual supplier selection model to select preferred suppliers based on two layers or more. They solved their model with fuzzy ANP, which was redesigned using a matrix manipulation method. Mafakheri et al. (2011) proposed a two-stage multiple criteria dynamic programming approach for two of the most critical tasks in supply chain management, namely, supplier selection and order allocation. In the first stage, AHP was used to address multiple decision criteria in supplier ranking. In the second stage, supplier ranks were fed into an order allocation model that aimed at maximizing a utility function for the firm and minimizing the total supply chain costs.

A supply chain may be defined as an integrated process wherein a number of various business entities (i.e., suppliers, manufacturers, distributors, and retailers) work together in an effort to: (1) acquire raw materials, (2) convert these raw materials into specified final products, and (3) deliver these final products to retailers. This chain is traditionally characterized by a forward flow of materials and a backward flow of information (Shi and Xiao, 2008; Xiao and Yan, 2011). For years, researchers and practitioners have primarily investigated the various processes of the supply chain individually. Recently, however, increasing attention has been placed on the performance, design, and analysis of the supply chain as a whole. From a practical standpoint, the supply chain concept arose from a number of changes in the manufacturing environment, including the rising costs of manufacturing, the shrinking resources of manufacturing bases, shortened product life cycles, the leveling of the playing field within manufacturing, and the globalization of market economies.

In spite of the extended research in supplier selection, Sevkli et al. (2008) have argued that "more research is definitely called for within the context of studying a more complex supply chain with multiple supply network and nodes. There is also a crucial need for investigating other hybrid methods to find the optimum supplier." In addition, most supplier selection models in the literature are intended to support DMs in the final selection phase and they have failed to consider a holistic view of the supplier selection process. Supplier evaluation and selection problems are inherently multi-criteria decision problems. Supply networks are now not only configured by suppliers, but also consist of manufacturers, retailers and customers. Therefore, a holistic and comprehensive approach for evaluating these elements in supply networks is required. We propose a multi-objective mathematical programming approach to select the most appropriate supply network elements. The process performance index (PPI) is used as an assessment tool for the supply network elements (i.e., supplier, distributor, retailer, and customer) and the AHP is used to integrate the objectives of the proposed mathematical program into a single one.

6.2 Problem Definition

As depicted in Figure 6.1, a supply network can be grouped into four distinctive but inter-related layers. The first layer is a group of suppliers, the second layer is a group of distributors, the third layer is a group of retailers, and the fourth layer is a group of customers.

We select a measure in each layer to represent the number of nonconformities in that level. The following measures are used in this study for the suppliers, distributors, retailers and customers layers:

Suppliers—the cost of transportation,

Distributors—the number of distribution services per month,

Retailers—the number of sale services per month, and

Customers—the number of purchasing services per month.

Then, the X-bar/S chart is configured to determine the control area and also the outliers. In this stage, each element of our network that is out of the control limit is omitted from the evaluation. After that, we apply another criterion entitled PPI for more precise

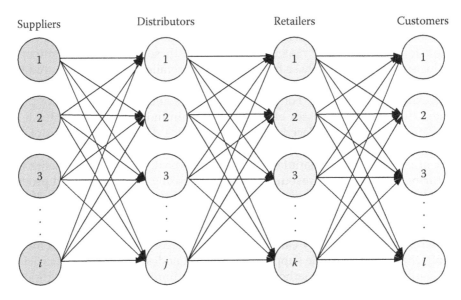

FIGURE 6.1
The proposed configuration of the supply network.

investigations. For each of the suppliers, the PPI is measured and the values are recorded. The same procedure is exerted for other layers. Now, we have the P_{pk}-based suppliers in each layer. But the problem is how to choose a set of supply network elements that are more appropriate to include in the development plan. Here, we apply a mathematical programming to gain the collection of optimal suppliers.

6.3 Integrated Performance Evaluation Model

6.3.1 X-bar/S Chart

There are two different methods for designing the control charts: the statistical methods, which aim at achieving the best statistical performance (Lucas, 1982; Reynolds et al., 1990; Castagliola et al., 2008; Costa et al., 2009), and the economic methods, which attempt to minimize the SPC cost (Duncan, 1956; Zhang et al., 2008; Ho and Trindade 2009; Torng et al., 2009). The potential poor statistical performance is a major drawback of economic designs because the calculated cost savings may be misleading (Woodall 1986). Consequently, the majority of the control charts used in practice are designed with statistical methods. Among them, an X-bar/S chart is a specific type of control chart that depicts the variability of average characteristics of a process over time when variables are collected in sub-groups.

The X-bar/S charts are generally employed for plotting variability of sub-groups with sizes greater than 10 while the X-bar/R charts are used for plotting variability when sub-group sizes are less than 10. X-bar/S charts plot the process mean (the X-bar chart) and process standard deviation (the S chart) over time for variables *within* sub-groups. Both the X-bar and S chart must be seen together to interpret the stability of the process. The

S chart must be examined first as the control limits of the X-bar chart is determined by considering both the process spread and center. Process variation, which is a characteristic of the spread, must be in control to correctly interpret the X-bar chart. If data-points in the S chart are outside the control limits, then the limits on the X-bar chart may be inaccurate and may falsely indicate an out-of-control condition. As in other types of control charts, data-points outside of control limits in an Xbar-S chart indicate special cases. In this chapter, as suggested by Wu et al. (2011), we employ the statistical method to design the control charts, because statistical design is more realistic and used almost exclusively in today's SPC practice.

6.3.2 Process Performance Index

The PPI is an estimate of the process capability during its initial setup, *before* it has been brought into a state of statistical control. Formally, if the upper and lower specifications of the process are USL and LSL, the estimated mean of the process is $\hat{\mu}$, and the estimated variability of the process (expressed as a standard deviation) is $\hat{\sigma}$, then the PPI is defined as:

$$\hat{P}_{pk} = \min\left[\frac{USL - \hat{\mu}}{3 \times \hat{\sigma}}, \frac{\hat{\mu} - LSL}{3 \times \hat{\sigma}}\right], \tag{6.1}$$

$\hat{\sigma}$ is estimated using the sample standard deviation. P_{pk} may be negative if the process mean falls outside the specification limits (because the process is producing a large proportion of defective output). Some specifications may only be one sided (for example, strength). For specifications that only have a lower limit, $\hat{P}_{p,lower} = (\hat{\mu} - LSL/3 \times \hat{\sigma})$; for those that only have an upper limit, $\hat{P}_{p,upper} = (USL - \hat{\mu}/3 \times \hat{\sigma})$.

Practitioners may also encounter $\hat{P}_p = (USL - LSL/6 \times \hat{\sigma})$, a metric that does not account for process performance that is not exactly centered between the specification limits, and therefore is interpreted as what the process would be capable of achieving if it could be centered and stabilized.

Let us introduce the following mathematical notations and definitions:

i	Number of suppliers; $i = 1, 2, ..., I$
j	Number of distributors; $j = 1, 2, ..., J$
k	Number of retailers; $k = 1, 2, ..., K$
l	Number of customers; $l = 1, 2, ..., L$
n	Index of the item in layer; $n = 1, ..., N$
m	Number of layers; $m = 1, 2, 3, 4$
D_m	Demand of the product in layer m
O_{mn}	Order cost for item n in layer m
Q_{mn}	Expected defect rate of item n in layer m
V_{mn}	Capacity of item n in layer m
C_{mn}	Purchasing price of the product from supply network item n in layer m
W_{mn}	The overall score of item n in layer m obtained from the P_{pk} process
X_{mn}	Total number of the product ordered to item n in layer m
Y_{mn}	$\begin{cases} 1 & \text{if an order is placed on item } n \text{ in layer } m \\ 0 & \text{otherwise} \end{cases}$

After the weights of the supply network elements are clarified, the most appropriate one in each layer is chosen. Here, we propose the following mathematical model for supply network element selection. The weight of the items in layers as specified using P_{pk} criterion are used as significant factors for the selection.

6.3.3 Objective Functions

Minimizing total cost—the total sum of the material cost and order cost is to be minimized:

$$\min Z_1 = \sum_{n=1}^{N} \left(\sum_{m=1}^{M} C_{mn} X_{mn} + \sum_{m=1}^{M} O_{mn} Y_{mn} \right) \tag{6.2}$$

Minimizing total defect rate—as Q_{mn} is the expected defect rate of the nth item in mth layer, the total defect rate to be minimized is:

$$\min Z_2 = \sum_{n=1}^{N} \sum_{m=1}^{M} Q_{mn} X_{mn}. \tag{6.3}$$

Maximization of total value of purchase—as W_{mn} and X_{mn} denote the normal weights of the item and the numbers of purchased units of nth item in mth layer, respectively, the following objective function is designed to maximize the total value of purchase:

$$\max Z_3 = \sum_{n=1}^{N} \sum_{m=1}^{M} W_{mn} X_{mn} \tag{6.4}$$

As illustrated above, the problem is a multi-objective one. Since the total cost and the total defect rate are independent, we can mix the first two objective functions yielding the following single cost minimization objective:

$$\min Z = \sum_{n=1}^{N} \left(\sum_{m=1}^{M} (C_{mn} + Q_{mn}) X_{mn} + \sum_{m=1}^{M} O_{mn} Y_{mn} \right) \tag{6.5}$$

Next, a bi-objective problem is configured to consider the third objective for maximizing the total value of purchase. We apply the AHP to differentiate the significance of the objectives based on the DMs' preferences.

6.3.4 The AHP

AHP is a multi-attribute decision making approach that simplifies complex and ill-structured problems by arranging the decision attributes and alternatives in a hierarchical structure with the help of a series of pair-wise comparisons. Dyer and Forman (1992) describe the advantages of AHP in a group setting as follows: (1) the discussion focuses on both tangibles and intangibles, individual and shared values; (2) the discussion can be focused on objectives rather than alternatives; (3) the discussion can be structured so that every attribute can be considered in turn; and (4) the discussion continues until all relevant information has been considered and a consensus choice of the decision alternative is achieved.

Saaty (2000) argues that a DM naturally finds it easier to compare two things than to compare all things together in a list. AHP also examines the consistency of the DMs and allows for the revision of their responses (Awasthi et al., 2008). AHP has been applied to many diverse decisions because of the intuitive nature of the process and its power in resolving the complexity in a judgmental problem. A comprehensive list of the major applications of AHP, along with a description of the method and its axioms, can be found in Saaty (1994, 2000), Weiss and Rao (1987) and Zahedi (1986). AHP has proven to be a popular technique for determining weights in multi-attribute problems (Zahedi 1986). The importance of AHP and the use of pairwise comparisons in decision-making are best illustrated in the more than 1000 references cited in Saaty (2000).

The main advantage of AHP is its ability to rank alternatives in the order of their effectiveness in meeting conflicting objectives. AHP calculations are not complex, and if the judgments made about the relative importance of the attributes have been made in good faith, then AHP calculations lead inexorably to the logical consequence of those judgments. AHP has been a controversial technique in the operations research community. Harker and Vargas (1990) show that AHP does have an axiomatic foundation, the cardinal measurement of preferences is fully represented by the eigenvector method, and the principles of hierarchical composition and rank reversal are valid. On the other hand, Dyer (1990a,b) has questioned the theoretical basis underlying AHP and argues that it can lead to preference reversals based on the alternative set being analyzed. In response, Saaty (1990) contends that rank reversal is a positive feature, when new reference points are introduced. AHP is based on three principles: (1) constructing the hierarchy, (2) priority setting, and (3) logical consistency.

6.3.5 Construction of the Hierarchy

A complex decision problem composed of multiple attributes is structured and decomposed into sub-problems (sub-objectives, criteria, alternatives, etc.), within the hierarchy.

6.3.6 Priority Setting

The relative "priority" given to each element in the hierarchy is determined by pair-wise comparison of the contributions of elements at a lower level in terms of the criteria (or elements) with a causal relationship (Macharis et al., 2004). In AHP, multiple paired comparisons are based on a standardized comparison scale of nine levels (see Table 6.1 of Saaty, 1980).

Let $C = \{C_j \mid j = 1, 2, ..., n\}$ be the set of criteria. The result of the pair-wise comparison on n criteria can be summarized in an $n \times n$ evaluation matrix A in which every element a_{ij} is the quotient of weights of the criteria, as shown in Equation 6.6 below:

$$A = (a_{ij}), \quad i, j = 1, ..., n.$$

(6.6)

The relative priorities are given by the right eigenvector (w) corresponding to the largest eigenvector ($\lambda_{\max}$) as:

$$Aw = \lambda_{\max} w$$

(6.7)

when the pair-wise comparisons are completely consistent, the matrix A has rank 1 and $\lambda_{max} = n$. In that case, weights can be obtained by normalizing any of the rows or columns of A. The procedure described above is repeated for all levels of the hierarchy. In order to synthesize the various priority vectors, these vectors are weighted with regards to the global priority of the parent criteria. This process starts at the top of the hierarchy. As a result, the overall relative priorities are obtained for the elements in the lowest level of the hierarchy. These overall relative priorities indicate the degree to which the alternatives contribute to the overall goal. These priorities represent a synthesis of the local priorities, and reflect an evaluation process that permits integration of the perspectives of the various stakeholders involved in the decision-making process (Macharis et al., 2004).

6.3.7 Logical Consistency

A measure of consistency of the given pair-wise comparison is needed. The consistency is defined by the relation between the entries of A; that is, we say A is consistent if $a_{ij} \cdot a_{jk} = a_{ik}$, for each i, j, k. The consistency index (CI) is

$$CI = \frac{(\lambda_{max} - n)}{(n - 1)}$$

(6.8)

The final consistency ratio (CR), calculated as the ratio of the CI and the random consistency index (RI), as indicated in Equation 6.9 below, can reveal the consistency and inconsistency of the pair-wise comparisons:

$$CR = \frac{CI}{RI}$$

(6.9)

The value *0.1* is the accepted upper limit for CR. If the final consistency ratio exceeds this value, the evaluation procedure needs to be repeated to improve consistency.

Notations and Definitions

n number of criteria
i number of items
p index for the items, $p = 1, ..., P$
b index for the sub-criteria, $b = 1, ..., B$
d index for the criteria, $d = 1, ..., D$
W'_{pb} the weight of the p-th item with respect to the b-th sub-criterion
W_{bd} the weight of the b-th sub-criterion with respect to the d-th criterion
R_{pd} the weight of the p-th item with respect to the d-th criterion
w_d the weight of the d-th criterion

6.3.8 Identify the Relationships and the Weights of Criteria with AHP

Step 1: Define the decision problem and goal.

Step 2: Structure the hierarchy from the top through the intermediate to the lowest level.

Step 3: Construct the supply network item-criteria matrix using steps 4 to 8 using the AHP.

Steps 4 to 6 are performed for all levels in the hierarchy.

Step 4: Construct pair-wise comparison matrices for each of the lower levels with one matrix for each element in the level immediately above by using a relative scale measurement. The DM has the option of expressing his or her intensity of preference on a nine-point scale. If two criteria are of equal importance, a value of 1, signifying equal importance, is given to a comparison, while a 9 indicates an absolute importance of one criterion over the other. Table 6.1 shows the measurement scale defined by Saaty (1980).

Step 5: In this step we compute the largest eigenvalue by the relative weights of the criteria and the sum taken over all weighted eigenvector entries corresponding to those in the next lower level of the hierarchy. We then analyze pair-wise comparison data using the eigenvalue technique. Using these pair-wise comparisons, estimate the parameters. The eigenvector of the largest eigenvalue of matrix A constitutes the estimation of relative importance of the attributes.

Step 6: In this step we evaluate the consistency of the judgments and perform consequence weights analysis as follows:

$$
A = (a_{ij}) =
\begin{bmatrix}
1 & w_1/w_2 & \cdots & w_1/w_n \\
w_2/w_1 & 1 & \cdots & w_2/w_n \\
\vdots & \vdots & \ddots & \\
w_n/w_1 & w_n/w_2 & & 1
\end{bmatrix}
$$

Note that if the matrix A is consistent (that is, $a_{ij} = a_{ik}\, a_{kj}$ for all $i, j, k = 1, 2, \ldots, n$), then A contains no error (the weights are already known) and we have,

$$
a_{ij} = \frac{w_i}{w_j}, \quad i, j = 1, 2, \ldots, n.
$$

If the pair-wise comparisons do not include any inconsistencies, then $\lambda_{max} = n$. The more consistent the comparisons are, the closer the value of computed λ_{max} is to n. A consistency index (CI), which measures the inconsistencies of pair-wise comparisons, is set to be:

$$
CI = \frac{(\lambda_{max} - n)}{(n-1)},
$$

and a consistency ratio (CR) is set to be:

$$
CR = 100\left(\frac{CI}{RI}\right),
$$

where n is the number of columns in A and RI is the random index, being the average of the CI obtained from a large number of randomly generated matrices. Note that RI depends on the order of the matrix, and a CR value of 10% or less is considered acceptable (Saaty, 1980).

Step 7: In this step we configure the item-sub-criteria and the sub-criteria-criteria matrices based on the preferences of the DM. Table 6.1 presents the relative importance scale used in AHP. The first column is the importance score and the second column is the verbal phrase used for making the pair-wise comparisons among the item-sub-criteria in Table 6.2 and the sub-criteria-criteria in Table 6.3.

Step 8: Next, we form the network item-criteria matrix as presented in Table 6.4, where

$$R_{pd} = \sum_{b=1}^{B} W'_{pb} \times W_{bd}, \quad \forall \quad p = 1, 2, \ldots, P, \quad d = 1, 2, \ldots, D$$

TABLE 6.1

The AHP Relative Importance Scale

Importance Score	Verbal Expression
1	Equal
2	Weak
3	Moderate
4	Moderate plus
5	Strong
6	Strong plus
7	Very strong or demonstrated
8	Very, very strong
9	Extreme

TABLE 6.2

The Network Item-Sub-Criteria Matrix

	SC_1	SC_2	...	SC_b
Item 1	$W'_{1,1}$	$W'_{1,2}$	...	$W'_{1,b}$
Item 2	$W'_{2,1}$	$W'_{2,2}$	...	$W'_{2,b}$
⋮	⋮	⋮	⋮	⋮
Item p	$W'_{p,1}$	$W'_{p,2}$	...	$W'_{p,b}$

TABLE 6.3

The Sub-Criteria-Criteria Matrix

	C_1	C_2	...	C_d
SC_1	$W_{1,1}$	$W_{1,2}$	...	$W_{1,d}$
SC_2	$W_{2,1}$	$W_{2,2}$	...	$W_{2,d}$
⋮	⋮	⋮	⋮	⋮
SC_b	$W_{b,1}$	$W_{b,2}$	...	$W_{b,d}$

TABLE 6.4

The Supply Network Item-Criteria Matrix

	C_1	C_2	...	C_d
Item 1	$R_{1,1}$	$R_{1,2}$	...	$R_{1,d}$
Item 2	$R_{2,1}$	$R_{2,2}$	...	$R_{2,d}$
⋮	⋮	⋮	⋮	⋮
Item p	$R_{p,1}$	$R_{p,2}$	...	$R_{p,d}$

TABLE 6.5

The Criteria-Criteria Pair-Wise Comparison Matrix

	C_1	C_2	...	C_d	w_d
C_1	1	$a_{1,2}$	...	$a_{1,d}$	w_1
C_2	$1/a_{1,2}$	1	...	$a_{2,d}$	w_2
⋮	⋮	⋮	⋮	⋮	⋮
C_d	$1/a_{1,d}$	$1/a_{2,d}$	...	1	w_d

Step 9: We then perform a pair-wise comparison among the criteria and configure the pair-wise comparison for criteria-criteria matrix presented in Table 6.5.

A normalization process is utilized to compute the w_ds presented in this table. Hence the w_ds are the criteria weights computed through the AHP.

Step 10: In this step we calculate the overall weights of the objective functions using Tables 6.3 and 6.4, as follows:

$$\text{Total weight for function } 1 = R_{11} \times w_1 + R_{12} \times w_2 + \ldots + R_{1d} \times w_d$$

$$\text{Total weight for function } p = R_{p1} \times w_1 + R_{p2} \times w_2 + \ldots + R_{pd} \times w_d$$

6.3.9 From Bi-Objective to Mono-Objective

The criteria considered here are

- Policy making
- Capability to control
- Strategic management

The process is similar to the one described above. As a result, assuming that ψ is the weight for the cost minimization and $\psi' = 1 - \psi$ is the weight for maximization of total value of purchase, the following single objective function is formed:

$$\min Z = \psi \left[\sum_{n=1}^{N} \left(\sum_{m=1}^{M} (C_{mn} + Q_{mn}) X_{mn} + \sum_{m=1}^{M} O_{mn} Y_{mn} \right) \right] - \psi' \left[\sum_{n=1}^{N} \sum_{m=1}^{M} W_{mn} X_{mn} \right]. \quad (6.10)$$

Next, we identify the capacity, demand, non-negativity and binary constraints as follows:

Capacity constraints—the corresponding constraints are used since item n can provide up to V_{mn} units of the product and its order quantity in layer m (X_{mn}) should be equal to or less than its capacity:

$$X_{mn} \leq V_{mn} Y_{mn}, \quad m = 1, 2, \dots, M, \quad n = 1, 2, \dots, N. \tag{6.11}$$

Demand constraints—the following constraints are imposed since the sum of the assigned order quantities to layer m should meet the buyer's demand:

$$\sum_{m=1}^{M} X_{mn} \geq D_m, \quad n = 1, 2, \dots, N. \tag{6.12}$$

Non-negativity and binary constraints—the following are the non-negativity and binary constraints imposed in the model:

$$X_{mn} \geq 0, \quad m = 1, 2, \dots, M, \quad n = 1, 2, \dots, N. \tag{6.13}$$

$$Y_{mn} = 0 \quad \text{or} \quad 1, \quad m = 1, 2, \dots, M, \quad n = 1, 2, \dots, N. \tag{6.14}$$

The proposed model can be applied as a decision aid tool for suppliers in a supply network. The advantages of the proposed model include the multi-objective structure helping DMs achieve multiple and conflicting objectives simultaneously, and the capability to design a multi-layer model which is beneficial for the long term planning purposes.

6.4 Numerical Illustrations

In this section we demonstrate the efficacy and applicability of the proposed methodology with a numerical example. Consider an example with three suppliers, four distributors, five retailers, and four customers. The parameter level of each of them is collected via 25 samples, separately. We applied MINITAB 14 package for simplicity in computations. Then, the X-bar/S chart is configured and analyzed whether it is under control or not. If the X-bar/S chart is not under control, we revise it and reconfigure the X-bar/S chart. The same procedure is continued to construct an under control X-bar/S chart. A process performance for the parameters of the layers is shown in Figure 6.2. The process performance indices for each supply network element in each layer are presented in Table 6.6.

TABLE 6.6

The PPI for Each Supply Network Element in Each Layer

Layer	Supplier			Distributor				Retailer					Customer			
Element	1	2	3	1	2	3	4	1	2	3	4	5	1	2	3	4
P_p	0.66	0.71	0.63	0.8	0.7	0.72	0.79	0.68	0.69	0.67	0.88	0.88	0.97	0.93	1.16	1.05
P_{pk}	0.57	0.58	0.41	0.7	0.62	0.69	0.71	0.65	0.65	0.62	0.88	0.7	0.83	0.89	1.11	0.92

TABLE 6.7

The Capacity of Each Supply Network Element (V_{mn})

		Item (m)				
		1	2	3	4	5
Layer (n)	1	63	72	89	0	0
	2	60	75	92	53	0
	3	62	71	90	51	64
	4	61	70	91	50	0

Figure 6.2 presents six charts concerning the supply network. The sample mean chart in the upper-left corner shows whether the mean for the observed samples is under control. In our example, the upper, lower, and center control limit are 5.924, 2.636, and 4.28, respectively. Also, all samples are within control limits. The sample standard deviation chart in the middle left illustrates the same results for the standard deviation of the observed samples. The last five subgroups of the observed samples are shown in the values chart in the lower-left corner. The capability histogram for the observed samples is shown in the upper-right corner. This chart certifies the normality of the samples distribution. The normal probabilities are plotted in the middle right chart. The P-value of this sample data set is very small showing that the distribution of the samples is normal. The capability plot and the process performance indices are all given in the last chart shown in the lower right corner.

Here, we need to determine the weights of the objective functions using AHP. Considering the stated criteria and applying the procedure described before, the following weights are computed.

$$\psi = 0.468, \quad \psi' = 0.532.$$

Some input data are required to configure the mathematical model. The capacity of each supply network element is presented in Table 6.7. The demand for each layer is indicated in Table 6.8. The order cost, purchasing price, and expected defect rate for each supplier in the corresponded time period are shown in Tables 6.9 and 6.10. The defect rate (Q) for all supply network element in all layers is equal to 0.2.

Now, we solve the model using the objective function (Equation 6.10) along with the constraints. We used LINGO 9 to facilitate the computations. The numerical results are summarized in Table 6.11. Note that the objective value is 207.8606.

As shown in Table 6.11, 60 units of item 3 must be ordered in layer 1, 50 units of item 1 must be ordered in layer 2. As for layers 3 and 4, we have a split order because 41 units of item 4 and 39 units of item 5 must be ordered in layer 3 and 70 units of item 2 and 20 units of item 3 must be ordered in layer 4.

TABLE 6.8

The Demand for Each Layer

Layer	D
1	60
2	80
3	50
4	90

TABLE 6.9

The Order Cost of Each Supply Network Item in Each
Layer (O_{mn})

		\multicolumn{5}{c}{Item (m)}				
		1	2	3	4	5
Layer (n)	1	5	4	4	0	0
	2	5	5	5	3	0
	3	5	5	5	4	2
	4	5	5	5	2	0

TABLE 6.10

The Purchasing Price of Each Supply Network Item in
Each Layer (C_{mn})

		\multicolumn{5}{c}{Item (m)}				
		1	2	3	4	5
Layer (n)	1	2	2	1	0	0
	2	1	4	2	5	0
	3	3	6	3	2	3
	4	5	3	4	4	0

TABLE 6.11

The Numerical Values for Decision Variables X_{mn} and Y_{mn}

X_{mn}		\multicolumn{5}{c}{Item (m)}				
		1	2	3	4	5
Layer (n)	1	0	0	60	0	0
	2	50	0	0	0	0
	3	0	0	0	41	39
	4	0	70	20	0	0

Y_{mn}		\multicolumn{5}{c}{Item (m)}				
		1	2	3	4	5
Layer (n)	1	0	0	1	0	0
	2	1	0	0	0	0
	3	0	0	0	1	1
	4	0	1	1	0	0

6.5 Discussions

Supplier selection process is one of the key operational tasks for sustainable supply chain
management. In spite of the extended research in supplier selection, researchers have
called for more research on: (1) complex supply chains with multiple supply network and
nodes; (2) hybrid supplier selection methods; and (3) holistic supplier selection processes.
We showed that supplier evaluation and selection problems are multi-criteria decision
problems and supply networks are not only configured by suppliers, but also consist of

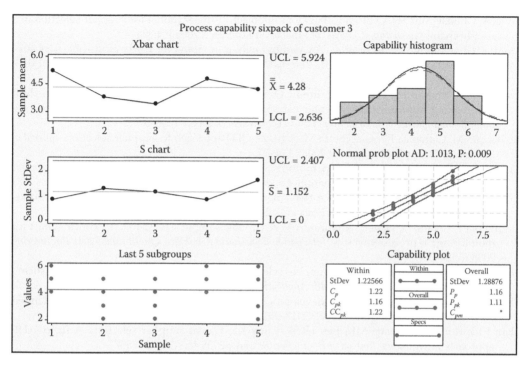

FIGURE 6.2
A process performance analyzer.

manufacturers, retailers and customers. We proposed a holistic and comprehensive multi-objective mathematical programming approach for evaluating the supply network elements. We presented the details of the proposed supplier selection approach and used PPI as an assessment tool for the supply network elements and the AHP to combine the objectives of the proposed mathematical program into a single one.

It is obvious that strategic sourcing and supplier selection topics have been mainly targeted and applied to manufacturing firms producing and marketing products. We suggest that researchers explore the application of strategic sourcing and supplier selection in non-manufacturing settings to determine its applicability and overall value. In addition, little research has been done on measuring purchasing performance. It would be beneficial to explore how purchasing views itself, how it is viewed by top management, how it is viewed by internal (i.e., other functional areas within the firm) and external stakeholders (i.e., suppliers to the firm). Last, as more and more companies around the world participate in global sourcing activities, international studies should be conducted to examine supplier selection strategies and practices across different countries and cultures.

References

Albino V, Garavelli C, Gorgoglione M. Fuzzy logic in vendor rating: A comparison between a fuzzy logic system and a neural network. *Fuzzy Economic Review* 1998;3:25–47.

Amid A, Ghodsypour SH, O'Brien C. Fuzzy multiobjective linear model for supplier selection in a supply chain. *International Journal of Production Economics* 2006;104:394–407.

Amid A, Ghodsypour SH, O'Brien C. A weighted max-min model for fuzzy multi-objective supplier selection in a supply chain. *International Journal of Production Economics* 2011;131(1): 139–145.

Awasthi A, Chauhan SS, Hurteau X, Breuil D. An Analytical Hierarchical Process-based decision-making approach for selecting car-sharing stations in medium size agglomerations. *International Journal of Information and Decision Sciences* 2008;1(1):66–97.

Bevilacqua M, Ciarapica FE, Giacchetta M. A fuzzy-QFD approach to supplier selection. *Journal of Purchasing & Supply Management* 2006;12:14–27.

Büyüközkan G, Çifçi G. A novel fuzzy multi-criteria decision framework for sustainable supplier selection with incomplete information. *Computers in Industry* 2011;62(2):164–174.

Cakravastia A, Toha IS, Nakamura N. A two-stage model for the design of supply chain networks. *International Journal of Production Economics* 2002;80(3):231–248.

Castagliola P, Celano G, Fichera S, Nunnari V. A variable sample size S2-EWMA control chart for monitoring the process variance. *International Journal of Reliability, Quality and Safety Engineering* 2008;15:181–201.

Chamodrakas I, Batis D, Martakos D. Supplier selection in electronic marketplaces using satisficing and fuzzy AHP. *Expert Systems with Applications* 2010;37:490–498.

Chan FTS, Kumar N. Global supplier development considering risk factors using fuzzy extended AHP-based approach. *Omega* 2007;35:417–431.

Chan FTS, Kumar N, Tiwari MK, Lau HCW, Choy KL. Global supplier selection: A fuzzy-AHP approach. *International Journal of Production Research* 2008;46(14):3825–3857.

Chen CT, Lin CT, Huang SF. A fuzzy approach for supplier evaluation and selection in supply chain management. *International Journal of Production Economics* 2006;102:289–301.

Choy KL, Lee WB, Lo V. An intelligent supplier management tool for benchmarking suppliers in outsource manufacturing. *Expert Systems with Applications* 2002;22:213–224.

Costa AFB, de Magalhães MS, Epprecht EK. Monitoring the process mean and variance using a synthetic control chart with two-stage testing. *International Journal of Production Research* 2009;47:5067–5086.

Dahel NE. Vendor selection and order quantity allocation in volume discount environments. *Supply Chain Management: An International Journal* 2003;8(4):335–342.

Degraeve Z, Labro E, Roodhooft F. An evaluation of supplier selection methods from a total cost of ownership perspective. *European Journal of Operational Research* 2000;125(1):34–59.

Duncan AJ. The economic design of Xbar charts used to maintain current control of a process. *Journal of the American Statistical Association* 1956;51:228–242.

Dyer JS. Remarks on the analytic hierarchy process. *Management Science* 1990a;36(3):249–258.

Dyer JS. A clarification of 'Remarks on the analytic hierarchy process.' *Management Science* 1990b;36(3): 274–275.

Dyer RF, Forman EH. Group decision support with the analytic hierarchy process, *Decision Support Systems* 1992;8(2):99–124.

England WB, Leenders MR. *Purchasing and Materials Management*. Homewood, IL: Richard Irwin, 1975.

Florez-Lopez R. Strategic supplier selection in the added-value perspective: A CI approach. *Information Sciences* 2007;177:1169–1179.

Ghodsypour SH, O'Brien C. A decision support system for supplier selection using an integrated analytical hierarchy process and linear programming. *International Journal of Production Economics* 1998;56/57:199–212.

Haq AN, Kannan G. Design of an integrated supplier selection and multi-echelon distribution inventory model in a built-to-order supply chain environment. *International Journal of Production Research* 2006;44(10):1963–1985.

Harker PT, Vargas LG. Reply to 'Remarks on the analytic hierarchy process' by J.S. Dyer. *Management Science* 1990;36(3):269–273.

Ho LL, Trindade ALG. Economic design of an X chart for short-run production. *International Journal of Production Economics* 2009;120:613–624.

Kannan G, Pokharel S, Kumar PS. A hybrid approach using ISM and fuzzy TOPSIS for the selection of reverse logistics provider. *Resources, Conservation and Recycling* 2009;54(1):28–36.

Kuo RJ, Lee LY, Hu TL. Developing a supplier selection system through integrating fuzzy AHP and fuzzy DEA: A case study on an auto lighting system company in Taiwan. *Production Planning & Control* 2010;21(5): 468–484.

Labib AW. A supplier selection model: A comparison of fuzzy logic and the analytic hierarchy process. *International Journal of Production Research* 2011;49.

Lee CC, Ou-Yang C. A neural networks approach for forecasting the supplier's bid prices in supplier selection negotiation process. *Expert Systems with Applications* 2009;36(2):2961–2970.

Lewis H. 1943. *Industrial Purchasing Principles and Practices*. Homewood, IL: Richard Irwin, 1943.

Liao CN, Kao HP. Supplier selection model using Taguchi loss function, analytical hierarchy process and multi-choice goal programming. *Computers & Industrial Engineering* 2010;58(4):571–577.

Lucas JM. Combined Shewhart-CUSUM quality control schemes. *Journal of Quality Technology* 1982;14: 51–59.

Macharis C, Springael J, De Brucher K, Verbeke A. PROMETHEE and AHP: The design of operational synergies in multicriteria-analysis. Strengthening PROMETHEE with ideas of AHP. *European Journal of Operational Research* 2004;153(2):307–317.

Mafakheri F, Breton M, Ghoniem A. Supplier selection-order allocation: A two-stage multiple criteria dynamic programming approach. *International Journal of Production Economics* 2011;132(1):52–57.

Mandal A, Deshmukh SG. Vendor selection using interpretive structural modelling (ISM). *International Journal of Operations & Production Management* 1994;14(6):52–59.

Mummalaneni V, Dubas KM, Chao C. Chinese purchasing managers' preferences and trade-offs in supplier selection and performance evaluation. *Industrial Marketing Management* 1996;25:115–124.

Nobar MN, Setak M, Tafti AF. Selecting Suppliers Considering Features of 2nd Layer Suppliers by Utilizing FANP Procedure. *International Journal of Business and Management* 2011;6(2):265–275.

Oliveria RC, Lourenço JC. A multicriteria model for assigning new orders to service suppliers. *European Journal of Operational Research* 2002;139:390–399.

Qu Z, Brocklehurst M. What will it take for China to become a competitive force in offshore outsourcing? An analysis of the role of transaction costs in supplier selection. *Journal of Information Technology* 2003;18(1):53–67.

Ramanathan R. Data envelopment analysis for weight derivation and aggregation in the analytic hierarchy process. *Computers & Operations Research* 2006;33:1289–1307.

Razmi J, Rafiei H. An integrated analytic network process with mixed-integer non-linear programming to supplier selection and order allocation. *The International Journal of Advanced Manufacturing Technology* 2010;49(9–12):1195–1208.

Reynolds MR Jr, Amin RW, Arnold JC. CUSUM charts with variable sampling intervals. *Technometrics* 1990;32:371–384.

Roodhooft F, Konings J. Vendor selection and evaluation: An activity based costing Approach. *European Journal of Operational Research* 1997;96(1):97–102.

Saaty TL. *The Analytic, Hierarchy Process*. New York, NY: McGraw-Hill, 1980.

Saaty TL. An exposition of the AHP in reply to the paper 'Remarks on the analytic hierarchy process' by J.S. Dyer. *Management Science* 1990;36(3):259–268.

Saaty TL. Highlights and critical points in the theory and application of the analytic hierarchy process. *European Journal of Operations Research* 1994;74:426–447.

Saaty TL. *Fundamentals of decision making and priority theory with the AHP*, 2nd edition. Pittsburgh, PA: RWS Publications, 2000.

Sen CG, Sen S, Basligil H. Pre-selection of suppliers through an integrated fuzzy analytic hierarchy process and max-min methodology. *International Journal of Production Research* 2010;48(6):1603–1625.

Sevkli M, Koh SCL, Zaim S, Demirbag M, Tatoglu E. An application of data envelopment analytic hierarchy process for supplier selection: A case study of BEKO in Turkey. *International Journal of Production Research* 2007;45(9):1973–2003.

Sevkli M, Koh SCL, Zaim S, Demirbag M, Tatoglu E. Hybrid analytical hierarchy process model for supplier selection. *Industrial Management & Data Systems* 2008;108(1):122–142.

Shi K, Xiao T. Coordination of a supply chain with a loss-averse retailer under two types of contracts. *International Journal of Information and Decision Sciences* 2008;1(1):5–25.

Ting SC, Cho DI. An integrated approach for supplier selection and purchasing decisions. *Supply Chain Management* 2008;13(2):116–127.

Torng CC, Lee PH, Liao NK. An economic-statistical design of double sampling control chart. *International Journal of Production Economics* 2009;120:495–500.

Tracey M, Tan CL. Empirical analysis of supplier selection and involvement, customer satisfaction and firm performance. *Supply Chain Management: An International Journal* 2001;6(4):174–188.

Tsai WH, Hung SJ. A fuzzy goal programming approach for green supply chain optimisation under activity-based costing and performance evaluation with a value-chain structure. *International Journal of Production Research* 2009;47(18):4991–5017.

Verma R, Pullman ME. An analysis of the supplier selection process. *Omega* 1998;26(6):739–750.

Wang HS, Che ZH, Wu C. Using analytic hierarchy process and particle swarm optimization algorithm for evaluating product plans. *Expert Systems with Applications* 2010;37:1023–1034.

Wang G, Huang SH, Dismukes JP. Manufacturing supply chain design and evaluation. *International Journal of Advanced Manufacturing Technology* 2005;25(1/2):93–100.

Wang T-Y, Yang Y-H. A fuzzy model for supplier selection in quantity discount environments. *Expert Systems with Applications* 2009;36:12179–12187.

Weber CA, Current JR, Benton WC. Vendor selection criteria and methods. *European Journal of Operational Research* 1991;50:2–18.

Weiss EN, Rao VR. AHP design issues for large-scale systems. *Decision Sciences* 1987;18:43–61.

Woodall WH. Weaknesses of the economic design of control charts. *Technometrics* 1986;28:408–410.

Wray B, Palmer A, Bejou D. Using neural network analysis to evaluate buyer-seller relationship. *European Journal of Marketing* 1994;28:32–48.

Wu Z, Yang M, Khoo MBC, Castagliola P. What are the best sample sizes for the Xbar and CUSUM charts? *International Journal of Production Economics* 2011;131(2):650–662.

Xiao T, Yan X. Coordinating a two-stage supply chain via a markdown money and advertising subsidy contract. *International Journal of Information and Decision Sciences* 2011;3(2):107–127.

Yan H, Yu Z, Cheng TCE. A strategic model for supply chain design with logical constraints: Formulation and solution. *Computers & Operations Research* 2003;30:2135–2155.

Zahedi F. The analytical hierarchy process: A survey of the method and its applications. *Interfaces* 1986;16(4):96–108.

Zhang CW, Xie M, Goh TN. Economic design of cumulative count of conforming charts under inspection by samples. *International Journal of Production Economics* 2008;111:93–104.

7

Supply Chain: Product Life Cycle Model

SUMMARY In this chapter, we consider a four-layer supply chain transferring raw material from supplier to manufacturer to produce a product and delivering it to the customers. Because life cycle consideration within the stages of the product life cycle is significant, it is crucial to include such considerations in modeling. We model the problem as a cost minimization formulation.

7.1 Introduction

Customer demand, severe competition and internal dynamics are major concerns for supply chain management (Faisal, 2012). The market originator must endure not only the substantial risk of whether the market would materialize or not, but also the difficulty of recovering major costs, such as research and development and advertisement (Prasanna Venkatesan and Kumanan, 2012).

Many durable products provide value only when used together with contingent services or consumable components, e.g., light fixtures (bulbs), printers (ink), electronics (batteries). Consumers need only have access to the contingent consumable components to continue to derive service from a durable. In fact, many firms rely primarily upon the revenues generated from the contingent services or consumables as the primary source of profitability, e.g., giving away the razors to make money on the blades. Such firms often invest considerable effort into making sure that consumers of their durables are held captive to their own branded consumables by impeding their access to generically available consumables. They do so by designing their products in such a way that they are not readily compatible with the generic consumables. Erzurumlu (2013) considered the implications of competition from third-party manufacturers that can provide generic consumables and the manufacturer's production decisions of a durable good under such contingencies. This allowed the decision makers to draw managerial insights about how a firm should decide on its product compatibility and production quantity when the generic contingent consumables enter the market.

Cai et al. (2013) considered a supply chain in which a producer supplies a fresh product, through a third-party logistics (3PL) provider, to a distant market where a distributor purchases and sells it to end customers. The product was perishable, both the quantity and quality of which may deteriorate during the process of transportation.

Jonrinaldi and Zhang (2013) proposed a model and solution method for coordinating integrated production and inventory cycles in a whole manufacturing supply chain involving reverse logistics for multiple items with a finite horizon period. A whole manufacturing supply chain involving reverse logistics consisted of tier-2 suppliers supplying raw materials to tier-1 suppliers; tier-1 suppliers producing parts; a manufacturer that manufactures and assembles parts from tier-1 suppliers into finished products; distributors distributing finished products to retailers; retailers selling products to end customers; and a third party that collected the used finished products from end customers, disassembled collected products into parts, and fed the parts back to the supply chain. In this system, a finite horizon period was considered.

Product life cycle (PLC) management as the integrated, information-driven approach to all aspects of a product's life, from concept to design, manufacturing, maintenance and removal from the market, has become a strategic priority in many company's boardrooms.

Bio-based materials have come increasingly into focus during the last years as alternatives to conventional materials such as fossil-based polymers, metals, and glass. LaRosa et al. (2012) presented an application of life cycle assessment (LCA) methodology in order to explore the possibility of improving the eco efficiency of glass fiber composite materials by replacing part of the glass fibers with hemp mats. The main purpose and contribution of the study was the exploration of the eco-efficiency of this new material. To a minor degree, it was also a contribution in the sense that it provides life cycle inventory data on composites, which as yet is scarce in the LCA community. Andersen et al. (2012) proposed an MILP multi-period formulation for the optimal design and planning of the Argentinean biodiesel supply chain, considering land competition and alternative raw materials. The model included intermediate and final products, i.e., seed, flour, pellets and expellers, oil, pure and blending biodiesel and glycerol. Given high variability of demands for short life cycle products, a retailer has to decide about the products' prices and order quantities from a manufacturer. In the meantime, the manufacturer has to determine an aggregate production plan involving for example, production, inventory and work force levels in a multi-period, multi-product environment. Due to the imprecise and fuzzy nature of products' parameters such as unit production and replenishment costs, a hybrid fuzzy multi-objective programming model including both quantitative and qualitative constraints and objectives was proposed to determine the optimal price markdown policy and aggregate production planning in a two-echelon supply chain (Ghasemy Yaghin et al., 2012).

Asset utilization is a major mid-term lever to increase shareholder value creation. Since rough-cut planning of capacity (dis)investments is performed at the long-term level, detailed timing of adjustments remains for the mid-term level. In combination with capacity control measures, capacity adjustment timing can be used to optimize asset utilization. Hahn and Kuhn (2012) provided a corresponding framework for value-based performance and risk optimization in supply chains covering investment, operations, and financial planning simultaneously.

Iglesias et al. (2012) performed a comparative Life Cycle Assessment with the aim of finding out how the environmental impact derived from biodiesel production (using raw sunflower oil or waste cooking oils) could be affected by the degree of decentralization of the production (number of production plants in a given territory). The decentralized production of biodiesel has been proposed for several reasons, such as the possibility of small scale production, the fact that there is no need to use high technology or make large investments, and because small plants do not need highly specialized technical staff.

Chen et al. (2012) presented a review of the issues associated with a manufacturer's pricing strategies in a two-echelon supply chain that comprises one manufacturer and two competing retailers, with warranty period-dependent demands. The manufacturer, as a Stackelberg leader, specifies wholesale prices to two competing retailers who face warranty period-dependent demand and have different sales costs. The manufacturer considered three pricing options: (1) setting the same price for both retailers, while disregarding their difference with regard to sales cost; (2) setting a different price to each retailer on the basis of their sales cost; and (3) setting the same price to both retailers according to the average sales cost of the industry.

Mass-customization production (MCP) companies must fight with shop-floor uncertainty and complexity caused by wide variety of product components. Zhong et al. (2013) was motivated by a typical MCP company that has experienced inefficient scheduling due to paper-based identification and manual data collection. Zhong et al. (2013) presented an RFID-enabled real-time manufacturing execution system (RT-MES). RFID devices were deployed systematically on the shop floor to track and trace manufacturing objects and collect real-time production data. Disturbances were identified and controlled within RT-MES. Planning and scheduling decisions were more practically and precisely made and executed. Online facilities were provided to visualize and manage real-time dynamics of shop-floor WIP (work-in-progress) items. A case study was reported in a collaborating company which manufactures large-scale and heavy-duty machineries. The efficiency and effectiveness of the proposed RT-MES were evaluated with real-life industrial data for shop-floor production management in terms of workers, machines and materials.

Product variety is one of the most important advantages in highly competitive markets. However, excessive product proliferation's reducing the profit margin has caused increased focus on developing a management method for maximal profit. In a closed-loop supply chain, product proliferation affects the reverse supply chain as well as the forward supply chain. Although increasing the number of product types can better satisfy diverse customer needs, complexity in the product recycling, remanufacturing, and resale processes may erode a firm's overall profits. Huang and Su (2013) developed a mathematical model for analyzing a capacitated reverse supply chain consisting of a single manufacturer and multiple retailers. They revealed closed-form solutions for the optimal batch size and maximal profit, and discussed managerial insights into how the number of products and other factors can affect both batch size and profit. Finally, the relationship between product proliferation and the choice of logistics strategy was investigated.

Qiang et al. (2013) investigated a closed-loop supply chain network with decentralized decision-makers consisting of raw material suppliers, retail outlets, and the manufacturers that collect the recycled product directly from the demand market. The optimal conditions of the various decision-makers were derived and the governing equilibrium conditions that can be formulated as a finite-dimensional variation inequality problem were established. The convergence of the proposed algorithm that can allow for the discussion of the effects of competition, distribution channel investment, yield and conversion rates, combined with uncertainties in demand, on equilibrium quantity transactions and prices was established. Numerical examples were provided for illustration.

Stadtler and Sahling (2013) presented a new model formulation for lot-sizing and scheduling of multi-stage flow lines which works without a fixed lead-time offset and still guarantees a feasible material flow. In the literature, multi-stage lot-sizing model formulations often use a fixed lead time offset of one period leading to increased planned lead times. Computational tests have shown that the total costs resulting from our new model formulation are at least 10% lower. Furthermore, they presented a solution approach based on Fix-and-Relax and Fix-and-Optimize. Numerical results showed that this solution approach generated high-quality solutions in moderate computational time.

Waste stemming from inappropriate quality control and excessive inventories is a major challenge for perishable food management in grocery retail chains. Improvement of visibility and traceability in food supply chains facilitated by tracking and tracing technologies has great potential to improve operations efficiency. Wang and Li (2012) aimed to reduce food spoilage waste and maximize food retailers' profit through a pricing approach based on dynamically identified food shelf life. The proposed model was evaluated through different pricing policies to exploit the benefits from utilizing accurate product shelf life information captured through innovated tracking and monitoring technologies. The main aim of Elgazzar et al. (2012) was to develop a performance measurement method which links supply chain (SC) processes' performance to a company's financial strategy through demonstrating and utilizing the relationship between SC processes' performance and a company's financial performance. The Dempster Shafer/Analytical Hierarchy Processes (DS/AHP) model was employed to link SC processes' performance to the company's financial performance through determining the relative importance weights of SC performance measures with respect to the priorities of financial performance.

7.2 Problem Definition

In this chapter, we consider a four-layer supply chain transferring raw material from supplier to manufacturer to produce a product and delivering it to the customers. Since life cycle consideration within the stages of the product life cycle is significant, it is crucial to include such considerations in modeling. We model the problem as a cost minimization formulation. Several cost factors are considerable in the different stages of product life cycle influencing the managers' investment decisions. Therefore, it is significant to provide a mathematical model for inclusion of life cycle in the process of decision making. The purpose of the mathematical model is to minimize the total costs and at the same time to maximize the process capability, as a quality measure of the activities performed in the life cycle stages. Mathematical model outputs are the amount of product i production and the occurrence of an operation in stage j for product i. The occurrence of an activity incurs costs and the amount of production provides benefits. The tradeoff between the objectives helps finding optimal values which are in compliance with the product life cycle.

7.3 Product Life Cycle Model

Here, we develop a mathematical model for product life cycle analyses in a multilayer supply chain. The mathematical model considers the costs incurred in different stages of product life cycle, namely, introduction, growth, maturity and decline. In each stage several operations are performed. The on-time occurrence of an operation is very significant, i.e., doing an operation earlier or later incurs costs. Only delivering within the due date is appropriate for avoiding product expiration date. The product transferring costs in the layers of supply chain also incur costs influencing the optimal product quantity decisions.

The indices, parameters and decision variables are defined as follows:

Indices

Counter for products $\qquad i = 1, 2, ..., I$
Counter for product life-cycle stages $\qquad j = 1, 2, ..., J$
Counter for operations $\qquad k = 1, 2, ..., K$

Parameters

E_{ijk} Earliness for product i in stage j for operation k
α Penalty for earliness
αE_{ijk} Earliness cost for product i in stage j for operation k
T_{ijk} Tardiness for product i in stage j for operation k
β Penalty for tardiness
βT_{ijk} Tardiness cost for product i in stage j for operation k
O_{ijk} Operation cost for product i in stage j for operation k
D_{ijk} Due date for product i in stage j for operation k
TR_{ijk} Transferring time from supplier to manufacturer for product i in stage j for operation k
KTR_{ijk} Transferring time from manufacturer to distributor for product i in stage j for operation k
TRC_{ijk} Transferring time from distributor to customer for product i in stage j for operation k
P_{ijk} Processing time for product i in stage j for operation k
γ Transferring unit cost from supplier to manufacturer
ρ Transferring unit cost from manufacturer to distributor
ε Transferring unit cost from distributor to customer
C_{ijk} Completion time for product i in stage j for operation k
Cp_{ik} Process capability for product i of operation k
θ_i Upper limit for total costs of earliness and tardiness for any products
λ_i Upper limit for total transferring costs for any products
ϑ_i Upper limit for total operation costs for any products

Decision variables

X_i The amount of production for product i
$Y_{ijk} = 1$ If operation k is done for product i in stage j
$Y_{ijk} = 0$ Otherwise

7.3.1 Objective Functions

$$Min\, Z : \sum_{i=1}^{I}\sum_{j=1}^{J}\sum_{k=1}^{K}((X_i + Y_{ijk})(\alpha E_{ijk} + \beta T_{ijk} + O_{ijk} + \gamma TR_{ijk} + \rho TRI_{ijk} + \varepsilon TRC_{ijk})) \qquad (7.1)$$

$$Max\, Z' : \sum_{i=1}^{I}\sum_{j=1}^{J}\sum_{k=1}^{K} X_i Y_{ijk}\, Cp_{ik} \qquad (7.2)$$

7.3.2 Constraints

$$\sum_{j=1}^{J}\sum_{k=1}^{K}((X_i + Y_{ijk})(\alpha E_{ijk} + \beta T_{ijk})) \leq \theta_i, \quad \forall i, \tag{7.3}$$

This constraint expresses an upper level for the total earliness and tardiness costs for each product.

$$\sum_{j=1}^{J}\sum_{k=1}^{K}((X_i + Y_{ijk})(\gamma TR_{ijk} + \rho TRI_{ijk} + \varepsilon TRC_{ijk})) \leq \lambda_i, \quad \forall i, \tag{7.4}$$

This constraint expresses an upper level for the total transferring cost for each product.

$$\sum_{j=1}^{J}\sum_{k=1}^{K}X_i O_{ijk} \leq \vartheta_i, \quad \forall i, \tag{7.5}$$

This constraint expresses an upper level for the total cost of operations for all products.

$$\sum_{i=1}^{I}\sum_{j=1}^{J}Y_{ijk} \geq 1, \quad \forall k. \tag{7.6}$$

This constraint expresses that for all products at least one operation should be chosen in all stages of life cycle.

The earliness and tardiness are computed as below:

$$T_{ijk} = \max\{0, C_{ijk} - D_{ijk}\}, \quad i = 1, ..., I, j = 1, ..., J, k = 1, ..., K, \tag{7.7}$$

$$E_{ijk} = \max\{0, D_{ijk} - C_{ijk}\}, \quad i = 1, ..., I, j = 1, ..., J, k = 1, ..., K, \tag{7.8}$$

To control the quality of various services, the concept of six sigma is considered as service weight. The six sigma approach is one of the most widely known best practices in providing a tolerance for a parameter. The concept of six sigma originates from statistical terminology, wherein sigma (σ) represents standard deviation. In the recent years, a few researchers have focused on the application of six sigma methodology in balancing process.

This approach assumes that the ideal value of the process mean is between specification intervals, i.e., $\varphi^L < \varphi < \varphi^U$ (with 1.5σ shift from the mean). It is implied that six sigma concept with a 1.5σ shift from the mean holds and the probability of conformance can be shown to be 0.9999966 (or 3.44 ppm). The level of assurance is targeted, but the terminology is also used to evaluate current level of φ with the following sigma level.

Analysis of the six sigma approach makes use of process capability indices C_p and C_{pk}. Process capability index C_p is defined as

$$C_p = \frac{\varphi^U - \varphi^L}{SL}. \tag{7.9}$$

The difference $\varphi^U - \varphi^L$ represents specification width. When $Cp = 2$ and service presence mean is centered at $(\varphi^U - \varphi^L)/2$ without any shift, then the probability of conformance is 99.9999998%.

Here, we make use of six sigma concept and the related equation to control the variation of the services offered by the company. Thus, we have

$$Cp_{ik} = \frac{\varphi_{ik}^U - \varphi_{ik}^L}{SL}, \tag{7.10}$$

where Cp_{ik} is the process capability for operation k performed on product i. Also, the decision variables' sign and type are

$$X_i \geq 0, \quad \forall i, \tag{7.11}$$

$$Y_{ijk} \in \{0,1\}, \quad \forall i, j, k. \tag{7.12}$$

We note that the nonlinear Equation 7.2 can be linearized by using the inequality

$$X_i \leq MY_{ijk}, \quad \forall i, j, k, \tag{7.13}$$

where M is a sufficiently large number; since Y_{ijk} is equal to 0 or 1, if Y_{ijk} is equal to 0 then X is also equal to 0, and if Y_{ijk} is equal to 1 then X gets value. The latter is imposed when M is large enough, and the former is guaranteed by the simultaneous incurred inequalities.

$$Y_{ijk} \geq 0 \tag{7.14}$$

$$Y_{ijk} \leq 0 \tag{7.15}$$

The earliness-tardiness relations are also nonlinear. We linearize them as follows:

$$E_{ijk} \geq 0, \quad i = 1, \ldots, I, j = 1, \ldots, J, k = 1, \ldots, K, \tag{7.16}$$

$$E_{ijk} \geq D_{ijk} - C_{ijk}, \quad i = 1, \ldots, I, j = 1, \ldots, J, k = 1, \ldots, K, \tag{7.17}$$

$$T_{ijk} \geq 0, \quad i = 1, \ldots, I, j = 1, \ldots, J, k = 1, \ldots, K, \tag{7.18}$$

$$T_{ijk} \geq C_{ijk} - D_{ijk}, \quad i = 1, \ldots, I, j = 1, \ldots, J, k = 1, \ldots, K. \tag{7.19}$$

7.4 Discussions

We proposed a mathematical model in a four layer supply chain using the product life cycle factors to satisfy customers' interests. Then, the mathematical model was employed to obtain the optimal number of products and the fulfillment of an activity in product life cycle stages. The problem was modeled to minimize the total cost of the system and maximize the process capability. An application example is illustrated to verify the effectiveness of the proposed methodology. The contributions of the work focused on the product life cycle management, product delivery within due dates causing customer satisfaction, and considering cost minimization and process capability maximization as quality measures at the same time. As future research, including uncertainty of the variables and parameters, and adding other variables such as pricing and other objectives such as benefit could be considered.

The results help managers to determine the operations in different stages of the life cycle corresponding to the optimal solutions for more attention. These operations are considered in two aspects: determination and investment. Of course, the determination ensues to investment, i.e., when an operation is determined to be optimal, then economic investment is concentrated on that operation.

References

Andersen F, Iturmendi F, Espinosa S, Diaz MS. Optimal design and planning of biodiesel supply chain with land competition. *Computers & Chemical Engineering* 2012;47(20):170–182.

Cai X, Chen J, Xiao Y, Xu X, Yu G. Fresh-product supply chain management with logistics outsourcing. *Omega* 2013;41(4):752–765.

Chen X, Li L, Zhou M. Manufacturer's pricing strategy for supply chain with warranty period-dependent demand. *Omega* 2012;40(6):807–816.

Elgazzar SH, Tipi NS, Hubbard NJ, Leach DZ. Linking supply chain processes' performance to a company's financial strategic objectives. *European Journal of Operational Research* 2012;223(1):276–289.

Erzurumlu SS. The compatibility of durable goods with contingent generic consumables. *Omega* 2013;41(3):574–585.

Faisal MN. Sustainability metrics for a supply chain: The case of small and medium enterprises. *International Journal of Services and Operations Management* 2012;13(3):392–414.

Ghasemy Yaghin R, Torabi SA, Fatemi Ghomi SMT. Integrated markdown pricing and aggregate production planning in a two echelon supply chain: A hybrid fuzzy multiple objective approach. *Applied Mathematical Modelling* 2012;36(12):6011–6030.

Hahn GJ, Kuhn H. Simultaneous investment, operations, and financial planning in supply chains: A value-based optimization approach. *International Journal of Production Economics* 2012;140(2):559–569.

Huang SM, Su JCP. Impact of product proliferation on the reverse supply chain. *Omega* 2013;41(3):626–639.

Iglesias L, Laca A, Herrero M, Díaz M. A life cycle assessment comparison between centralized and decentralized biodiesel production from raw sunflower oil and waste cooking oils. *Journal of Cleaner Production* 2012;37:162–171.

Jonrinaldi J, Zhang DZ. An integrated production and inventory model for a whole manufacturing supply chain involving reverse logistics with finite horizon period. *Omega* 2013;41(3):598–620.

LaRosa AD, Cozzo G, Latter A, Recca A, Björklund A, Parrinello E, Cicala G. Life cycle assessment of a novel hybrid glass-hemp/thermoset composite. *Journal of Cleaner Production*, 2012.

Prasanna Venkatesan S, Kumanan S. Supply chain risk prioritization using a hybrid AHP and PROMETHEE approach. *International Journal of Services and Operations Management* 2012;13(1):19–41.

Qiang Q, Ke K, Anderson T, Dong J. The closed-loop supply chain network with competition, distribution channel investment, and uncertainties. *Omega* 2013;41(2):186–194.

Stadtler H, Sahling F. A lot-sizing and scheduling model for multi-stage flow lines with zero lead times. *European Journal of Operational Research* 2013;225(3):404–419.

Wang X, Li D. A dynamic product quality evaluation based pricing model for perishable food supply chains. *Omega* 2012;40(6):906–917.

Zhong RY, Dai QY, Qu T, Hu GJ, Huang GQ. RFID-enabled real-time manufacturing execution system for mass-customization production. *Robotics and Computer-Integrated Manufacturing* 2013;29(2):283–292.

8

Multi Echelon Supply Chain: CRM Model

SUMMARY In this chapter, we consider a three-layer supply chain transferring raw material from supplier to manufacturer. Since life cycle consideration lead to customer satisfaction, it is crucial to include such considerations in modeling. For efficiency and customer satisfaction, a CRM system is conducted to collect customers" opinion about the life cycle parameters of the products. The customer relationship is worked out through online and media technologies. A mathematical model is formulated to integrate customer views of product life cycle.

8.1 Introduction

The need for tailored logistic channels in the product delivery process (PDP) is well recognized. Certainly "one size does not fit all" (Shewchuck, 1998) is in our experience a reasonable summary of both theory and practice. For example, in automotive spares supply chains at least three distinctive delivery channels are required. These are typified by the availability needs and holding cost requirements for (say) light bulbs, oil pumps, and bumpers (Towill, 2001). Each of the three resultant channels thus identified is then designed according to the optimal choice of location, mode and frequency of transport, inventory levels, and degree of postponement appropriate to that particular product. However an innovative manufacturer must do more than recognize the need for tailored logistics (Kumar et al., 2012). Not only is the operating scenario likely to be more complex due to the wide range of both his customers and his suppliers forming an extensive interactive network, but also there is considerable interaction between the PDP and the new product introduction process (PIP).

Tompkins (Bradley et al., 1999) introduced the concept of Supply Chain Synthesis. It is a holistic, continuous improvement process of ensuring customer satisfaction and is all about using partnerships and communication to integrate the supply chain. Advertisement through the mass media and the development of the internet has speeded up the diffusion of new products. At the same time, technical innovation and severe competition in the market promote rapid obsolescence of existing products and technologies. When a company succeeds in developing a new product category, other competitors may soon emerge. The market originator must endure not only the substantial risk of whether the market would materialize or not, but also the difficulty of recovering major costs, such as research and development and advertisement (Prasanna Venkatesan and Kumanan, 2012). Increasingly, the supply chain becomes the mechanism for coping with these problems because it is often inefficient for any single company to produce a whole product. Hence, modern business is essentially the competition of one supply chain with another (Tompkins, 2000).

Many durable products provide value only when used together with contingent services or consumable components, for example, light fixtures (bulbs), printers (ink), electronics (batteries). Consumers need only have access to the contingent consumable components to continue to derive service from a durable. In fact, many firms rely primarily upon the revenues generated from the contingent services or consumables as the primary source of profitability, for example, giving away the razors to make money on the blades. Such firms often invest considerable effort into making sure that consumers of their durables are held captive to their own branded consumables by impeding their access to generically available consumables. They do so by designing their products in such a way that they are not readily compatible with the generic consumables. Erzurumlu (2013) considered the implications of competition from third-party manufacturers that can provide generic consumables and the manufacturer's production decisions of a durable good under such contingencies. This allowed the decision makers to draw managerial insights about how a firm should decide on his product compatibility and production quantity when the generic contingent consumables enter the market.

The origins of strategic supply chain management were founded in the 1970s, when Geoffrion and Graves (1974) developed a distribution model, long before supply chain management was invented. Although supply chain management has been on the top of research in recent years, the advances in the past decade for integrating strategic planning tasks to supply chain management are limited. The latest available reviews from Vidal and Goetschalckx (1997) and Thomas and Griffin (1996) have identified many opportunities for research and investigation in strategic supply chain management. Supply chain management covers the short- and long-term collaboration of a company with other companies to develop and manufacture products with the required internal and inter-company organization, planning and control of the flows of materials, financial value and information along the business processes (Schonsleben, 1998; Stadtler, 2000).

The extended supply chain network consists of several business processes and elements. In strategic supply chain management, materials are the most important process-overlapping elements. They are defined by their physical properties such as weight, size and volume. Other important process-overlapping elements of the material flow are transport paths and warehouses. Transport times can be considered for the means of transport and the transport paths between the different elements (Kasilingam, 1998). Warehouses are necessary to synchronize consumption and manufacturing of the materials through stockpiling (Schonsleben, 1998).

Product life cycle (PLC) management as the integrated, information-driven approach to all aspects of a product's life, from concept to design, manufacturing, maintenance and removal from the market, has become a strategic priority in many company's boardrooms (Teresko, 2004). For example, in the pharmaceutical industry, the development time for new drugs has almost doubled over the last 30 years and the average drug development costs exceed US $ 800 million. Reshaping the life cycle curve so that profitability starts earlier and maturity ends later is seen as a matter of survival (Daly and Kolassa, 2004). The automotive industry is another vivid example of where success or failure is strongly influenced by the company's ability to proactively manage product life cycles (Korth, 2003). Increased product complexity, greater reliance on outsourcing and a growing need for collaboration with a rapidly expanding list of business partners are the specific PLC management challenges the industry faces (Teresko, 2004). Furthermore, in high-tech or fashion industries, accelerated technological and design changes explain why PLC management is at the forefront. PLC management confronts the need to balance fast response to changing consumer demands with competitive pressure to seek cost reductions in sourcing, manufacturing and distribution. It needs to be based on a close alignment between customer-facing functions (e.g., marketing, sales, customer service) and supply functions (e.g., purchasing, manufacturing, logistics) (Hughes, 1990; O'Marah, 2003; Combs, 2004; Conner, 2004).

First, life-cycle design seeks to maximize the life-cycle value of a product at the early stages of design, while minimizing cost and environmental impact. Ishii et al., (1994) introduced the concept of the life-cycle value and illustrated a prototype computer tool of Design for Product Retirement (DFPR). Their paper focused on product retirement and advanced planning for material recycling. For the issue of designing for remanufacturing or recycling, Klausne and Wolfgang (1999) outlined a concept to integrate product repair and product take-back. They showed that the replacement of a large share of conventional repairs with remanufacturing and reconditioning would result in a higher service level in product repair.

Bio-based materials have come increasingly into focus during the last years, as alternatives to conventional materials such as fossil-based polymers, metals, and glass. LaRosa et al. (2012) presented an application of life cycle assessment (LCA) methodology in order to explore the possibility of improving the eco efficiency of glass fiber composite materials by replacing part of the glass fibers with hemp mats. The main purpose and contribution of the study was the exploration of the eco-efficiency of this new material. To a minor degree, it was also a contribution in the sense that it provides life cycle inventory data on composites, which as yet is scarce in the LCA community. Andersen et al. (2012) proposed an Mixed integer linear programming (MILP) multi-period formulation for the optimal design and planning of the Argentinean biodiesel supply chain, considering land competition and alternative raw materials. The model included intermediate and final products, that is, seed, flour, pellets and expellers, oil, pure and blending biodiesel and glycerol. Given high variability of demands for short life cycle products, a retailer has to decide about the products' prices and order quantities from a manufacturer. In the meantime, the manufacturer has to determine an aggregate production plan involving for example, production, inventory and work force levels in a multi period, multi product environment. Due to imprecise and fuzzy nature of products' parameters such as unit production and replenishment costs, a hybrid fuzzy multi-objective programming model including both quantitative and qualitative constraints and objectives was proposed to determine the optimal price markdown policy and aggregate production planning in a two echelon supply chain (Ghasemy Yaghin et al., 2012).

Asset utilization is a major mid-term lever to increase shareholder value creation. Since rough-cut planning of capacity (dis)investments is performed at the long-term level, detailed timing of adjustments remains for the mid-term level. In combination with capacity control measures, capacity adjustment timing can be used to optimize asset utilization. Hahn and Kuhn (2012) provided a corresponding framework for value-based performance and risk optimization in supply chains covering investment, operations, and financial planning simultaneously.

Iglesias et al. (2012) performed a comparative Life Cycle Assessment with the aim of finding out how the environmental impact derived from biodiesel production (using raw sunflower oil or waste cooking oils) could be affected by the degree of decentralization of the production (number of production plants in a given territory). The decentralized production of biodiesel has been proposed for several reasons, such as the possibility of small scale production, the fact that there is no need to use high technology or make large investments, and because small plants do not need highly specialized technical staff.

Chen et al. (2012) presented a review of the issues associated with a manufacturer's pricing strategies in a two-echelon supply chain that comprises one manufacturer and two competing retailers, with warranty period-dependent demands. The manufacturer, as a Stackelberg leader, specifies wholesale prices to two competing retailers who face warranty period-dependent demand and have different sales costs. The manufacturer considered three pricing options: (1) setting the same price for both retailers, while disregarding their difference with regard to sales cost; (2) setting a different price to each retailer on the basis of their sales cost; and (3) setting the same price to both retailers according to the average sales cost of the industry.

Well-integrated supply chain is one of the primary business strategies to improve supply chain performance. Real-time information exchange with suppliers in the upstream and with customers in the downstream will create an opportunity where optimization can take place. Linkage, which helps reducing lead-times, undoubtedly will reduce the adverse effect (i.e., bullwhip effects) and contribute to enhancing performance. Theoretically it has been well-known that supply chain integration creates strategic advantages. However, there has been a lack of research to actually measure such total integration and link it to performance metrics in real-world supply chain strategy situations (Pagh and Cooper, 1998). For example, it has been argued that a well-connected business process improves supply chain management (SCM) performance through lowering cost, shortening delivery time, providing appropriate feedback, maintaining low inventory levels, and improving reliability (Davis, 1993; Mason-Jones and Towill, 1997; Krajewski et al., 2005).

Mass-customization production (MCP) companies must fight with shop-floor uncertainty and complexity caused by wide variety of product components. Zhong et al. (2013) motivated by a typical MCP company that has experienced inefficient scheduling due to paper-based identification and manual data collection. Zhong et al. (2013) presented an RFID-enabled real-time manufacturing execution system (RT-MES). RFID devices were deployed systematically on the shop-floor to track and trace manufacturing objects and collect real-time production data. Disturbances were identified and controlled within RT-MES. Planning and scheduling decisions were more practically and precisely made and executed. Online facilities were provided to visualize and manage real-time dynamics of shop-floor WIP (work-in-progress) items. A case study was reported in a collaborating company which manufactures large-scale and heavy-duty machineries. The efficiency and effectiveness of the proposed RT-MES were evaluated with real-life industrial data for shop-floor production management in terms of workers, machines and materials.

Product variety is one of the most important advantages in highly competitive markets. However, excessive product proliferation's reducing the profit margin has caused increased focus on developing a management method for maximal profit. In a closed-loop supply chain, product proliferation affects the reverse supply chain as well as the forward supply chain. Although increasing the number of product types can better satisfy diverse customer needs, complexity in the product recycling, remanufacturing, and resale processes may erode a firm's overall profits. Huang and Su (2013) developed a mathematical model for analyzing a capacitated reverse supply chain consisting of a single manufacturer and multiple retailers. They revealed closed-form solutions for the optimal batch size and maximal profit, and discussed managerial insights into how the number of products and other factors can affect both batch size and profit. Finally, the relationship between product proliferation and the choice of logistics strategy was investigated.

Qiang et al. (2013) investigated a closed-loop supply chain network with decentralized decision-makers consisting of raw material suppliers, retail outlets, and the manufacturers that collect the recycled product directly from the demand market. The optimality conditions of the various decision-makers were derived and the governing equilibrium conditions that can be formulated as a finite-dimensional variation inequality problem was established. The convergence of the proposed algorithm that can allow for the discussion of the effects of competition, distribution channel investment, yield and conversion rates, combined with uncertainties in demand, on equilibrium quantity transactions and prices was established. Numerical examples were provided for illustration.

Stadtler and Sahling (2013) presented a new model formulation for lot-sizing and scheduling of multi-stage flow lines which works without a fixed lead-time offset and still guarantees a feasible material flow. In the literature, multi-stage lot-sizing model formulations often use a fixed lead time offset of one period leading to increased planned lead times. Computational tests have shown that the total costs resulting from our new model formulation are at least 10% lower. Furthermore, they presented a solution approach based on Fix-and-Relax and Fix-and-Optimize. Numerical results showed that this solution approach generated high-quality solutions in moderate computational time.

Waste stemmed from inappropriate quality control and excessive inventories is a major challenge for perishable food management in grocery retail chains. Improvement of visibility and traceability in food supply chains facilitated by tracking and tracing technologies has great potential to improve operations efficiency. Wang and Li (2012) aimed to reduce food spoilage waste and maximize food retailer's profit through a pricing approach based on dynamically identified food shelf life. The proposed model was evaluated through different pricing policies to exploit the benefits from utilizing accurate product shelf life information captured through innovated tracking and monitoring technologies. The main aim of Elgazzar et al. (2012) was to develop a performance measurement method which links supply chain (SC) processes' performance to a company's financial strategy through demonstrating and utilizing the relationship between SC processes' performance and a company's financial performance.

8.2 Problem Definition

In this chapter, we consider a three-layer supply chain transferring raw material from supplier to manufacturer to produce a product and delivering it to the customers. Since life cycle consideration within the stages of the product life cycle is significant in customer satisfaction, it is crucial to include such considerations in modeling. For efficiency and customer satisfaction, a CRM system is conducted to collect customers' opinion about the life cycle parameters of the products. The customer relationship is worked out through online and media technologies. Thus, the availability of the media and technology and the human resource required to service and collect the customers' opinions. We model the problem as a cost minimization formulation. Several cost factors are considerable in the different stages of product life cycle influencing the managers' investment decisions. Therefore, it is significant to provide a decision aid mathematical model for inclusion of life cycle in the process of decision making. The purpose of the mathematical model is to minimize the total costs and at the same time to maximize the process capability, as a quality measure of the activities performed in the life cycle stages to keep the customer satisfaction quality measure. Mathematical model outputs are the amount of product ith production and the occurrence of an operation in stage jth for product ith. The occurrence of an activity incurs costs and the amount of production provides benefits. The tradeoff between the objectives helps optimal values keeping the product life cycle with respect its costs.

8.3 Integrated Customer Related Life Cycle Model

Here, we develop a mathematical model for product life cycle analyses in a multilayer supply chain. The mathematical model considers the costs incurred in different stages of product life cycle namely, introduction, growth, maturity and decline. In each stage several operations are performed. The on time occurrence of an operation is very significant, that is, doing an operation earlier or later cause costs. Only delivering within the due date is appropriate for the satisfaction purpose. The product transferring costs in the layers of supply chain also cause costs influencing the optimal product quantity decisions.

The indices, parameters and decision variables are defined as follows:

Indices

$i = 1, 2, ..., I$ Counter for Products
$j = 1, 2, ..., J$ Counter for product life-cycle stages
$k = 1, 2, ..., K$ Counter for operations

Parameters

E_{ijk}	Earliness for product ith in stage jth for operation kth
α	Penalty for earliness
αE_{ijk}	Earliness cost for product ith in stage jth for operation kth
T_{ijk}	Tardiness for product ith in stage jth for operation kth
β	Penalty for tardiness
βT_{ijk}	Tardiness cost for product ith in stage jth for operation kth
O_{ijk}	Operation cost for product ith in stage jth for operation kth
D_{ijk}	Due date for product ith in stage jth for operation kth
TR_{ijk}	Transferring time from supplier to manufacturer for product ith in stage jth for operation kth
TRC_{ijk}	Transferring time from manufacturer to customer for product ith in stage jth for operation kth
P_{ijk}	Processing time for product ith in stage jth for operation kth
γ	Transferring unit cost from supplier to manufacturer
ε	Transferring unit cost from manufacturer to customer
C_{ijk}	Completion time for product ith in stage jth for operation kth
Cp_{ik}	Process capability for product ith of operation kth
θ_i	Upper limit for total costs of earliness and tardiness for any products
λ_i	Upper limit for total transferring costs for any products
ν	Upper limit for total customers' opinion collection costs
ϑ_i	Upper limit for total operation costs for any products
CRH_i	Human resource costs for product ith
CRT_i	Technology costs for product ith

Decision Variables

X_i The amount of production for product ith
$Y_{ijk} = 1$ If operation kth is done for product ith in stage jth
$Y_{ijk} = 0$ Otherwise

8.3.1 Objective Functions

$$Min\,Z \sum_{i=1}^{I}\sum_{j=1}^{J}\sum_{k=1}^{K}((X_{ijk} + Y_{ijk})(\alpha E_{ijk} + \beta T_{ijk} + O_{ijk} + \gamma TR_{ijk} + \varepsilon TRC_{ijk}))$$

$$+ X_i(CRH_i + CRT_i)$$

$$Max\,Z' \sum_{i=1}^{I}\sum_{j=1}^{J}\sum_{k=1}^{K} X_{ijk} Y_{ijk}\, Cp_{ik}$$

8.3.2 Constraints

$$\sum_{j=1}^{J}\sum_{k=1}^{K}((X_i + Y_{ijk})(\alpha E_{ijk} + \beta T_{ijk})) \le \theta_i, \quad \forall i,$$

This constraint expresses an upper level for the total earliness and tardiness costs for each product.

$$\sum_{j=1}^{J}\sum_{k=1}^{K}((X_i + Y_{ijk})(\gamma TR_{ijk} + \varepsilon TRC_{ijk})) \le \lambda_i, \quad \forall i,$$

This constraint expresses an upper level for the total transferring cost for each product.

$$\sum_{i=1}^{I} X_i(CRH_i + CRT_i) \le \nu$$

This constraint expresses an upper level for the total cost of collecting customers' opinions for all products.

$$\sum_{j=1}^{J}\sum_{k=1}^{K} X_i O_{ijk} \le \vartheta_i, \quad \forall i,$$

This constraint expresses an upper level for the total cost of operations for all products.

$$\sum_{i=1}^{I}\sum_{j=1}^{J} Y_{ijk} \ge 1, \quad \forall k.$$

This constraint expresses that for all products at least one operation should be chosen in all stages of life cycle.

The earliness and tardiness are computed as below:

$$T_{ijk} = \max\left\{0, C_{ijk} - D_{ijk}\right\}, \quad i = 1, ..., I, j = 1, ..., J, k = 1, ..., K,$$

$$E_{ijk} = \max\left\{0, D_{ijk} - C_{ijk}\right\}, \& \ i = 1, ..., I, j = 1, ..., J, k = 1, ..., K$$

To control the quality of various services, the concept of six sigma is considered as service weight. The six sigma approach is one of the most widely known best practices in providing a tolerance for a parameter. The concept of six sigma originates from statistical terminology, wherein sigma (σ) represents standard deviation. In the recent years, a few researchers have focused on the application of six sigma methodology in balancing process.

This approach assumes that the ideal value of the process mean is between specification intervals, that is, $\varphi^L < \varphi < \varphi^U$ (with 1.5σ shift from the mean). It is implied that six sigma concept with a 1.5σ shift from the mean holds and the probability of conformance can be shown to be 0.9999966 (or 3.44ppm). The level of assurance is targeted, but the terminology is also used to evaluate current level of φ with the following sigma level.

Analysis of the six sigma approach makes use of process capability indices C_p and C_{pk}. Process capability index C_p is defined as,

$$C_p = \frac{\varphi^U - \varphi^L}{SL},$$

The difference $\varphi^U - \varphi^L$ represents specification width. When $Cp = 2$ and service presence mean is centered at $(\varphi^U - \varphi^L)/2$ without any shift, then the probability of conformance is 99.9999998%.

Here, we make use of six sigma concept and the related Equation to control the variation of the services offered by the company. Thus, we have,

$$Cp_{ik} = \frac{\varphi_{ik}^U - \varphi_{ik}^L}{SL},$$

where Cp_{ik} is the process capability for operation kth performed on product ith. Also, the decision variables" sign and type are,

$$X_i \geq 0, \quad \forall i,$$

$$Y_{ijk} \in \{0, 1\}, \quad \forall i, j, k.$$

8.3.3 Linearization

We note that the nonlinear equation can be linearized by using the inequality,

$$X_i \leq MY_{ijk}, \quad \forall i, j, k,$$

where M is a sufficiently large number; since Y_{ijk} is equal to 0 or 1, then if Y_{ijk} is equal to 0 then X is is also equal to 0 and if Y_{ijk} is equal to 1 then X gets value. The latter is imposed when M is large enough, and the former is guaranteed by the simultaneous incurred inequalities.

$$Y_{ijk} \geq 0$$

$$Y_{ijk} \leq 0$$

The earliness and tardiness relations are also nonlinear. We linearize them as follows:

$$E_{ijk} \geq 0, \quad i=1,...,I, j=1,...,J, k=1,...,K,$$

$$E_{ijk} \geq D_{ijk} - C_{ijk}, \quad i=1,...,I, j=1,...J, k=1,...,K,$$

$$T_{ijk} \geq 0, \quad i=1,...,I, j=1,...,J, k=1,...,K,$$

$$T_{ijk} \geq C_{ijk} - D_{ijk}, \quad i=1,...,I, j=1,...,J, k=1,...,K.$$

8.3.4 Weighing the Objectives by AHP

Since the mathematical model is a bi-objective one, we weigh the objectives to integrate and optimize the proposed model. To weight the objectives, we take a multi-criteria decision-making approach. Multi-criteria decision-making (MCDM), dealing primarily with problems of evaluation or selection (Keeney and Raiffa, 1976; Teng, 2002), is a rapidly developing area in operations research and management science. AHP, developed by Saaty (1980), is a technique of considering data or information for a decision in a systematic manner (Schniederjans and Garvin, 1997). It is mainly concerned with a way of solving decision problems with uncertainties in multiple criteria characterization. It is based on three principles: construction of the hierarchy, priority setting, and logical consistency. We apply AHP to weight the objectives.

8.3.5 Construction of the Hierarchy

A complicated decision problem, composed of various attributes of an objective, is structured and decomposed into sub-problems (sub-objectives, criteria, alternatives, etc.), within a hierarchy.

8.3.6 Priority Setting

The relative "priority" given to each element in the hierarchy is determined by pair-wise comparisons of the contributions of elements at a lower level in terms of the criteria (or elements) with a causal relationship. In AHP, multiple paired comparisons are based on a standardized comparison scale of nine levels (see Table 8.1, from Saaty, 1980).

Let $C = \{c_1, ..., c_n\}$ be the set of criteria. The result of the pair-wise comparisons on n criteria can be summarized in an $n \times n$ evaluation matrix A in which every element a_{ij} is the quotient of weights of the criteria, as shown below:

$$A = (a_{ij}), \quad i,j = 1, ..., n.$$

The relative priorities are given by the eigenvector (w) corresponding to the largest eigenvalue (λ_{max}) as:

$$Aw = \lambda_{max} w.$$

When pair-wise comparisons are completely consistent, the matrix A has rank 1 and $\lambda_{max} = n$. In that case, weights can be obtained by normalizing any of the rows or columns of A.

The procedure described above is repeated for all subsystems in the hierarchy. In order to synthesize the various priority vectors, these vectors are weighted with the global priority of the parent criteria and synthesized. This process starts at the top of the hierarchy. As a result, the overall relative priorities to be given to the lowest level elements are obtained. These overall, relative priorities indicate the degree to which the alternatives contribute to the objective. These priorities represent a synthesis of the local priorities, and reflect an evaluation process that permits integration of the perspectives of the various stakeholders involved.

8.3.7 Consistency Check

A measure of consistency of the given pair-wise comparison is needed. The consistency is defined by the relation between the entries of A; that is, we say A is consistent if $a_{ik} = a_{ij} \cdot a_{jk}$, for all i,j,k. The consistency index (CI) is:

$$CI = \frac{(\lambda_{max} - n)}{(n-1)}.$$

The final consistency ratio (CR), on the basis of which one can conclude whether the evaluations are sufficiently consistent, is calculated to be the ratio of the CI and the random consistency index (RI):

$$CR = \frac{CI}{RI}.$$

The value 0.1 is the accepted upper limit for CR. If the final consistency ratio exceeds this value, the evaluation procedure needs to be repeated to improve consistency. The measurement of consistency can be used to evaluate the consistency of decision-makers as well as the consistency of all the hierarchies.

We are now ready to give an algorithm for computing objective weights using the AHP. The following notations and definitions are used.

n number of criteria
i number of objectives

p index for objectives, $p = 1$ or 2
d index for criteria, $1 \leq d \leq D$
R_{pd} the weight of pth item with respect to dth criterion
w_d the weight of dth criterion

Algorithm 8.1: OWAHP (compute objective weights using the AHP)

Step 1: Define the decision problem and the goal.

Step 2: Structure the hierarchy from the top through the intermediate to the lowest level.

Step 3: Construct the objective-criteria matrix using steps 4–8 using the AHP.

(Steps 4–6 are performed for all levels in the hierarchy.)

Step 4: Construct pair-wise comparison matrices for each of the lower levels for each element in the level immediately above by using a relative scale measurement. The decision-maker has the option of expressing his or her intensity of preference on a nine-point scale. If two criteria are of equal importance, a value of 1 is set for the corresponding component in the comparison matrix, while a 9 indicates an absolute importance of one criterion over the other (Table 8.1 shows the measurement scale defined by Saaty, 1980).

Step 5: Compute the largest eigenvalue by the relative weights of the criteria and the sum taken over all weighted eigenvector entries corresponding to those in the next lower level of the hierarchy.

Analyze pair-wise comparison data using the eigenvalue technique. Using these pair-wise comparisons, estimate the objectives. The eigenvector of the largest eigenvalue of matrix A constitutes the estimation of relative importance of the attributes.

Step 6: Construct the consistency check and perform consequence weights analysis as follows:

$$A = (a_{ij}) = \begin{bmatrix} 1 & w_1/w_2 & \cdots & w_1/w_n \\ w_2/w_1 & 1 & \cdots & w_2/w_n \\ \vdots & \vdots & \ddots & \\ w_n/w_1 & w_n/w_2 & & 1 \end{bmatrix}.$$

TABLE 8.1

Scale of Relative Importance

Intensity of Importance	Definition of Importance
1	Equal
2	Weak
3	Moderate
4	Moderate plus
5	Strong
6	Strong plus
7	Very strong or demonstrated
8	Very, very strong
9	Extreme

Note that if the matrix A is consistent (that is, $a_{ik} = a_{ij} \cdot a_{jk}$, for all $i, j, k = 1, 2, \ldots, n$), then we have (the weights are already known),

$$a_{ij} = \frac{w_i}{w_j}, \quad i, j = 1, 2, \ldots, n.$$

If the pair-wise comparisons do not include any inconsistencies, then $\lambda_{\max} = n$. The more consistent the comparisons are, the closer the value of computed $\lambda_{\max}$ is to n. Set the consistency index (CI), which measures the inconsistencies of pair-wise comparisons, to be:

$$CI = \frac{(\lambda_{\max} - n)}{(n - 1)},$$

and let the consistency ratio (CR) be:

$$CR = 100 \left(\frac{CI}{RI} \right),$$

where n is the number of columns in A and RI is the random index, being the average of the CI obtained from a large number of randomly generated matrices.

Note that RI depends on the order of the matrix, and a CR value of 10% or less is considered acceptable (Saaty, 1980).

Step 7: Form the objective-criteria matrix as specified in Table 8.2:

Step 8: As a result, configure the pair-wise comparison for criteria-criteria matrix as in Table 8.3:
The w_d are gained by a normalization process. The w_d are the weights for criteria.

Step 9: Compute the overall weights for the objectives, using Tables 8.2 and 8.3

$$\psi = \text{Total weight for objective } 1 = R_{11} \times w_1 + R_{12} \times w_2 + \ldots + R_{1d} \times w_d,$$
$$\psi' = \text{Total weight for objective } 2 = R_{21} \times w_1 + R_{22} \times w_2 + \ldots + R_{2d} \times w_d,$$

where $\psi + \psi' = 1$.

TABLE 8.2

The Objective-Criteria Matrix

	C_1	C_2	...	Cd
Objective 1	R_{11}	R_{12}	...	R_{1d}
Objective 2	R_{21}	R_{22}	...	R_{2d}

TABLE 8.3

The Criteria-Criteria Pair-Wise Comparison Matrix

	C_1	C_2	...	C_d	w_d
Criteria 1	1	a_{12}	...	a_{1d}	w_1
Criteria 2	$1/a_{12}$	1	...	a_{2d}	w_2
⋮	⋮	⋮	⋮	⋮	⋮
Criteria d	$1/a_{1d}$	$1/a_{2d}$	...	1	w_d

8.4 Numerical Example

Product life cycle is defined in four stages: introduction, growth, maturity and decline. Introduction is the stage of low growth rate of sales as the product is newly launched in the market. Monopoly can be created, depending upon the efficiency and need of the product to the customers. A firm usually incurs losses rather than profit. If the product is in the new product class, the users may not be aware of its true potential. In order to achieve that place in the market, extra information about the product should be transferred to consumers through various media. This stage has the following characteristics: (1) low competition, (2) firm mostly incurs losses and not profit. Growth comes with the acceptance of the innovation in the market and profit starts to flow. As the monopoly still exists manufacturer can experiment with its new ideas and innovation in order to maintain the sales growth. It is the best time to introduce new effective product in the market thus creating an image in the product class in the presence of its competitors who tries to copy or improve the product and present it as a substitute the growth of a product is determined on the country of a product. Maturity is the end stage of the growth rate, sales slowdown as the product have already achieved it acceptance in the market. So, new firms start experimenting in order to compete by innovating new models of the product. With many companies in the market, competition for customers becomes fierce, even though the increase in the growth rate of sales at the initial part of this stage. Aggressive competition in the market results the profit to acme at the end of the growth stage thus beginning the maturity stage. In addition to this the maturity section of the development process is the most vital. Decline is the stage where most of the product class usually dies due to the low growth rate in sales. As number of companies starts dominating the market, makes it difficult for the existing company to maintain its sale. Not only the efficiency of the company play an important factor in the decline, but also the product category itself becomes a factor, as market may perceive the product as 'OLD' and may not be in demand.

In each of the stages some effective cost parameters related to better service provision to customers are considered as shown in Table 8.4.

The CRM system provides eight factors: Marketing and trust in introduction stage, quality and advertisement in growth stage, quality and guaranty in puberty stage, redesign and economic evaluation in decline stage, are the actions that should be performed to satisfy life cycle purposes, calling them mathematically $k1$, $k2$, $k3$, $k4$, $k5$, $k6$, $k7$ and $k8$. Here, we consider four products ($i = 1, 2, 3, 4$) the life cycle of the product is four stages ($j = 1, 2, 3, 4$) the operations are eight. The coefficients for earliness and tardiness are 3 and 6, respectively. The operations cost for the products in different stages are reported in Table 8.5.

TABLE 8.4

The Cost Parameters in Different Life Cycle Stages

Introduction	Growth	Maturity	Decline
Design	Capacity increase	Delay in service	New product development
Marketing	Lost sale	Sale guarantee	Research and development
Advertisement	Back order	Trust	Redesign
Trust	Quality	Quality	Reverse engineering
–	Sale guarantee	–	Economic evaluation
–	Advertisement	–	–

TABLE 8.5

The Operations Costs for the Products in Different Stages

$k=1$	$i=1$	$i=2$	$i=3$	$i=4$	$k=5$	$i=1$	$i=2$	$i=3$	$i=4$
$j=1$	13	12	14	11	$j=1$	18	11	18	14
$j=2$	15	18	13	17	$j=2$	17	16	13	18
$j=3$	17	13	16	15	$j=3$	15	15	12	13
$j=4$	11	17	18	14	$j=4$	13	18	17	16
$k=2$	$i=1$	$i=2$	$i=3$	$i=4$	$k=6$	$i=1$	$i=2$	$i=3$	$i=4$
$j=1$	13	18	12	15	$j=1$	17	14	16	18
$j=2$	18	13	14	13	$j=2$	18	13	17	14
$j=3$	16	14	17	18	$j=3$	13	12	15	16
$j=4$	14	16	11	12	$j=4$	15	18	17	13
$k=3$	$i=1$	$i=2$	$i=3$	$i=4$	$k=7$	$i=1$	$i=2$	$i=3$	$i=4$
$j=1$	16	13	15	18	$j=1$	17	12	13	16
$j=2$	18	14	17	14	$j=2$	16	15	18	14
$j=3$	16	13	11	12	$j=3$	15	13	14	17
$j=4$	15	18	12	17	$j=4$	18	11	16	13
$k=4$	$i=1$	$i=2$	$i=3$	$i=4$	$k=8$	$i=1$	$i=2$	$i=3$	$i=4$
$j=1$	14	11	18	16	$j=1$	11	13	17	12
$j=2$	13	15	17	12	$j=2$	17	15	14	16
$j=3$	17	13	12	14	$j=3$	16	12	18	13
$j=4$	16	18	13	17	$j=4$	18	16	11	17

The due date, the processing time, the transferring time from supplier to producer, the transferring time from producer to customer and the completion time are given in Tables 8.6 through 8.10.

The transferring cost from supplier to manufacturer is 5 unit of money and the transferring cost from manufacturer to customer is 11 unit of money. The upper bounds for total earliness and tardiness for various products are 17, 21, 14, and 26 unit of money. The upper bounds for total transferring cost for different products are 35, 29, 43, and 32 unit of money. The upper bounds for collecting customers' opinions total cost for all products is 45 unit of money. The upper bounds for operation total costs for different products are 47, 53, 39, and 42 unit of money. The human resource costs for different products are 6, 6, 9 and 4 unit of money. Technology costs for different products are 11, 14, 10 and 14 unit of money. The process capabilities for different operations for different products that are extracted from several quality control statistical samples and tests are given in Table 8.11.

To optimize the proposed bi-objective mathematical model, AHP is employed. Here, the objectives are considered to be the alternatives and the criteria are strategic viewpoints, competitive advantage and macroeconomic capability. Doing the AHP as stated in Algorithm 8.1, the weights are obtained to be 0.42 and 0.58, respectively. Therefore, the integrated weighted objective function is as below:

$$Min\, Z\left[0.42\sum_{i=1}^{I}\sum_{j=1}^{J}\sum_{k=1}^{K}((X_{ijk}+Y_{ijk})(\alpha E_{ijk}+\beta T_{ijk}+O_{ijk}+\gamma TR_{ijk}+\varepsilon TRC_{ijk}))+X_i(CRH_i+CRT_i)\right)$$

$$-\left(0.58\sum_{i=1}^{I}\sum_{j=1}^{J}\sum_{k=1}^{K}X_{ijk}Y_{ijk}Cp_{ik}\right)$$

TABLE 8.6

The Due Dates

$k=1$	$i=1$	$i=2$	$i=3$	$i=4$	$k=5$	$i=1$	$i=2$	$i=3$	$i=4$
$j=1$	5	8	6	9	$j=1$	7	10	7	10
$j=2$	7	10	5	7	$j=2$	8	8	6	5
$j=3$	6	5	8	10	$j=3$	6	6	8	5
$j=4$	9	6	10	8	$j=4$	10	9	5	6
$k=2$	$i=1$	$i=2$	$i=3$	$i=4$	$k=6$	$i=1$	$i=2$	$i=3$	$i=4$
$j=1$	10	8	5	10	$j=1$	6	10	9	9
$j=2$	7	7	7	7	$j=2$	7	5	7	5
$j=3$	6	6	9	6	$j=3$	10	5	5	7
$j=4$	9	6	5	8	$j=4$	9	7	10	8
$k=3$	$i=1$	$i=2$	$i=3$	$i=4$	$k=7$	$i=1$	$i=2$	$i=3$	$i=4$
$j=1$	8	7	10	9	$j=1$	7	10	6	10
$j=2$	6	5	6	8	$j=2$	9	7	8	7
$j=3$	10	9	5	6	$j=3$	10	5	5	5
$j=4$	5	8	9	9	$j=4$	6	5	10	7
$k=4$	$i=1$	$i=2$	$i=3$	$i=4$	$k=8$	$i=1$	$i=2$	$i=3$	$i=4$
$j=1$	6	10	9	8	$j=1$	7	10	5	6
$j=2$	7	8	5	5	$j=2$	5	5	8	7
$j=3$	5	8	6	7	$j=3$	8	5	6	9
$j=4$	10	7	10	9	$j=4$	10	6	9	10

TABLE 8.7

The Processing Time

$k=1$	$i=1$	$i=2$	$i=3$	$i=4$	$k=5$	$i=1$	$i=2$	$i=3$	$i=4$
$j=1$	2	2	4	7	$j=1$	3	7	4	2
$j=2$	7	7	6	3	$j=2$	5	6	5	4
$j=3$	4	4	2	5	$j=3$	6	2	2	6
$j=4$	7	3	7	2	$j=4$	4	5	7	2
$k=2$	$i=1$	$i=2$	$i=3$	$i=4$	$k=6$	$i=1$	$i=2$	$i=3$	$i=4$
$j=1$	5	4	7	7	$j=1$	2	6	4	7
$j=2$	4	5	3	2	$j=2$	4	2	3	2
$j=3$	7	2	5	5	$j=3$	7	7	7	5
$j=4$	2	6	4	3	$j=4$	5	3	2	4
$k=3$	$i=1$	$i=2$	$i=3$	$i=4$	$k=7$	$i=1$	$i=2$	$i=3$	$i=4$
$j=1$	4	7	5	4	$j=1$	3	7	7	5
$j=2$	2	2	5	5	$j=2$	5	2	4	3
$j=3$	7	5	7	3	$j=3$	7	4	3	7
$j=4$	5	6	2	7	$j=4$	2	5	6	3
$k=4$	$i=1$	$i=2$	$i=3$	$i=4$	$k=8$	$i=1$	$i=2$	$i=3$	$i=4$
$j=1$	2	4	2	6	$j=1$	6	5	2	2
$j=2$	5	7	6	3	$j=2$	2	3	7	5
$j=3$	3	6	5	5	$j=3$	5	3	3	7
$j=4$	2	2	4	7	$j=4$	7	2	5	4

TABLE 8.8

The Transferring Time from Supplier to Producer

$k=1$	$i=1$	$i=2$	$i=3$	$i=4$	$k=5$	$i=1$	$i=2$	$i=3$	$i=4$
$j=1$	10	12	11	11	$j=1$	11	10	11	12
$j=2$	10	11	12	12	$j=2$	10	12	12	11
$j=3$	12	10	10	10	$j=3$	10	12	11	12
$j=4$	11	12	12	10	$j=4$	12	10	11	10
$k=2$	$i=1$	$i=2$	$i=3$	$i=4$	$k=6$	$i=1$	$i=2$	$i=3$	$i=4$
$j=1$	10	10	12	11	$j=1$	10	12	11	10
$j=2$	12	11	11	10	$j=2$	12	10	10	11
$j=3$	12	11	10	10	$j=3$	11	11	12	12
$j=4$	11	12	11	12	$j=4$	11	10	12	11
$k=3$	$i=1$	$i=2$	$i=3$	$i=4$	$k=7$	$i=1$	$i=2$	$i=3$	$i=4$
$j=1$	10	12	11	10	$j=1$	11	10	12	10
$j=2$	11	10	12	11	$j=2$	10	11	10	11
$j=3$	10	10	10	12	$j=3$	12	12	11	12
$j=4$	12	11	10	10	$j=4$	12	11	10	10
$k=4$	$i=1$	$i=2$	$i=3$	$i=4$	$k=8$	$i=1$	$i=2$	$i=3$	$i=4$
$j=1$	10	12	11	10	$j=1$	10	11	12	10
$j=2$	12	11	10	12	$j=2$	10	12	12	11
$j=3$	10	12	10	11	$j=3$	12	10	10	11
$j=4$	10	11	12	10	$j=4$	10	11	11	12

TABLE 8.9

The Transferring Time from Producer to Customer

$k=1$	$i=1$	$i=2$	$i=3$	$i=4$	$k=5$	$i=1$	$i=2$	$i=3$	$i=4$
$j=1$	7	17	12	11	$j=1$	13	12	10	12
$j=2$	16	7	10	16	$j=2$	12	8	9	16
$j=3$	12	11	9	8	$j=3$	8	13	16	15
$j=4$	15	13	15	10	$j=4$	7	11	14	17
$k=2$	$i=1$	$i=2$	$i=3$	$i=4$	$k=6$	$i=1$	$i=2$	$i=3$	$i=4$
$j=1$	17	16	9	10	$j=1$	12	17	12	10
$j=2$	8	17	11	15	$j=2$	9	12	8	17
$j=3$	10	11	13	17	$j=3$	17	17	10	14
$j=4$	7	9	10	4	$j=4$	11	10	15	13
$k=3$	$i=1$	$i=2$	$i=3$	$i=4$	$k=7$	$i=1$	$i=2$	$i=3$	$i=4$
$j=1$	16	9	10	10	$j=1$	12	16	13	10
$j=2$	14	11	12	14	$j=2$	15	14	10	11
$j=3$	7	16	16	17	$j=3$	8	13	7	17
$j=4$	10	8	15	9	$j=4$	10	8	9	13
$k=4$	$i=1$	$i=2$	$i=3$	$i=4$	$k=8$	$i=1$	$i=2$	$i=3$	$i=4$
$j=1$	14	12	7	11	$j=1$	7	10	13	15
$j=2$	17	10	8	13	$j=2$	11	15	7	16
$j=3$	12	14	16	10	$j=3$	12	13	16	10
$j=4$	10	13	9	17	$j=4$	14	7	9	17

TABLE 8.10

The Completion Time

k = 1	i = 1	i = 2	i = 3	i = 4	k = 5	i = 1	i = 2	i = 3	i = 4
j = 1	11	12	8	10	j = 1	6	12	8	10
j = 2	6	11	12	8	j = 2	8	9	11	7
j = 3	7	6	9	9	j = 3	12	6	9	11
j = 4	10	9	11	12	j = 4	10	7	6	9
k = 2	i = 1	i = 2	i = 3	i = 4	k = 6	i = 1	i = 2	i = 3	i = 4
j = 1	9	12	8	10	j = 1	8	9	7	10
j = 2	7	9	12	8	j = 2	7	8	9	12
j = 3	12	10	8	9	j = 3	10	11	6	8
j = 4	9	11	10	6	j = 4	6	10	12	9
k = 3	i = 1	i = 2	i = 3	i = 4	k = 7	i = 1	i = 2	i = 3	i = 4
j = 1	9	10	8	6	j = 1	12	8	11	10
j = 2	9	12	11	10	j = 2	7	11	8	9
j = 3	11	8	10	12	j = 3	10	12	6	7
j = 4	11	6	8	7	j = 4	6	7	11	12
k = 4	i = 1	i = 2	i = 3	i = 4	k = 8	i = 1	i = 2	i = 3	i = 4
j = 1	7	11	10	9	j = 1	11	8	12	10
j = 2	8	12	9	6	j = 2	12	7	9	8
j = 3	12	6	8	10	j = 3	9	10	6	12
j = 4	9	11	10	12	j = 4	6	9	10	9

TABLE 8.11

The Process Capabilities for Different Operations for Different Products

Cp_{ik}	i = 1	i = 2	i = 3	i = 4
k = 1	2.8	2.93	3.1	3.3
k = 2	3.2	3.3	2.95	2.85
k = 3	2.7	3.2	2.75	2.78
k = 4	2.81	2.96	3.25	3.05
k = 5	2.75	2.95	3	3.15
k = 6	2.98	3.1	2.93	2.7
k = 7	3.3	2.89	2.83	2.9
k = 8	2.95	2.83	3.24	3.18

Using the input data, the model optimization is fulfilled using LINGO software and the optimal solutions are as follows:

The objective function value: 4575.25 unit of money,

$X1 = 27$

$X2 = 35$

$X3 = 31$

$X4 = 38.$

where $X1 = 27$ implies that the optimal value for product one should be produced 27 unit to satisfy the objective function being the minimal costs and maximal process capability. The binary decision variable's values are reported in Table 8.12.

TABLE 8.12

The Results of Binary Variable

$k=1$	$i=1$	$i=2$	$i=3$	$i=4$	$k=5$	$i=1$	$i=2$	$i=3$	$i=4$
$j=1$	1	1	1	1	$j=1$	0	0	0	0
$j=2$	0	0	0	0	$j=2$	1	0	1	1
$j=3$	0	0	0	0	$j=3$	1	1	1	0
$j=4$	0	0	0	0	$j=4$	0	0	0	0
$k=2$	$i=1$	$i=2$	$i=3$	$i=4$	$k=6$	$i=1$	$i=2$	$i=3$	$i=4$
$j=1$	1	1	0	1	$j=1$	0	0	0	0
$j=2$	0	0	0	0	$j=2$	0	0	0	0
$j=3$	0	0	0	0	$j=3$	1	1	1	1
$j=4$	0	0	0	0	$j=4$	0	0	0	0
$k=3$	$i=1$	$i=2$	$i=3$	$i=4$	$k=7$	$i=1$	$i=2$	$i=3$	$i=4$
$j=1$	0	0	0	0	$j=1$	0	0	0	0
$j=2$	0	1	1	0	$j=2$	0	0	0	0
$j=3$	1	1	1	1	$j=3$	0	0	0	0
$j=4$	0	0	0	0	$j=4$	1	0	1	0
$k=4$	$i=1$	$i=2$	$i=3$	$i=4$	$k=8$	$i=1$	$i=2$	$i=3$	$i=4$
$j=1$	0	0	0	0	$j=1$	0	0	0	0
$j=2$	0	0	0	0	$j=2$	0	0	0	0
$j=3$	1	1	1	1	$j=3$	0	0	0	0
$j=4$	0	0	0	0	$j=4$	1	0	1	1

Table 8.12 shows the operations being performed in each stage of product life cycle for different products. For instance, operation 3 (quality) should be considered for product 2 in the second stage (growth) leading to customer satisfaction and at the same time cost minimization and process capability maximization. The results help the managers to determine the operations in different stages of life cycle corresponding to the optimal solutions for more attention. These operations are considered in two aspects—determination and investment. Of course, the determination ensues to investment, that is, when an operation determined to be optimal, then economic investment is concentrated on that operation.

8.5 Discussions

We proposed a mathematical model in a three layer supply chain using customer relationship management extracting the product life cycle factors to satisfy customers' interests. Then, the mathematical model was employed to obtain the optimal number of products and the fulfillment of an activity in a product life cycle stages. Customers' opinions about the product life cycle were collected via a customer relationship system. The problem was modeled to minimize the total cost of the system and maximize the process capability. An application example is illustrated to verify the effectiveness of the proposed methodology. The contributions of the work focused on the customer satisfaction via product life cycle operations, product delivery within due dates causing customer satisfaction, and considering cost minimization and process capability maximization as quality measure at the same time. As future research, including uncertainty of the variables and parameters, adding other variables such as pricing and other objectives such as benefit could be considered.

References

Andersen F, Iturmendi F, Espinosa S, Diaz MS. Optimal design and planning of biodiesel supply chain with land competition. *Computers & Chemical Engineering* 2012;47(20):170–182.

Bradley PJ, Thomas TG, Cooke J. Future competition: Supply chain vs. supply chain. *Logistics Management and Distribution Report* 1999;39(3):20–21.

Chen X, Li L, Zhou M. Manufacturer's pricing strategy for supply chain with warranty period-dependent demand. *Omega* 2012;40(6):807–816.

Combs L. The right channel at the right time. *Industrial Management* 2004;46(4):8–16.

Conner M. The supply chain's role in leveraging PLM. *Supply Chain Management Review* 2004;8(2):36–43.

Daly R, Kolassa M. Start earlier, sell more, sell longer. *Parmaceutical Executive* 2004;1(8):30–38.

Davis T. Effective supply chain management. *Sloan Management Review* 1993;34(4):35–46.

Elgazzar SH, Tipi NS, Hubbard NJ, Leach DZ. Linking supply chain processes" performance to a company's financial strategic objectives. *European Journal of Operational Research* 2012;223(1):276–289.

Erzurumlu SS. The compatibility of durable goods with contingent generic consumables. *Omega* 2013;41(3): 574–585.

Geoffrion AM, Graves GW. Multicommodity distribution system design by Benders decomposition. *Management Science* 1974;20(5):822–844.

Ghasemy Yaghin R, Torabi SA, Fatemi Ghomi SMT. Integrated markdown pricing and aggregate production planning in a two echelon supply chain: A hybrid fuzzy multiple objective approach. *Applied Mathematical Modelling* 2012;36(12):6011–6030.

Hahn GJ, Kuhn H. Simultaneous investment, operations, and financial planning in supply chains: A value-based optimization approach. *International Journal of Production Economics* 2012;140(2):559–569.

Huang SM, Su JCP. Impact of product proliferation on the reverse supply chain. *Omega* 2013;41(3):626–639.

Hughes D. Managing high-tech product cycles. *The Executive* 1990;4(2):44–56.

Iglesias L, Laca A, Herrero M, Díaz M. A life cycle assessment comparison between centralized and decentralized biodiesel production from raw sunflower oil and waste cooking oils. *Journal of Cleaner Production* 2012;37:162–171.

Ishii K, Eubanks CF, Marco PD. Design for product retirement and material life-cycle. *Journal of Materials & Design* 1994;15(4):225–233.

Kasilingam RG. *Logistics and Transportation—Design and Planning*. Dordrecht: Kluwer Academic, 1998.

Keeney R, Raiffa H. *Decision with multiple objectives: Preference and value tradeoffs*. New York, NY: Wiley and Sons, 1976.

Klausne M, Wolfgang MG. Integrating product take-back and technical service. In: *Proceedings from the IEEE International Symposium on Electronics & the Environment*, Danvers, MA, 1999:48–53.

Korth K. Learning to compete in the current environment. *Automat Design and Production* 2003.

Krajewski L, Wei JR, Tang LL. Responding to schedule changes in build-to-order supply chains. *Journal of Operations Management* 2005;23(5):452–469.

Kumar N, Singh SR, Kumari R. Three echelon supply chain inventory model for deteriorating items with limited storage facility and lead-time under inflation. *International Journal of Services and Operations Management* 2012;13(1):98–118.

LaRosa AD, Cozzo G, Latter A, Recca A, Björklund A, Parrinello E, Cicala G. Life cycle assessment of a novel hybrid glass-hemp/thermoset composite. *Journal of Cleaner Production*, 2012.

Mason-Jones R, Towill DR. Information enrichment designing the supply chain for competitive advantage. *Supply Chain Management: An International Journal* 1997;2(4):137-148.

O'Marah K. The business case for PLM. *Supply Chain Management Review* 2003;7(6):16–18.

Pagh JD, Cooper MC. Supply chain postponement and speculation strategies how to choose the right strategy. *Journal of Logistics Management* 1998;19(2):13–33.

Prasanna Venkatesan S, Kumanan S. Supply chain risk prioritization using a hybrid AHP and PROMETHEE approach. *International Journal of Services and Operations Management* 2012;13(1):19–41.

Qiang Q, Ke K, Anderson T, Dong J. The closed-loop supply chain network with competition, distribution channel investment, and uncertainties. *Omega* 2013;41(2):186–194.

Saaty TL. *The Analytic Hierarchy Process*. McGraw-Hill, New York, 1980.

Schniederjans MJ, Garvin T. Using the analytic hierarchy process and multi-objective programming for the selection of cost drivers in activity based costing. *European Journal of Operational Research* 1997;100: 72–80.

Schonsleben P. *Integrales Logistikmanagement: Planung and Steuerung von Umfassenden Geschäftsprozessen*, Berlin, 1998.

Shewchuck P. Agile manufacturing: One size does not fit all. *Proceedings of International Conference on Manufacturing Value Chains*, Troon, 1998:143–150.

Stadtler H. Supply chain management—an overview. In: Stadtler H, Kilger C (ed.), *Supply Chain Management and Advanced Planning—Concepts, Models, Software and Case Studies*. Berlin, 2000:7–29.

Stadtler H, Sahling F. A lot-sizing and scheduling model for multi-stage flow lines with zero lead times. *European Journal of Operational Research* 2013;225(3):404–419.

Teng J-Y. *Project Evaluation: Method and Applications*. National Taiwan Ocean University, Keelung, Taiwan, 2002.

Teresko J. The PLM Revolution. *Industry Week* 2004;253(2):32–36.

Thomas DJ, Griffin PM. Coordinated supply chain management. *European Journal of Operational Research* 1996;94:1–15.

Tompkins JA. *No Boundaries*. North Carolina: Tompkins Press, 2000:50.

Towill DR. Engineering the agile supply chain. In: Gunesekaran A (ed.), *Agile Manufacturing: 21st Century Manufacturing Strategy Chapter 8*. Oxford: Elsevier Science, 2001:377–396.

Vidal CJ, Goetschalckx M. Strategic production—distribution models: A critical review with emphasis on global supply chain models. *European Journal of Operational Research* 1997;98:1–18.

Wang X, Li D. A dynamic product quality evaluation based pricing model for perishable food supply chains. *Omega* 2012;40(6):906–917.

Zhong RY, Dai QY, Qu T, Hu GJ, Huang GQ. RFID-enabled real-time manufacturing execution system for mass-customization production. *Robotics and Computer-Integrated Manufacturing* 2013;29(2):283–292.

9

Supply Chain: Activity-Based Costing, Pricing and Earned Value

SUMMARY This chapter proposed an integrated methodology in a supply chain to consider the costs in pricing. The costs were obtained via activity based costing due to different activities performed in the supply chain. The customer willingness to pay was a merit effective on the total price of a product. Also, earned value technique is employed to analyze the effectiveness of pricing model.

9.1 Introduction

For many years, members of supply chains have been separated by organization and philosophy. Interactions between them have often been adversarial, with each trying to gain at the other's expense. Today, this long-established pattern is rapidly giving way to system integration due to increasing external competitive threat. The advocates argue that all of the subsystems of a supply chain are connected. The outputs from one system are the inputs of the other systems. Thus, integration of the complete scope of the supply chain from the supplier, through the manufacturer, to the retailer needs to be considered so that fully transparent information is shared freely among members, and collective strategies can be designed to optimize the system's joint objectives. While the importance of achieving integration in the supply chain is generally well recognized, for real-world applications designing a sophisticated integrated system is an arduous task. Few firms are so powerful that they can manage the entire supply chain so as to drive individual members to a superimposed integrated objective (Lee, 2007).

A fundamental change in the global competitive landscape is driving prices to levels that in real terms are as low as they have ever been. A number of causal factors have contributed to this new market environment. First, there are new global competitors who have entered the marketplace supported by low-cost manufacturing bases. The dramatic rise of China as a major producer of quality consumer products is evidence of this. Second, the removal of barriers to trade and the de-regulation of many markets have accelerated this trend enabling new players to rapidly gain ground. One result of this has been overcapacity in many industries (Greider, 1998). Over capacity implies an excess of supply against demand and hence leads to further downward pressure on price. A further cause of price deflation, it has been suggested (Marn et al., 2003), is the Internet, which makes price comparison so much easier. The Internet has also enabled auctions and exchanges to be established at industry wide levels that have also tended to drive down prices.

Changes in competition (globalization, standardization in production and so on) have recently led to many businesses cutting production in order to focus on key competencies. Thus, an even larger portion of value added is subcontracted resulting in significant expansion in the supply chain in many industrial markets. While this trend has brought benefits and businesses have been able to concentrate on their strengths and focus their main assets in specific areas, this strategic orientation also has increased the need to collaborate and integrate activities between the different companies in the supply chain. Therefore, most companies today try to establish relationships with their partners in the supply chain rather than concentrating on purchasing (Narayandas and Rangan, 2004).

This development is further supported by today's business relationships offering one of the most effective remaining opportunities for significant cost reduction and value improvement (Christopher and Gattorna, 2005). However, Frazier et al. (1988) observed that these opportunities mainly depend on the closeness of the relationship. In this sense, suppliers in particular have cultivated business relationships for years by investing in their customers with a view to safeguarding subsequent business dealings from outsuppliers (Jackson, 1985). However, there comes a point where making business relationships closer is only possible when both the supplier and the customer are prepared to invest in this special type of collaboration, as relationships in which the reason for staying in are solely determined by investments made on the part of the supplier are unstable by their very nature. As soon as competitors offer comprehensive benefits in alternative business transactions, there is an economic reason for customers to switch suppliers (Bonner and Calantone, 2005). This means that further investments will only become financially viable from the supplier's point of view if the customer is also prepared to put himself into a position of some dependence on the supplier. But transaction partners may devolve their economic welfare, at least in part, to the conduct of the other partner. Companies must be aware that supply chain pricing (SCP) will only provide a clear competitive advantage for the period of time when the competitors not yet have adapted to the new perspective. Taking the situation into consideration where a market or branch has completely switched into SCP, the use of our concept will no longer dispose of our stated overall advantage. In this situation, it can surely amount to nothing more than the prevention of competitive disadvantage (Rokkan et al., 2003).

9.2 Problem Definition

Globalization of the markets and competition between enterprises as well as increasing the customers' expectations for achieving high quality products and better services has led enterprises to try for their survival, and increase their efficiency in the supply chain. Since a reasonable competitive appraisal with other enterprises is one of the important goals of any organization, applying some methods to allocate costs based on the activity can be a good way to reduce costs and be competitive appraisal in the market. The recognition of the present costs, collecting of the costs in the related centers, and applying multiple factors are three main components to allocate costs based on activity. The activities evaluated in the current research include human resources, production, maintenance, and transportation. Using costs, customers willing to pay, and earned value we can do pricing.

9.2.1 Activity Based Costing

Activity based costing method is that activities spent products, resources spent activities, costs spent resources. This study considers activity costs for a three-layer supply chain including supplier, manufacturer, and customer.

9.2.2 Costs of Manufacturing

Cost of raw material production: First, raw materials (that are meats in our study) are provided from renowned slaughter houses. To save time, the received materials are prepared for final processing in a special salon.

Cost of human resources: Skilled manpower and the knowledge are assets of an organization and a key competitive advantage and scarce resource in today's knowledge-based economy. Therefore, the business strategy of today's organizations is essentially focused on human resources.

Cost of manpower = (Variable cost of manpower per person × number of hours) + Fixed costs

Cost of production: Production costs include costs are that created during the production.

Cost of production = (Variable cost of production × each unit of output) + Fixed costs

Cost of maintaining products: Maintenance costs comprise the major part of production costs. According to type of industry will take about 15%–60% of the cost of production.

Cost of maintaining products = (Variable cost of maintenance × each unit of output) + Fixed costs

Cost of transportation: Activities will be created between the manufacturer and the customer, including the transportation.

Cost of transportation = (Variable cost of transportation × each unit of output) + Fixed costs

9.2.3 The Earned Value

The percentage deviation of the cost difference between the real cost of work performed and the earned value represents the amount of deviation costs.

$$CV = EV - AC; CV\% = CV/EV$$

$$CPI = EV/AC \text{ Cost performance Index}$$

9.3 Pricing Model

The scope of pricing and revenue optimization (PRO) is to set and update the prices for each combination of product, customer segment and channel. The goal in PRO is to provide the appropriate price for any products, any customer segments and any channels. Due to market condition changes over time, the PRO is responsible to update the prices. The basic element of a PRO is the price-response function or the price-response curve showing by $d(P)$. The following specifications are considered for price-response curve:

- Non-negativity
- Continuous
- Differentiable

The price-response curve is pursuing the demand changes considering the variations in price for a product. The demand function is linear and increasing the price causes reduction in demand until for the maximum demand the price gets to zero. This type of demand function is called demand curve for monopoly market in economy.

In the real world, the producer determines the price and customers decide to buy or not. The price-response function identifies the number of potential customers transforming to active ones while the producer lowers the price or how many active customers are missed if the producer raises the prices. Thus, customers' behavior configures the price-response function. Let us investigate the concept of willingness to pay (WTP). Any potential customer has a maximum WTP which is called reservation price interchangeably. Therefore, a customer would buy a product only if its price is lower than his maximum WTP. We can compute the WTP of different customers for a given price interval $[p1, p2]$,

$$\text{WTP} = \int_{P2}^{P1} W(X)\,dx$$

Also $W(x)$ is the WTP function. As stated, the maximum demand (D) is obtained to be $D = d(p)$, that is, the maximum demand is when the price is zero. Here, we can obtain the price-response function using WTP,

$$d(x) = D \cdot \int w(x)d(x)$$

In our proposed model, assuming WTP is a uniform probability distribution function, the demand function is considered to be linear using the proposed integral. As a result, the following linear demand function is obtained

$$d(p) = D - m \cdot P$$

$D = d(0)$ is the maximum demand, m is the gradient of the demand curve and P is the price.

Also for making the maximum profit, obtained prices are multiplied by the number of products.

9.3.1 Calculate the Costs of Supply Chain Based on Activity Based Costing

Total production cost, manpower cost, transportation cost and maintenance cost in the supply chain can be formulated based on activity based costing for each product as follows:

$$F(C) = F(CI) + F(CH) + F(CP) + F(CM) + F(CTr)$$

9.3.2 Customer Demands

Each of the products made by a manufacturer has specific demands from customers. First, we get the costs of the products. As a result, the total costs of production divided by the number and multiplied by the production rate provide the cost per product. Now, we can get breakeven point price for the products using the following formula:

$$(Price * number) - costs = 0$$

Breakeven point prices are the lower boundary for the customers' willingness to pay integral; the upper limit of integral is a certain amount of profit multiplied by the breakeven point value to obtain expected profit. After obtaining the price of each product in the breakeven point and an ideal profit, and the computed willingness to pay for different customers, now we get to the pricing using the following equation:

$$Price = (breakeven\ price + willingness\ to\ pay) * number$$

In order to obtain the earned value of different pricing technique, the following relations are used:

$$CV = EV - AC; CV\% = CV/EV$$

9.4 Discussions

This chapter proposed an integrated methodology in a supply chain to consider the costs in pricing. The costs were obtained via activity based costing due to different activities performed in the supply chain. The customer willingness to pay was a merit effective on the total price of a product. Breakeven point (via activity based costing), favorite expected profit for company and willingness to pay for customer were three significant elements in forming the optimal price. Earned value was an analytical method to compare the current prices and prices obtained by the proposed approach confirming the proposed approach's efficiency.

References

Bonner JM, Calantone RJ. Buyer attentiveness in buyer–supplier relationships. *Industrial Marketing Management* 2005;34(1):53–62.

Christopher M, Gattorna J. Supply chain cost management and value-based pricing. *Industrial Marketing Management* 2005;34:115–121.

Frazier GL, Spekman R, O'Neal C. Just-in-time exchange relationshipsin industrial markets. *Journal of Marketing* 1988;52(4):52–67.

Greider W. *One world, ready or not*. New York: Simon & Schuster, 1998.

Lee CH. Coordination on stocking and progressive pricing policies for a supply chain. *International Journal of Production Economics* 2007;106:307–319.

Jackson B. *Winning and Keeping Industrial Customers*. Toronto: Lexington Books, 1985.

Marn MV, Roegner EV, Zawada CC. The power of pricing. *McKinsey Quarterly* 2003;1:27–39.

Narayandas D, Rangan VK. Building and sustaining buyer–seller relationships immature industrial markets. *Journal of Marketing* 2004;68(3):63–77.

Rokkan AI, Heide JB, Wathne KH. Specific investments in marketing relationships: Expropriation and bonding effects. *Journal of Marketing Research* 2003;40(2):210–225.

10

Multi-Product Supply Chain: Customer Utility and Risk Model

SUMMARY   In this chapter, we are looking for pricing in a multi-product supply chain including suppliers, producers, and customers based on the utility of the customers. First, by using cost functions that include fixed and variable costs of human resources, and transportation, inventory and production costs, we obtain the supply chain costs, and then we investigate customer willingness to pay for function as utility. Using the utility model, the pricing for products considering the risk of decision-making is investigated.

10.1 Introduction

Today, with the increasing complexity of manufacturing processes and also the need to create a variety of goods and services, one organization alone, without the assistance and cooperation of other organizations, will not be able to produce goods or provide suitable services anymore. For this purpose, creating a supply chain that consists of suppliers, producers, etc., to work together along with producing and providing goods and services acceptable to the customer is a necessity for companies to survive and thrive in today's world.

On the other hand, customers' expectations of high quality, prompt service and reasonable prices for goods and services have also increased pressure on organizations. Therefore, organizations seek to reduce costs and increase quality so that they can satisfy customers' needs as the most important factor toward creating value for organizations and increasing their utility. In recent years, organizations have considered customer satisfaction as the most important factor in gaining competitive edge. To determine and modify costs and also to obtain the cost of the products, activity based costing method can be used. Activity based costing method is one of the new systems in costing products and services, which satisfies needs such as accurate calculation of the cost of production, improvement of production processes, elimination of redundant activities, and identification of cost drivers, planning operation and determination of business strategies for economic units (Roudposhti, 2009).

After calculating the cost of the supply chain, to start the pricing process, considering customer satisfaction and utility is of great importance. Some scholars have defined the concept of customer satisfaction as feelings of pleasure or disappointment resulting from comparing their perception of performance and the outcome of a service or a particular product with their own expectations of that product or service (Brady and Robertson, 2001). Also, customer willingness to pay expresses the maximum amount of money that the customer is willing to pay for a particular product or service.

By calculating the cost of the supply chain and the price of products, the chain profit can be achieved. Profit is under the influence of risk factors; these factors can reduce the total profit of the supply chain. Supply chain risk is the uncertainties or events that are unpredictable which affect one or more of the members of the supply chain and influence its undertaking position in achieving its objectives (Fischhoff et al., 1981).

Thus, decision makers in a supply chain can take actions to process pricing products, calculating their costs and considering utility and the maximum of customer willingness to pay. Also, with regard to risk factors, profit is calculated considering supply chain risk and accordingly prices are modified.

In a study done based on how the focus of supply chain processes is on the client, it was noted that integrated supply chain is able to improve processes within the company and supply chain. This study compares New Zealand and the UK automotive industries and also examines the two countries' supply chain of the automotive industry. Finally, this study ends with understanding value, making flows for clients and considering the uncertainty in these flows (Childerhouse et al., 2002).

Costing based on activity and its application in the production environment has been widely discussed. There are several examples of applications using activity based costing in making decisions related to the management of production. Some of these studies include Berling's research (2008) in terms of storage costs and inventory, Baykasoglu and Kaplanoglu's surveys (2008) in relation to the cost of transportation, Pirttila and Hautaniemi's study (1995) of the distribution of logistics, and Tatsiopoulos and Panayiotou's research (2000) on re-engineering and its costs.

In a study, a framework was presented as "Activity-based costing model for supply chain management" (Figure 10.1). It was expressed in this model that the traditional tools for costing and accounting are not useful in the field of supply chain management. Additionally, there are not desired standards for the definition and composition of costs. These obstacles and deficits make the conversion and comparison of the cost data different and sometimes conflicting among members of a supply chain.

Accordingly, reviewing different models develops a conceptual and perceptual framework for activity based costing in a supply chain. Using an appropriate model, cost reduction opportunities both within the company and across the supply chain are identified (Schulze et al., 2012).

Zhang and Xia (2010) began their studies in relation to a pricing model based on cost of a manufacturing and operational unit. This was formed by studying cost function and showed that as the average cost is reduced, efficiency (output) increases. Then they extended their studies and presented a costing model based on short-term pricing in a supply chain with multiple operation units. To price products in the supply chain, microeconomic methods can be used. The main methods used are

- Exclusive pricing: In which the maximum profit of the whole supply chain network is searched.
- Exclusive-competitive pricing: In the network of supply chain, every manufacturing and operational unit looks for maximizing profit, so we get the exclusive-competitive equilibrium price (Dong and Zhang, 2004; Li et al., 2007; Lee and Wilhelm, 2010).
- Pricing based on costs: For each manufacturing and operational unit in the supply chain, it is considered that cost equals income and demand equals supply (Zhang and Xia, 2010).

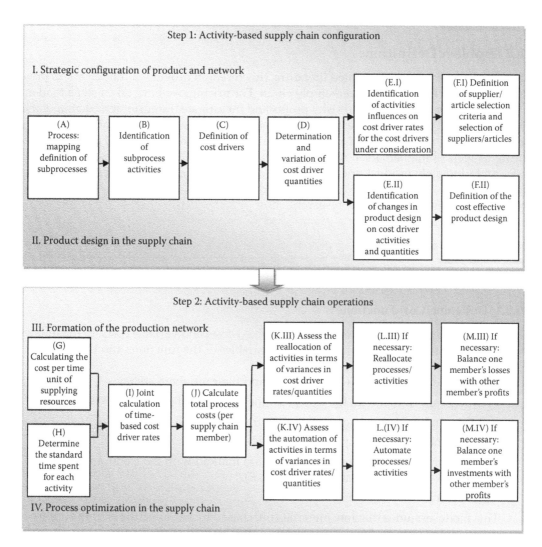

FIGURE 10.1
Activity-based costing model for supply chain management.

In a study on the pricing of new products, a comprehensive methodology was presented based on analysis of the costs to develop, produce and deliver a new product from its idea to market, and then using the model, a case study related to a new product, carbon nanotubes, was investigated. This study continued with determining the minimum and maximum price and also using cost-based, market-based and value-based approaches in pricing and ultimately the cost-based pricing process in which a percentage is added to the whole cost.

The most important advantage of this model is that despite the use of cost-based pricing, a percentage is added to the cost and thereby the selling price is obtained, but this pricing method does not neglect studying market and demand and assesses the competitive position in the market through market considerations in the product life cycle (Bandariyan et al., 2010).

10.2 Problem Definition

Studying and reviewing the related literature, this research provides a framework to price different products in a three-layer supply chain. For pricing, we have taken steps to identify the costs of supply chain and risky profits and investigate the customers' demand and the maximum of their willingness to pay. This task is done using mathematical modeling of cost function in the areas of manpower, maintenance, transportation, and manufacturing as well as mathematical modeling of customers' willingness to pay, benefits and risks, and ultimately product pricing. In this study, the four factors causing risks in profits include inflation, exchange rates, lost sales and competitors. In this study, the analytical methods are used to conduct the research.

10.3 Utility-Based Model

10.3.1 Designing Cost Functions

Each of the four cost functions consists of both fixed costs and variable costs. In the following, we attempt to define the parameters and variables of the functions and their design.

- The suppliers are marked with the icon i and their numbers are I ($i = 1, 2, ..., I$).
- Producers are shown with the icon j and the model has J producer ($j = 1, 2, ..., J$).
- K is the number of customers and is shown with k ($k = 1, 2, ..., K$).
- Q is the kinds of products produced, that is, shown with q ($q = 1, 2, ..., Q$).
- The raw materials that suppliers provide are marked with the icon r and R is the kind of raw material available ($r = 1, 2, ..., R$).
- Costs are marked with the letter c.
- This model assumes that raw materials purchased by manufacturers from suppliers are not maintained and immediately enter the flow of production.

10.3.2 Functions of Manpower Costs

Two functions of manpower costs are considered, one for suppliers and the other for manufacturers. Manpower will be represented by the index m. The functions of manpower costs consist of fixed and variable costs. The main fixed costs of manpower are the salaries, benefits, and fixed wages that are paid periodically and the services offered by human resources. Other fixed costs that can be noted are the costs of personnel insurance, hedging and training manpower. The main variable cost of manpower is overtime. Variable cost of human resources is obtained and calculated by multiplying the cost of per hour of overtime by overtime hours.

10.3.3 The Function of Suppliers' Manpower Cost

$$f(Cm_i) = \sum_{i=1}^{I} [M_i + (c'm_i \times hm_i)] \tag{10.1}$$

Cm_i: The cost of the supplier's manpower of i

M_i: Fixed cost of the supplier's manpower of i

$c'm_i$: The cost of per hour overtime of suppliers' manpower of i

hm_i: Overtime of suppliers' manpower of i

10.3.4 The Function of Manufacturers' Manpower

$$f(Cm_j) = \sum_{j=1}^{J} [M_j + (c'm_j \times hm_j)]$$
(10.2)

Cm_j: The cost of manufacturers' manpower of j

M_j: The fixed cost of manufacturers' manpower of j

$c'm_j$: The cost of per hour of overtime of manufacturers' manpower of j

hm_j: The hours of the overtime of manufacturers' manpower of j

10.3.5 Functions of Transportation Costs

In this model, two functions of transportation costs are considered: one, raw material transportation from suppliers to manufacturers and the other, product transportation from manufacturers to customers. Each one of these functions consists of fixed and variable costs of transportation. This model is marked with the symbol of transportation activities. In this model, transportation activities are shown with t. The cost of buying or renting machinery and transportation equipment, including means of transport by road, rail, air and sea, as well as insurance of displaced items and depreciation of machinery, equipment and transport facilities and generally all the costs which are not under the influence of weight, size and number of displaced products, are parts of transportation fixed costs.

Fuel cost of transportation equipment, the cost of repairing it, customs fees and border duties—generally all costs that change due to the increase or decrease of the amount of material moved—are included in the variable costs of transportation.

10.3.6 Transportation Costs of Raw Materials from Suppliers to Manufacturers

$$f(Ct_{rij}) = \sum_{j=1}^{J} \sum_{r=1}^{R} [T_{rij} + (c't_{rij} \times nt_{rij})]; \forall i$$
(10.3)

Ct_{rij}: The cost of raw material transportation r from suppliers i to manufacturers j

T_{rij}: The fixed cost of raw material transportation r from suppliers i to manufacturers j

$c't_{rij}$: The cost of every unit of raw material r from suppliers i to manufacturers j

nt_{rij}: The number of raw material transported of the kind r from suppliers i to manufacturers j

10.3.7 The Function of Goods' Transportation Cost from Manufacturers to Consumers

$$f(Ct_{qjk}) = \sum_{k=1}^{K} \sum_{q=1}^{Q} [T_{qjk} + (c't_{qjk} + nt_{qjk})]; \; \forall j \tag{10.4}$$

Ct_{qjk}: Goods' transportation cost q from manufacturer j to customer k

T_{qjk}: Transfer fixed cost q of goods from manufacturer j to customer k

$c't_{qjk}$: Transportation cost of every raw material unit q from manufacturer j to customer k

nt_{qjk}: The number of transported goods q from manufacturer j to customer k

10.3.8 Functions of Inventory Holding Costs

In this model, like the manpower and transportation costs, two functions of inventory holding costs are considered: one, the holding cost of suppliers' raw material and the other, the holding cost of manufacturers' products. The inventory holding activity is marked with the index S. Each one of the above mentioned functions consists of the fixed and variable costs of holding. The fixed costs of inventory holding that can be mentioned are the cost of purchasing, building or leasing a warehouse or any place for inventory holding, the cost of storage and item insurance, cost of equipment, machinery and holding costs, software, depreciation of storage and goods, costs of warehouse side facilities and generally expenses that are not under the influence of inventory levels. The variable costs of inventory holding are the ones that change due to the increase or decrease of the inventory levels. Variable costs can be obtained and calculated by multiplying holding cost by the number of held inventory.

10.3.9 The Function of Raw Material Holding Costs by Suppliers

$$f(Cs_{ri}) = \sum_{i=1}^{I} \sum_{r=1}^{R} [S_{ri} + (c's_{ri} \times ns_{ri})] \tag{10.5}$$

Cs_{ri}: Raw material holding cost r by suppliers i

S_{ri}: Fixed holding cost of raw material r by suppliers i

$c's_{ri}$: Holding cost of each unit of raw material r by suppliers i

ns_{ri}: The amount of held raw material r by suppliers i

10.3.10 The Function of the Holding Cost of Products by Manufacturers

$$f(Cs_{qj}) = \sum_{j=1}^{J} \sum_{q=1}^{Q} [S_{qj} + (c's_{qj} \times ns_{qj})] \tag{10.6}$$

Cs_{qj}: Products' holding cost q by manufacturer j

S_{qj}: Fixed holding cost of q product by manufacturer j

$c's_{qj}$: Holding cost of every unit of products q by manufacturer j

ns_{qj}: The number of held products q by manufacturer j

10.3.11 Function of Production Cost

In this model, there is a function of product cost specific to the manufacturer. The function consists of the number of manufacturer (J) produce product type Q. As an example, the production cost of q product can be calculated by j manufacturer. Production activity is marked with the icon p.

Production costs like all the mentioned costs consist of two parts: fixed and variable costs. Fixed costs are the ones like costs of machinery and manufacturing equipment, depreciation cost of machinery, equipment insurance costs, cost of production energy such as water, electricity, fuel, and generally all the costs that have nothing to do with the level of production and always exist. Variable costs are the ones whose level of fluctuation is directly related to the level of production fluctuation. For example, the cost of providing raw materials as the most important one can be noted.

$$f(Cp_{qj}) = \sum_{j=1}^{J} \sum_{q=1}^{Q} [P_{qj} + (c'p_{qj} \times np_{qj})] \tag{10.7}$$

Cp_{qj}: Production cost of products q by manufacturer j

P_{qj}: Fixed production cost q products by manufacturer j

$c'p_{qj}$: Production cost of every unit of product q type by manufacturer j

np_{qj}: Number of produced product by manufacturer j

10.3.12 Product Pricing Based on Customers' Demand and Willingness to Pay

Customer demand in time periods is considered as fixed amount. In this model, customer demand is shown by D. Each one of the number of K customers has a fixed demand of Q type of product, also each one of Q type of product is made by each one of the manufacturers. Thus, D_{qjk} is the product demand q of manufacturer j from customer k. Each K customer for each Q type of product has a maximum willingness to pay. The maximum willingness to pay represents the maximum amount of money that the customer is willing to pay for a specific product. Different customers' willingness to pay in price mileage [p_1, p_2] is calculated using the following equation:

$$\text{WTP} = \int_{p_1}^{p_2} W(x)dx \tag{10.8}$$

In this model, the p_1 price is considered in breakeven point and p_2 price considered with ideal profit. $W(x)$ is the function of customer's willingness to pay which is a Uniform Probability Distribution Function, so we will have:

$$W(x) = \begin{cases} \dfrac{1}{b-a} & a \le x \le b \\ 0 & \text{o.w} \end{cases} \tag{10.9}$$

Customers' maximum willingness to pay for each product is calculated as WTP_{qjk} (Rezaie et al., 2012). In other words, willingness to pay for product q of manufacturer j on behalf of customer k.

Then, adding customers' willingness to pay to the lower level of integration that is usually the breakeven price of products, the price can be calculated.

10.3.13 The Function of Profit

The profit can be calculated in hedging networks by making a relation among the amount of product sales, price of sales and supply chain costs.

$$\Pi = \sum_{j=1}^{J} \sum_{k=1}^{K} \left(P_{qjk} \times N_{qjk} - \sum f(C); \forall q \right) \tag{10.10}$$

P_{qjk}: The price of product sale q of manufacturer j for customer k

N_{qjk}: The amount of product sales q of manufacturer j for customer k

$f(C)$: Supply chain costs

10.3.14 Supply Chain Risk

There are a lot of factors affecting supply chain profit. In this model, four risk factors that are examined include inflation, exchange rates, competitors and lost sales. Through the influence of these factors on the amount of sale and also supply chain costs, the hedging network is affected by benefits.

10.3.15 Inflation Rate Risk

Inflation rate is considered as a factor influencing profit of manufacturers and supply chain. With changes in the general level of prices, raw material prices, manpower costs, transportation, maintenance and many other items are subject to changes that increase the costs and reduce sales rate and consequently supply chain profit. In this model, inflation risk is considered in accordance with prevailing distribution, with uniform probability distribution function of a random variable and parameters that are inflation risk:

$$f(x) = \begin{cases} \dfrac{1}{b-a} & \text{for } a \le x \le b \\ 0 & \text{for } x < a \text{ or } x > b \end{cases} \tag{10.11}$$

10.3.16 Exchange Rate Risk

Supply chains that operate internationally are exposed to the risk of exchange rate fluctuations. Providing foreign raw materials or working with foreign suppliers, purchasing machinery and equipment from international resources, international shipping costs, exporting supply chain goods to other countries and many other issues by impacting the cost of sales, the amount of supply chain sales, network profits can be at risk.

Prevailing distribution of exchange rate risk is Exponential Probability Distribution with the parameter λ and random variable λ that are indicative of exchange rate:

$$f(x) = \lambda e^{-\lambda x}, \quad x \geq 0, \lambda > 0 \tag{10.12}$$

10.3.17 Competitors Risk

In an environment where several manufacturing or service companies are competing with each other, one or more companies' decisions can influence other companies. There is the same situation in supply chain. In the same business environment, decisions and actions that are taken by the members of another supply chain influence the situation of investigated supply chain and can reduce supply chain profit. Prevailing distribution of competitor risk like exchange risk, probability distribution is indicated by parameter λ and random variable λ.

$$f(x) = \lambda e^{-\lambda x}, \quad x \geq 0, \lambda > 0 \tag{10.13}$$

10.3.18 The Risk of Lost Sales

The lost sales are poor responses to customer demand or, in other words, the amount manufacturers can gain by selling products to customers. Due to the lack of product for any reason, it is lost. In prevailing distribution of this risk, Normal Probability Distribution is with the mean μ, standard deviation σ and random variable x:

$$f(x) = \frac{1}{\sigma\sqrt{2\pi}} \exp\left(-\frac{(x-\mu)^2}{2\sigma^2}\right) \tag{10.14}$$

10.3.19 The Function of Risk

In this model, the function of risk consists of the mentioned risk factors, that is, the inflation rate risk, exchange rate risk, competitor risk and the risk of lost sales. Each one of these risks has a Loss Function. The function of risk is as follows:

$$\Lambda = \int_{\infty} \lambda(x) f(x) dx \tag{10.15}$$

$\lambda(x)$: Loss function

x: Continuous random variable associated with each of the risk factors

$f(x)$: The probability density function associated with each risk factor

Equation 10.16 is indicative of Loss Function, in which the constant value is usually 1, t is the objective value and x is continuous random variable.

$$\lambda(x) = C\,|t - x|^2 \tag{10.16}$$

In the end, we can compare the obtained prices from the function of customer willingness to pay, that is, customers' favorable price with supply chain favorable prices in which the risks and benefits are considered taking risks in consideration (Fazlollahtabar and Abbasi, 2012).

10.4 Numerical Study

In a supply chain consisting of three suppliers, two manufacturers and two customers, five kinds of raw materials will be sent to manufacturers by suppliers. The five types of raw materials are used in the construction of two types of products, and eventually the two products are sent to the customer. Each one of the suppliers gives all five raw materials to manufacturers. Also, any two manufacturers make two products and send them to each of two customers. In the construction of the first and second products, all five types of raw materials are used in equal proportions.

10.4.1 Supply Chain Costs

The model has three suppliers, thus, by using Equation 10.1 the total cost of manpower of three suppliers is calculated. In this example, there are also two manufacturers, thus by using Equation 10.2 the total manpower cost of two manufacturers can be achieved. Finally, adding the two numbers together, the total cost of the supply chain manpower is calculated as follows:

$$f(Cm) = 1,260,150,000 \quad \text{Monetary unit}$$

Each of the suppliers gives all five raw materials to two manufacturers. The manufacturers also produce two types of products and send them to the two customers. Thus, by using Equations 10.3 and 10.4, the cost of transporting raw materials from suppliers to manufacturers and also products from manufacturers to customers can be acquired. As a result, the total cost of supply chain transportation equals the following amount.

$$f(Ct) = 3,275,570,000 \quad \text{Monetary unit}$$

Suppliers attempt to hold the five types of raw material to a certain level. The manufacturer also has two holding costs of two types of products. Consequently, by using Equations 10.5 and 10.6 the total cost of supply chain inventory holding can be achieved.

$$f(Cs) = 1,310,080,000 \quad \text{Monetary unit}$$

By using Equation 10.7, the cost of producing two products by two manufacturers is also calculated as follows:

$$f(Cp) = 1,732,800,000 \quad \text{Monetary unit}$$

Finally, by calculating manpower, transportation, inventory holding and production costs, the entire supply chain costs can be achieved.

$$f(C) = f(Cm) + f(Ct) + f(Cs) + f(Cp) = 1,260,150,000$$
$$+ 3,275,570,000 + 1,310,080,000 + 1,732,800,000$$
$$= 7,578,600,000 \quad \text{Monetary unit}$$

10.4.2 Calculation of the Customers' Favorable Prices

To do the product pricing process in a supply chain, the equation of customer willingness to pay, Equation 10.8, is used. Having the customers' demand of a manufacturers' product and also calculating the cost of each one of a manufacturers' products, breakeven price can be achieved. Considering a specific profit by supply chain decision makers, the ideal profit costs can be gained with an amount that represents a high level of integral of customer willingness to pay. Finally, by using Equations 10.8 and 10.9 the customers' favorable price is calculated. The results are shown in Table 10.1.

10.4.3 Calculating the Supply Chain Profit and the Risks

By calculating the demand and the customers' favorable price and by using Equation 10.10, supply chain profit is calculated as follows:

$$\Pi = 347,500,000 \quad \text{Monetary unit}$$

In this model, four factors: inflation rate, exchange rate, competitors and lost sales are considered as factors causing risks in the supply chain profit. Considering the prevailing probability distribution in the field of each risk factor and by using Equations 10.11 through 10.14, the distribution amount of each one of the factors can be calculated. Then, by using loss equation Equation 10.16 and the risk function, total amount of supply chain risk is calculated. The gained results are shown in Table 10.2 (in risk function, what is considered is the low level of integral 0 and its high level of objective value).

By adding the number 1 to risk percentage and multiplying it by supply chain profit, supply chain profit with considering risk can be calculated.

$$\Pi r = 347,500,000 \times 1.01961 = 354,314,475 \quad \text{Monetary unit}$$

TABLE 10.1

Customers' Favorable Prices Based on the Function of Customers' Willingness to Pay

(Product, manufacturer, customer)	(1,1,1)	(2,1,1)	(1,2,1)	(2,2,1)	(1,1,2)	(2,1,2)	(1,2,2)	(2,2,2)
The breakeven price	313,904	313,904	290,628	290,628	361,504	361,504	318,993	318,993
Price with ideal profit	351,404	351,404	328,128	328,128	399,004	399,004	356,493	356,493
Parameters of Equation 10.9 (a,b)	(6,11)	(7,13)	(7,9)	(7.5,10)	(7,11)	(8,13)	(7,9)	(9,11)
Customer willingness to pay	7500	6250	18,750	15,000	9375	7500	18,750	18,750
Customer favorable price	321,404	320,154	309,378	305,628	370,879	369,004	337,743	337,743

TABLE 10.2

Supply Chain Risks

	Inflation Rate	Exchange Rate	Competitors	Lost Sales	Supply Chain Risk
Prevailing probability distribution	Uniform	Exponential	Exponential	Normal	–
Amount of parameters	7,11	0.4	0.1	0.02, 0.1	–
Objective level	0.2	0.3	0.12	0.09	–
Percentage of risk	0.000667	0.00375	0.015	0.0001938	0.01961

10.4.4 Calculating Supply Chain Favorable Price according to Supply Chain Profit Considering Risk

In order to obtain the supply chain favorable prices, first the profit of each one of the manufacturers' products must be calculated. This can be done by having the demand and customers' favorable price and also the cost of manufacturers' products. Then by multiplying this amount by 0.01961 the profit of each of different products is obtained. The gained results are represented in Table 10.3.

In the end, supply chain favorable prices are computed as follows:

The price of first manufacturer's first product for two customers

$$\Pi r_{11} = (P_{111} \times N_{111}) + (P_{112} \times N_{112}) - f(C_{11})$$
$$29,314,807 = (P_{111} \times 3000) + (P_{112} \times 1000) - 1,255,615,000$$
$$3P_{111} + P_{112} = 1,284,930$$

The price of second manufacturer's first product for two customers

$$\Pi r_{12} = (P_{121} \times N_{121}) + (P_{122} \times N_{122}) - f(C_{12})$$
$$137,646,330 = (P_{121} \times 4000) + (P_{122} \times 4000) - 2,325,025,000$$
$$P_{121} + P_{122} = 615,667$$

The price of first manufacturer's second product for two customers

$$\Pi r_{21} = (P_{211} \times N_{211}) + (P_{212} \times N_{212}) - f(C_{21})$$
$$34,412,858 = (P_{211} \times 2000) + (P_{212} \times 2000) - 1,446,015,000$$
$$P_{211} + P_{212} = 740,214$$

TABLE 10.3

Profit Considering the Risk of Each One of Manufacturers' Product

(Product, manufacturer)	(1,1)	(2,1)	(1,2)	(2,2)
Profit	28,751,000	134,999,000	33,751,000	149,999,000
Profit considering risk	29,314,807	137,646,330	34,412,857	152,940,480

The price of second manufacturer's second product for two customers

$$\Pi r_{22} = (P_{221} \times N_{221}) + (P_{222} \times N_{222}) - f(C_{22})$$
$$152,94,0480 = (P_{221} \times 5000) + (P_{222} \times 3000) - 2,551,945,000$$
$$5P_{221} + 3P_{222} = 2,704,885$$

10.5 Discussions

Due to increasing supply chain costs and the need to reduce the prices on the other hand, using this model is recommended to investigate the costs and their reduction and taking the process of appropriate pricing. Additionally, with regard to the competitiveness of the industrial markets and the element of customer as the most important factor in creating competitive advantage, using the equation of customers' willingness to pay presented in this model can be beneficial in determining the maximum amount that a variety of customers can pay for supply chain products causing utility in customers. Also, due to the increasing unpredictability in terms of the competitive environment, considering the risk factors causing risk in supply chain profit and their calculation are recommended. The basis of the research is taking the process of pricing and calculating the supply chain risky profit. It means obtaining the prices of different products so that not only the costs are covered but also the chain gets some profit and more importantly product prices cause customer satisfaction and utility. In this regard, reducing costs to a small amount can be effective in product prices and supply chain profit. With obtaining different prices based on customers' willingness to pay and also obtaining supply chain favorable prices, that is, those achieved after calculating profit considering risk and comparing these two values, it was determined that product prices based on risky profit are slightly more than prices based on customers willingness to pay. The reason is that supply chain decision makers are aware of the various risks while customers do not know these risks exactly; therefore, supply chain decision makers adopt prices commensurate with the level of risk more than prices based on customer willingness to pay so that they cover existing risks. The price increases for a manufacturer's various products and for different customers can be different.

References

Bandariyan R, Safavi F, Rashidi A. Presenting a comprehensive and coherent methodology for the pricing of new products based on the idea to market fundamentals. *Business Surveys*, SNo. 43, 2010:43–57.

Baykasoglu A, Kaplanoglu V. Application of activity-based costing to a land transportation company: A case study. *International Journal of Production Economics* 2008;116(2):308–324.

Berling P. Holding cost determination: An activity-based cost approach. *International Journal of Production Economics* 2008;112(2):829–840.

Brady MK, Robertson CJ. Searching for a consensus on the antecedent role of service quality and satisfaction: An exploratory cross-national study. *Journal of Business Research* 2001;51(1):53–60.

Childerhouse P, Aitken J, Towill DR. Analysis and design of focused demand chains. *Journal of Operations Management* 2002;20:675–689.

Dong J, Zhang D. A supply chain network equilibrium model with random demands. *European Journal of Operational Research* 2004;156(1):194–212.

Fazlollahtabar H, Abbasi A. Heuristic probabilistic approach for prioritizing optimal course delivery policies in e-learning systems. *Annals of Spiru Haret University, Mathematics-Informatics Series* 2012;7(2):37–51.

Fischhoff B, Lichtenstein S, Slovic P, Derby S, Keeney R. *Acceptable risk*. New York: Cambridge University Press, 1981.

Lee C, Wilhelm W. On integrating theories of international economics in the strategic planning of global supply chains and facility location. *International Journal of Production Economics* 2010;124:225–240.

Li J, Cheng TCE, Wang, SY. Analysis of postponement strategy for perishable items by EOQ-based models. *International Journal of Production Economics* 2007;107:31–38.

Pirttila T, Hautaniemi P. Activity-based costing and distribution logistics management. *International Journal of Production Economics* 1995;41(1–3):327–333.

Rezaie B, Esmaeili F, Fazlollahtabar H. Developing the concept of pricing in a deterministic homogenous vehicle routing problem with comprehensive sensitivity analysis. *International Journal for Services and Operations Management* 2012;12(1):20–34.

Roudposhti F. *Activity-based costing and activity-based management*, Tehran, Iran: Termeh Publication. 2009.

Schulze M, Seuring S, Ewering CH. Applying activity-based costing in a supply chain environment. *International Journal of Production Economics* 2012;135(2):716–725.

Tatsiopoulos IP, Panayiotou N. The integration of activity based costing and enterprise modeling for reengineering purposes. *International Journal of Production Economics* 2000;66(1):33–44.

Zhang Y, Xia G. Short-run cost-based pricing model for a supply chain network. *International Journal of Production Economics* 2010;128(1):167–174.

11

Flexible Supply Network: VRP Model

SUMMARY In this chapter, we propose a supply chain that considers multiple depots, multiple vehicles, multiple products, and multiple customers, with multi-time periods. The supplier receives the order and forwards it to depots of multiple products. The depots investigate the capacity level and accept/refuse supplying the order. Considering the location of the customers, the depots decide upon sending the suitable vehicles. Each vehicle has its specific traveling time and cost. We present a mathematical model for the allocation of orders to depots and vehicles minimizing the total cost.

11.1 Introduction

The rapid industrialization and economic growth of many countries around the world have spurred the development of various supply chains reaching around the world. This has provided opportunities for manufacturers to cut costs and be closer to emerging and highly grown markets, but it has also created new risks. As supply chains become increasingly dependent on the efficient movement of materials among geographically dispersed facilities, there is more opportunity for disruption.

Supply chain coordination has gained considerable notice lately from both practitioners and researchers. In monopolistic markets with a single chain or markets with perfect competing retailers, a vertically integrated supply chain maximizes the profit of the chain; for example, Jeuland and Shugan (1983), Cachon (2003), and Bernstein and Federgruen (2005).

A market with two competing supply chains was investigated in the seminal work of McGuire and Staelin (1983). They considered a price (i.e., Bertrand) competition between two suppliers selling through independent retailers. They concluded that, for highly substitutable products, a decentralized supply chain Nash Equilibrium was preferred by both manufacturers. Coughlan (1985) applied this research to the electronics industry and Moorthy (1988) further explained why the decentralized chains could lead to higher profits for the manufacturer and the entire chains. Bonanno and Vickers (1988) investigated a similar model and used geometric insights to show that with franchise fees there were some settings in which the manufacturers' optimal strategy was to sell its products using an independent retailer. Thus, in these cases the manufacturer prefers a decentralized supply chain irrespective of the decision in other supply chains. Neither of these works considers demand uncertainty.

In the operations literature, Wu and Chen (2003) considered a quantity (i.e., Cournot) competition of a duopoly facing newsvendor demand, but they ignored pricing decisions. There are a few common features in all these papers, namely that they ignore the important interaction between price and quantity decisions and they also consider competition only over a single period.

A few supply chain coordination mechanisms that induce the chain to act as if they were vertically integrated (VI) were investigated; for example, buyback (Pasternack, 1985), quantity flexibility (Tsay, 1999), and revenue sharing (Cachon and Lariviere, 2005). See also Cachon (2003) for a survey of this literature. Two more reviews are found in Kouvelis et al. (2006) that focus on supply chain coordination literature published in Production and Operations Management journals during 1992–2006, and in Tang (2005) which covers much literature on supply chain coordination. Lin and Kong (2002) consider a duopoly that has no demand uncertainty and investigate a symmetric Nash Bargaining model. Similar to McGuire and Staelin (1983), they show that Nash Bargaining can lead to higher supply chain profits than a vertically integrated chain.

In a recent work, Baron et al. (2008) investigate the Nash Equilibrium of an industry with two supply chains by extending the seminal work of McGuire and Staelin (1983). Baron et al. (2008) show that both the traditional Manufacturer Stackleberg (MS) and the VI strategies are special cases of Nash Bargaining on the wholesale price when the demand is deterministic. They warn that the supply chain coordination mechanisms, which focus on inducing supply chains to act as if they were vertically integrated, should be treated with caution.

The *vehicle routing problem* (VRP) is one of the most studied problems in operations research. It consists of finding least cost routes for a set of homogeneous vehicles located at a depot to geographically scattered customers. Each customer has a known demand and service duration. The routes have to be designed such that each customer is visited only once by exactly one vehicle, each vehicle route starts and ends at the depot and the total capacity of a vehicle may not be exceeded. The VRP is computationally hard to solve and is usually tackled by heuristic approaches; see, for example, Toth and Vigo (2002), Laporte et al. (2000) and Cordeau et al. (2002).

Finding efficient vehicle routes is an important logistics problem that has been studied for several decades. When a firm is able to reduce the length of its delivery routes or is able to decrease its number of vehicles, then it is able to provide better service to its customers, operate in a more efficient manner and possibly increase its market share. A typical vehicle routing problem includes simultaneously determining the routes for several vehicles from a central supply depot to a number of customers and returning to the depot without exceeding the capacity constraints of each vehicle.

This problem is of economic importance to businesses because of the time and cost associated with providing a fleet of delivery vehicles to transport products to a set of geographically dispersed customers. Additionally, such problems are also significant in the public sector where vehicle routes must be determined for bus systems, postal carriers, and other public service vehicles. In each of these instances, the problem typically involves finding the minimum cost of the combined routes for a number of vehicles in order to facilitate delivery from a supply location to a number of customer locations. Because cost is closely related to distance, a company might attempt to find the minimum distance traveled by a number of vehicles in order to satisfy its customer demands. In doing so, the firm attempts to minimize costs while increasing or at least maintaining an expected level of customer service.

The transportation problem we tackle can be described as a multi-depot pickup and delivery problem with time windows and side constraints (Desrosiers et al., 1995), and is regarded as one of the richest within the class of time constraint vehicle routing and scheduling problems in terms of scope and complexity. The earliest pickup time for shipments corresponds to one-sided time window constraints. In addition, operating time restrictions at some locations impose delivery time windows. The coexistence of consolidation (and of less-than-truckload (LTL) shipments), relaying, and trailer availability requirements in our problem context makes it a unique and even more complicated problem than the ones studied before. Early major work on pickup and delivery problems with time windows has been reported by Savelsbergh and Sol (1995). Variants of the basic problem with context-specific characteristics have been reported by Currie and Salhi (2003), Liu et al. (2003), and Sigurd et al. (2004), to name a few.

The notion of transportation network equilibrium, on the other hand, has a much longer history than that of supply chain networks, and appears at the earliest in the work of Kohl (1841) and Pigou (1920), with the first rigorous mathematical treatment given by Beckmann et al. (1956) in their classic book. Other seminal work in terms of transportation network equilibrium modeling and methodological contributions include those of Smith (1979), Dafermos (1980, 1982), and Boyce et al. (1983). For additional research highlights in transportation network equilibrium, see the paper by Florian and Hearn (1995) and the books by Patriksson (1994) and Nagurney (1999, 2000).

In supply chain modeling and analysis (Lee and Billington, 1993; Slats et al., 1995; Anupindi and Bassok, 1996), one typically associates the decision-makers with the nodes of the multi-tiered supply chain network. In transportation networks, on the other hand, the nodes represent origins and destinations as well as intersections. Travelers or users of the transportation networks seek, in the case of user-optimization, to determine their cost-minimizing routes of travel.

11.2 Problem Definition

We consider different customers being serviced with one supplier. The supplier provides various products and keeps them in different depots. Each depot uses different types of vehicles to carry out the orders. All depots are already stationed at the related locations. Here, we consider a multi-echelon supply chain network (one supplier, multiple depots and customers, multi-commodity with deterministic demands). A set of vehicles exists at each depot. Each depot can store a set of products. The received order list from a customer can be handled by one or several depots at each time. Each selected vehicle for delivery can transfer only one product and after delivering the product, the vehicle returns to its corresponding depot. A penalty is assigned when a delivery time exceeds the predetermined time for transferring the products from depots to customers. A configuration of the proposed model is shown in Figure 11.1.

11.3 Vehicle Routing Model

The mathematical model for this problem is as follows:

Notations

P Set of products

I Set of depots stationed

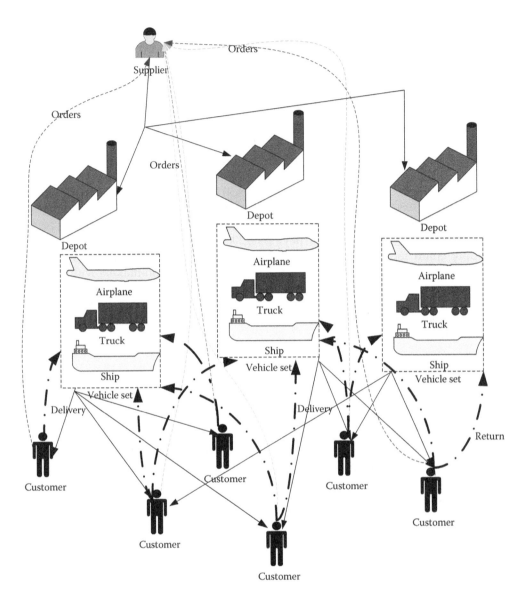

FIGURE 11.1
A configuration of the proposed model.

J	Set of customers
T	Set of time periods
V	Set of vehicles

Parameters

D_{jpt}	Demand of customer j for product p at time t
TH_{ipt}	Maximum throughput of depot i for product p at time t
CA_i	Total capacity of depot i

N_{ivt} The number of existing vehicles v in depot i at time t

VL_{vp} Capacity of vehicle v for product p

d_{ij} Distance between depot i and customer j

r_{ijpvt} Number of return vehicles of type v from customer j to depot i at time t that have already received product p

M A large number

CT_{ijv} Traveling fixed cost per mile from depot i to customer j using vehicle v

TRT_{ijv} Traveling time from depot i to customer j using vehicle v

C The fixed cost for the whole planning horizon

Pen The fixed cost as a penalty

$|t|$ The size of period t

α_{ijv} 1, if the time of delivery from depot i to customer j using vehicle v exceeds a pre-specified limit;
          0, otherwise

Decision Variables

x_{ijpvt} : $\begin{cases} 1, \text{if depot } i \text{ is selected to deliver product } p \text{ to customer } j \text{ by vehicle } v \text{ at time } t \\ 0, \text{otherwise} \end{cases}$

f_{ijpvt}:   Number of transferred vehicle type v from depot i to customer j at time t

QP_{ijpt}:   Quantity of product p that can be satisfied by depot i to customer j at time t

11.3.1 Objective Function

$$\text{Minimize } F = Min(f_1 + f_2 + f_3)$$

$$f_1 = \sum_{i \in I} \sum_{j \in J} \sum_{p \in P} \sum_{v \in V} \sum_{t \in T} f_{ijpvt} \cdot d_{ij} \cdot CT_{ijv} \cdot x_{ijpvt} \tag{11.1}$$

$$f_2 = \sum_{i \in I} \sum_{j \in J} \sum_{p \in P} \sum_{v \in V} \sum_{t \in T} TRT_{ijv} \cdot C \cdot x_{ijpvt} \tag{11.2}$$

$$f_3 = \sum_{i \in I} \sum_{j \in J} \sum_{p \in P} \sum_{v \in V} \sum_{t \in T} \alpha_{ijv} \cdot pen \cdot x_{ijpvt} \tag{11.3}$$

11.3.2 Constraints

$$\sum_{i \in I} QP_{ijpt} = D_{jpt}, \quad \forall j \in J, \forall p \in P, \forall t \in T, \tag{11.4}$$

$$QP_{ijpt} \left(1 - \sum_{v \in V} x_{ijpvt}\right) = 0, \quad \forall i \in I, \forall j \in J, \forall p \in P, \forall t \in T, \tag{11.5}$$

$$QP_{ijpt} \geq \sum_{v \in V} x_{ijpvt}, \quad \forall i \in I, \forall j \in J, \forall p \in P, \forall t \in T, \tag{11.6}$$

$$\left\lceil \left(\frac{QP_{ijpt}}{VL_{vp}}\right) \cdot x_{ijpvt} + 0.999 \right\rceil = f_{ijpvt}, \quad \forall i \in I, \forall j \in J, \forall p \in P, v \in V, t \in T, \tag{11.7}$$

$$r_{ijpv\left(t-1+\left\lceil\frac{2 \cdot TRT_{ijv}}{|t|}+0.999\right\rceil\right)} = f_{ijpvt}, \quad \forall i \in I, \forall j \in J, \forall p \in P, \forall v \in V, \forall t \in T, \tag{11.8}$$

$$\sum_{t \in T}\sum_{i \in I}\sum_{j \in J}\sum_{p \in P}\sum_{v \in V} r_{ijpvt} - \sum_{t \in T}\sum_{i \in I}\sum_{j \in J}\sum_{p \in P}\sum_{v \in V} r_{ijpv\left(t-1+\left\lceil\frac{2 \cdot TRT_{i,j,v}}{|t|}+0.999\right\rceil\right)} = 0, \tag{11.9}$$

$$N_{ivt-1} - \sum_{j \in J}\sum_{p \in P} f_{ijpvt} + \sum_{j \in J}\sum_{p \in P} r_{ijpvt} = N_{ivt}, \quad \forall i \in I, \forall v \in V, \forall t \in T, \tag{11.10}$$

$$\sum_{j \in J}\sum_{p \in P} f_{ijpvt} \leq N_{ivt-1}, \quad \forall i \in I, \forall v \in V, \forall t \in T, \tag{11.11}$$

$$TH_{ipt-1} - \sum_{j \in J} QP_{ijpt} = TH_{ipt}, \quad \forall i \in I, \forall p \in P, \forall t \in T, \tag{11.12}$$

$$\sum_{j \in J} QP_{ijpt} \leq TH_{ipt-1}, \quad \forall i \in I, \forall p \in P, \forall t \in T, \tag{11.13}$$

$$x_{ijpvt} \in \{0,1\}, \quad \forall i \in I, \forall j \in J, \forall p \in P, \forall v \in V, \forall t \in T, \tag{11.14}$$

$$f_{ijpvt} \geq 0, \quad Integer, \quad \forall i \in I, \forall j \in J, \forall p \in P, \forall v \in V, \forall t \in T, \tag{11.15}$$

$$QP_{ijpt} \geq 0, \quad Integer, \quad \forall i \in I, \forall j \in J, \forall p \in P, t \in T. \tag{11.16}$$

Formulas 11.1 and 11.2 are the objective functions which minimize the total cost and time, respectively. Formula 11.3 considers penalty for delivery times exceeding a pre-specified time limit. The constraints 11.4 guarantee that all customer demands are met for all products required at each period. The constraints 11.5 and 11.6 ensure that delivery is accomplished by only one vehicle. The number of each traveling vehicle between the depots and customers is shown by constraints 11.7. The constraints 11.8 and 11.9 identify return times of only the remaining vehicles. The constraints 11.10 represent the numbers of remaining vehicles at the end of the period. The constraints 11.11 ensure that the number of traveled vehicles from depot would not exceed the existing vehicles. The amounts of remaining product in depots at the end of the period are shown by constraints 11.12. The constraints 11.13 represent the capacity constraint of each depot for each product at the corresponding time. They must receive enough products from the supplier in order to meet all the demands. The constraints 11.14 impose that the variables be binary. The last constraints 11.15 and 11.16 show the non-negativity requirements for all the other variables.

TABLE 11.1

The Orders for Products in Different Periods

Order	Product 1	Product 2	Product 3
First Period			
Customer 1	40	45	60
Customer 2	70	30	50
Customer 3	0	20	30
Second Period			
Customer 1	19	0	18
Customer 2	0	0	13
Customer 3	13	15	17
Third Period			
Customer 1	30	25	17
Customer 2	16	20	18
Customer 3	26	25	20

TABLE 11.2

The Distance from Depots to Customers

Distance	Customer 1	Customer 2	Customer 3
Depot 1	20	25	10
Depot 2	10	15	17

TABLE 11.3

The Capacity of Vehicles for Different Products

Capacity of Vehicle	Product 1	Product 2	Product 3
Vehicle 1	15	9	12
Vehicle 2	10	15	8

TABLE 11.4

The Capacity of Depots for Different Products

Depot Capacity	Product 1	Product 2	Product 3
Depot 1	100	200	100
Depot 2	200	100	200

TABLE 11.5

The Transferring Time for Vehicles from
Depots to Customers

Transferring Time	Vehicle 1	Vehicle 2
Depot 1		
Customer 1	10	15
Customer 2	12	17
Customer 3	5	10
Depot 2		
Customer 1	5	10
Customer 2	7	12
Customer 3	9	14

TABLE 11.6

The Quantity of Products that Can Be Satisfied by Depots to Customers at Different Time Periods

Q_p	Depot (i)	Customer (j)	Product (p)	Time Period (t)	X	Vehicle	F
10	1	2	1	1	1	2	1
15	1	2	2	1	1	2	1
20	1	3	2	1	1	2	2
30	1	3	3	1	1	2	4
40	2	1	1	1	1	2	4
45	2	1	2	. 1	1	2	3
60	2	1	3	1	1	1	5
60	2	2	1	1	1	1	4
15	2	2	2	1	1	2	1
50	2	2	3	1	1	1	5
5	1	3	1	2	1	2	1
17	1	3	3	2	1	2	3
19	2	1	1	2	1	1	2
18	2	1	3	2	1	1	2
13	2	2	3	2	1	2	2
8	2	3	1	2	1	2	1
15	2	3	2	2	1	2	1
15	1	1	2	3	1	2	1
20	1	2	2	3	1	2	2
25	1	3	2	3	1	2	2
20	1	3	3	3	1	1	2
30	2	1	1	3	1	1	2
10	2	1	2	3	1	2	1
17	2	1	3	3	1	1	2
16	2	2	1	3	1	1	2
18	2	2	3	3	1	2	3
26	2	3	1	3	1	2	3

TABLE 11.7

The Number of Return Vehicles from Customers to Depots at Different Time Periods

r	Depot (*i*)	Customer (*j*)	Product (*p*)	Vehicle (*v*)	Time Period (*t*)
1	1	2	1	2	4
1	1	2	2	2	4
2	1	3	2	2	2
4	1	3	3	2	2
4	2	1	1	2	2
3	2	1	2	2	2
5	2	1	3	1	1
4	2	2	1	1	2
1	2	2	2	2	3
5	2	2	3	1	2
1	1	3	1	2	3
3	1	3	3	2	3
2	2	1	1	1	2
2	2	1	3	1	2
2	2	2	3	2	4
1	2	3	1	2	4
1	2	3	2	2	4
1	1	1	2	2	5
2	1	2	2	2	6
2	1	3	2	2	4
2	1	3	3	1	3
2	2	1	1	1	3
1	2	1	2	2	4
2	2	1	3	1	3
2	2	2	1	1	4
3	2	2	3	2	5
3	2	3	1	2	5

11.4 Numerical Illustrations

We present a numerical example to show the effectiveness of the proposed mathematical model. The number of customers is three, number of products is three, number of depots is two, and number of vehicles is two. We consider a six-period supply chain which receives the order list in periods one, two and three with the size of time period $|t| = 10$. The orders for products in different periods are given in Table 11.1.

The distance from depots to customers, the capacity of vehicles for different products, and the capacity of depots for different products are given in Tables 11.2 through 11.4, respectively.

The maximum capacity of both depots 1 and 2 are equal to 600. The transferring cost per unit of distance for vehicles 1 and 2 are 50 and 30, respectively. The transferring times for vehicles from depots to customers are given in Table 11.5.

The number of vehicle 1 in both depots 1 and 2 is 14 and the number of vehicle 2 in both depots 1 and 2 is 12. To facilitate the computation, LINGO 8 package is applied. The output for the decision variables are summarized in Tables 11.6 and 11.7. The quantity of products (Qp) that can be satisfied by depots to customers at different time periods, selected route (X), type of vehicle and number of transferred vehicle (F) are presented in Table 11.6.

The number of return vehicles from customers to depots at different time periods (r) is shown in Table 11.7.

The number of remaining vehicles at the end of each period is given in Table 11.8.

The number of remaining capacity at the end of each period is given in Table 11.9.

The best objective value for the problem is 34,350.

TABLE 11.8

The Number of Remaining Vehicles at the End of the Periods

Number of Remaining Vehicles	Vehicle 1	Vehicle 2
At the End of Period 1		
Depot 1	14	4
Depot 2	5	4
At the End of Period 2		
Depot 1	14	6
Depot 2	14	7
At the End of Period 3		
Depot 1	14	5
Depot 2	12	1
At the End of Period 4		
Depot 1	14	9
Depot 2	14	6
At the End of Period 5		
Depot 1	14	10
Depot 2	14	12
At the End of Period 6		
Depot 1	14	12
Depot 2	14	12

TABLE 11.9

The Amount of Remaining Capacity at the End of the Periods

Depot Capacity	Product 1	Product 2	Product 3
At the End of Period 1			
Depot 1	90	165	70
Depot 2	100	40	90
At the End of Period 2			
Depot 1	85	165	53
Depot 2	73	25	59
At the End of Period 3			
Depot 1	85	105	33
Depot 2	1	15	24

11.5 Discussions

We proposed a supply network model in which one supplier would provide various products for customers in different time periods. The contribution of the proposed model is in its flexibility with respect to vehicles and depots. The aim was to minimize the total cost and time of the orders' delivery process. Furthermore, deliveries needing times longer than the pre-specified limits are penalized. The effectiveness and validity of the proposed mathematical model can be illustrated by working out numerical examples.

References

Anupindi R, Bassok Y. Distribution channels, information systems and virtual centralization. In: *Proceedings of the Manufacturing and Service Operations Management Society Conference*, USA, 1996:87–92.

Baron O, Berman O, Wu D. Bargaining in the supply chain and its implication to coordination of supply chains in an industry. In: *Working Paper*, Joseph L. Rotman School of Management, University of Toronto, 2008. <http:// www.rotman.utoronto.ca/bicpapers/workingpapers/ bargainingNRL.pdf>.

Beckmann MJ, McGuire CB, Winsten CB. *Studies in the Economics of Transportation*. New Haven, CT: Yale University Press, 1956.

Bernstein F, Federgruen A. Decentralized supply chains with competing retailers under demand uncertainty. *Management Science* 2005;51(1):18–29.

Bonanno G, Vickers J. Vertical separation. *Journal of Industrial Economics* 1988;36:257–265.

Boyce DE, Chon KS, Lee YJ, Lin KT, LeBlanc LJ. Implementation and computational issues for combined models of location, destination, mode, and route choice. *Environment and Planning A* 1983;15:1219–1230.

Cachon G. Supply chain coordination with contracts. In: Graves S, Ton de K. (eds.), *Handbooks in operations research and management science*. Supply Chain Management, North Holland, 2003.

Cachon G, Lariviere, M. Supply chain coordination with revenue sharing contracts. *Management Science* 2005;51(1):30–44.

Cordeau JF, Gendreau M, Laporte G, Potvin JY, Semet F. A guide to vehicle routing heuristics. *Journal of the Operational Research Society* 2002;53:512–522.

Coughlan AT. Competition and cooperation in marketing channel choice: Theory and application. *Marketing Science* 1985;4(Spring):110–129.

Currie RH, Salhi S. Exact and heuristic methods for a full-load, multi-terminal, vehicle scheduling problem with backhauling and time windows. *The Journal of the Operational Research Society* 2003;54(4):390.

Dafermos S. Traffic equilibrium and variational inequalities. *Transportation Science* 1980;14:42–54.

Dafermos S. The general multimodal network equilibrium problem with elastic demand. *Networks* 1982;12:57–72.

Desrosiers J, Dumas Y, Solomon M, Soumis F. 1995. Time constrained routing and scheduling. In: Ball MB Magnanti TL, Monma CL, Nemhauser GL (eds.), *Network Routing, Handbooks in Operations Research and Management Science*, Amsterdam: Elsevier Science 1995;8:35–139.

Florian M, Hearn D. Network equilibrium models and algorithms. In: Ball MO, Magnanti TL, Monma CL, Nemhauser GL. (eds.), *Network Routing, Handbooks in Operations Research and Management Science*, Amsterdam: Elsevier Science, 1995;8:485–550

Jeuland AP, Shugan SM. Managing channel profits. *Marketing Science* 1983;2(3):239–272.

Kohl JE. *Der Verkehr Und Die Ansiedelungen Der Menschen In Ihrer Abhangigkeit Von Der Gestaltung Der Erdorberflache*. Dresden, Leipzig, Germany, 1841.

Kouvelis P, Chambers C, Wang H. Supply chain management research and production and operations management: Review, trends, and opportunities. *Production and Operations Management* 2006;15(3):449–470.

Laporte, G., Gendreau, M., Potvin JY, Semet F. Classical and modern heuristics for the vehicle routing problem. *International Transactions in Operational Research* 2000;7:285–300.

Lee L, Billington C. Material management in decentralized supply chains. *Operations Research* 1993;41:835–847.

Lin RY, Kong LZ. The impact of channel power of symmetric competing channels on the profit segmentation. In: *Proceeding of National Conference of Chinese National Academy*, Beijing, China, 2002;Dec 2:352–361.

Liu J, Li C-L, Chan C-Y. Mixed truck delivery systems with both hub-and-spoke and direct shipment. *Transportation Research Part E: Logistics and Transportation Review* 2003;39(4):325–339.

McGuire TW, Staelin R. An industry equilibrium analysis of downstream vertical integration. *Marketing Science* 1983;2(2):161–191.

Moorthy K. Strategic decentralization in channels. *Marketing Science* 1988;7(4):335–355.

Nagurney A. *Network economics: A variational inequality approach*, second and revised ed., Dordrecht, The Netherlands: Kluwer Academic Publishers, 1999.

Nagurney, A. *Sustainable transportation networks*. Cheltenham, England: Edward Elgar Publishing, 2000.

Pasternack BA. Optimal pricing and returns policies for perishable commodities. *Marketing Science* 1985;4(2):166–176.

Patriksson M. *The traffic assignment problems—models and methods*. Utrecht, The Netherlands: VSP, 1994.

Pigou, AC. *The economics of welfare*. London: Macmillan, 1920.

Savelsbergh MWP, Sol M. The general pickup and delivery problem. *Transportation Science* 1995;29(1):17–29.

Sigurd M, Pisinger D, Sig M. Scheduling transportation of live animals to avoid the spread of diseases. *Transportation Science* 2004;38(2):197–209.

Slats PA, Bhola B, Evers JJ, Dijkhuizen G. Logistic chain modelling. *European Journal of Operations Research* 1995;87:1–20.

Smith MJ. Existence, uniqueness, and stability of traffic equilibria. *Transportation Research* 1979;13B:259–304.

Tang SC. Perspectives in supply chain risk management. *International Journal of Production Economics* 2005;103(3):451–488.

Toth P, Vigo D. (eds.). *The vehicle routing problem*. Philadelphia, PA: SIAM Monographs on Discrete Mathematics and Applications, 2002.

Tsay A. Quantity–flexibility contract and supplier–customer incentives. *Management Science* 1999;45(10):1339–1358.

Wu O, Chen H. Chain-to-chain competition under demand uncertainty. In: *Working paper*, University of Michigan, 2003. <http://webuser.bus.umich.edu/ owenwu/academic/Chain-to-Chain_2. pdf>.

12

Multi-Aspect Supply Chain: Depot-Customer-Depot Model

SUMMARY This chapter concerns with a supply network that includes supplier, depots and customers. We consider multiple depots, multiple vehicles, multiple products, multiple customers, and different time periods. The supplier receives the order and forwards it to depots of multiple products. A set of depots should be selected among candidate depots. Considering the location of the customers, the depots decide about sending the suitable vehicles. Also, when the vehicles deliver the order to the customers, another allocation for the returning vehicles to depots is set. The aim is to identify the allocation of orders to depots, vehicles, and returning vehicles to depots to minimize the total cost.

12.1 Introduction

Over the last decade or so, supply chain management has emerged as a key area of research among the practitioners of operations research. A lot of research is being carried out to make the supply chain more efficient and economic. The smooth and efficient functioning of business involves the smooth and efficient functioning of the principal areas of the supply chain (SC).

A supply chain is a network comprised of a set of geographically dispersed facilities (suppliers, plants, and warehouses or distribution centers). It is often regarded as the art of bringing the right amount of the right product to the right place at the right time. If the facilities are to distribute product directly to customers, then single-stage model is appropriate. On the other hand, if several facilities are to be sited between the suppliers to the customers in order to produce product or act as regional warehouses or distribution centers, then the multi-stage model is the appropriate model (Bidhandi et al., 2009).

Mathematical programming models have proven their usefulness as analytical tools to optimize complex decision-making problems such as those encountered in supply chain planning. Geoffrion and Graves (1974) described a multi-commodity distribution system design problem and solved it by Benders Decomposition. This is probably the first paper that presents a comprehensive mixed integer programming (MIP) model for the strategic design of supply chain networks. After that, a diversity of deterministic mathematical programming models dealing with the design of supply chain networks can be found in the literature. See, for example, Aikens (1985), Goetschalckx et al. (2002), Geoffrion and Powers (1995), Yan et al. (2003), and Amiri (2004).

A crucial component of the planning activities of a manufacturing firm is the efficient design and operation of its supply chain logistics network. A supply chain is a network of suppliers, manufacturing plants, warehouses, and distribution channels organized to acquire raw materials, convert these raw materials to finished products, and distribute these products to customers. These decisions can be classified into three categories according to their importance and the length of the planning horizon considered.

First, choices regarding the location, capacity and technology of plants and warehouses are generally seen as strategic with a planning horizon of several years. Second, supplier selection, product range assignment as well as distribution channel and transportation mode selection belong to the tactical level and can be revised every few months. Finally, raw material, semi-finished and finished product flows in the network are operational decisions that are easily modified in the short term (Cordeau et al., 2006).

One of the supply chain process models is often represented as a resource network. The nodes in the network represent facilities, which are connected by links that represent direct transportation connections permitted by the company in managing its supply chain (Farahani and Elahipanah, 2008). Supply chain modeling has to configure this network and program the flows within the configuration according to a specific objective function based on algorithms (Tayur et al., 1999). Therefore, supply chain can be modeled as a configurable and flow-programmable resource network. The network employs a completely different and very selective view of what is going on in the supply chain (Wang et al., 2008).

Supply chain modeling offers short-, medium- or long-term optimization potentials. Elements within the optimization scope may be plants, distribution centers, suppliers, customers, orders, products, or inventories (Yao and Chu, 2008). The standard problems for supply chain modeling are formulated in the following manner. A set of goals should be achieved by minimizing the costs of transfer and transformation. In partial solutions, particular goals are selected, such as securing a certain service level to minimize the lead time and maximize capacity utilization, or to secure the availability of resources (Otto and Kotzab, 2003). Supply chain models can also be classified into various frameworks with respect to their problem scopes or application areas. Min and Zhou (2002) viewed the problem scope as a criterion for measuring the realistic dimensions of the model.

Considering the inherent nature of supply chain problems that cut across functional boundaries, supply chain models involve making tradeoffs between more than one business process (function) within the supply chain (Pan et al., 2009). Therefore, only models that attempt to integrate different functions of the supply chain are regarded as supply chain models. Such models deal with the multi-functional problems of location/routing, production/distribution, supplier selection/inventory control, and scheduling/transportation. Recently, Kerbache and Smith (2004) classified optimization problems associated with queuing networks as follows: optimal topological problem (OTOP), optimal routing problem (OROP) and optimal resource allocation problem (ORAP).

Meixell and Gargeya (2005) reviewed decision-support models for the design of global supply chains, and assess the fit between the research literature in this area and the practical issues of global supply chain design. Zhao et al. (2005) proposed a fuzzy linear programming model for bi-level distribution network design in supply chain management, in which both customer demands for products and production capacity of branch plants are treated as fuzzy parameters. Javid and Parikh (2006) discussed scanning location-specific barcodes as a possible way of localizing transactions to individual villages and customers. They presented the high-level design of this system and enumerate the possible technologies that can be used to determine a user's location via a mobile device. Li et al. (2004) constructed the military product supplier selection index system based on military supply chain. Nagumey (2006) considered the relationship between supply chain network equilibrium and transportation network equilibrium. This equivalence allows us to transfer the wealth of methodological tools developed for transportation network equilibrium modeling, analysis, and computation to the study of supply chain networks. Wang et al. (2006) proposed methods for modeling service reliability in a supply chain. The logistics system in a supply chain typically consists of thousands of retail stores along with multiple distribution centers (DC). Products are transported between DCs and stores through multiple routes. Wu et al. (2006) discussed a framework for supplier selection process and set up an evaluation model of supplier selection in terms of cost, quality, service, manufacture and technological capability, reputation and information system. Cárdenas-Barrón (2007) proposed an *n*-stage-multi-customer supply chain inventory model, where there is a company that can supply products to several customers. It concluded that it is possible to use an algebraic approach to optimize the supply chain model without the use of differential calculus.

This chapter concerns with a supply network that includes supplier, depots and customers. Here, we propose a supply chain that considers multiple depots, multiple vehicles, multiple products, multiple customers, and different time periods. The supplier receives the order and forwards it to depots of multiple products. A set of depots should be selected among candidate depots. The depots investigate the capacity level and accept/refuse supplying the order. Considering the location of the customers, the depots decide about sending the suitable vehicles. Each vehicle has its corresponded traveling time and cost. Also when the vehicles deliver the order to the customers, another allocation for the returning vehicles to depots is set. The aim is to identify the allocation of orders to depots, vehicles, and returning vehicles to depots to minimize the total cost. The main decision taken is the vehicle routing to optimize the cost and satisfy the time.

12.2 Problem Definition

The proposed problem of this chapter considers different customers that should be serviced with one supplier. The supplier provides various products and keeps them in different depots. The initial problem is choosing the appropriate depots among a set of candidate depots. Each depot uses different types of vehicle to satisfy the orders. All of the depots are already stationed at the related locations. Here, we consider a multi-echelon supply chain network (one supplier, multiple depots, and customers), multi-commodity, deterministic demand. Sets of vehicles are stationed at each depot. Each depot can store set of products. The received order list from customer can be responded by one or multiple depots at each time. Each selected vehicle to deliver can transfer only one product. The returning vehicles are allocated to the depots when depots may not have specific vehicles in a period and should respond to an order. A configuration of the proposed supply chain network is shown in Figure 12.1.

The novel contribution of the work is related to the return of the vehicles to the depots which need the vehicle in that period due to satisfy the customer's demand. This kind of decision-making certifying the flexibility of the proposed supply network is interesting.

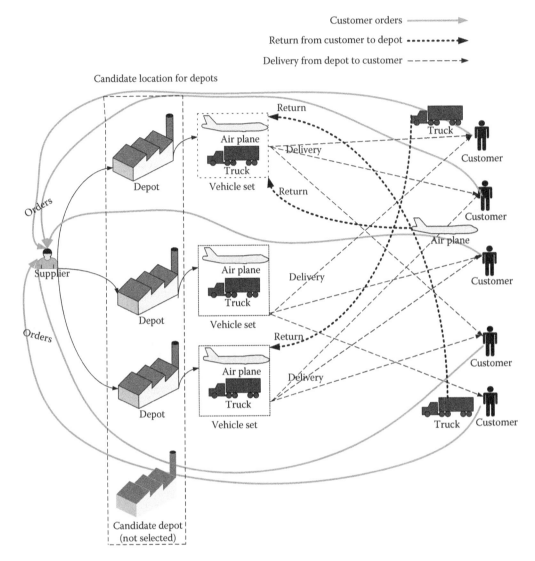

FIGURE 12.1
A configuration of the proposed supply network.

12.3 Returning Vehicle Model

The mathematical model for this problem is as follows:

Notations

P	Set of all products
I	Set of all depots stationed
J	Set of all customers
T	Unit of time

V Set of all vehicles

D_{jpt} Demand of product p for customer j at time t

ND_{ivt} The number of existing vehicles v in depot i at time t

NC_{jvt} The number of existing vehicles v in customer j at the end of time t

VL_{vp} Capacity of vehicle v for product p

CT_{ijv} Traveling cost per mile from depot i to customer j using vehicle v

d_{ij} Distance between depot i and customer j

M A large number

Ch_{ip} The holding cost for product p in depot i

Cs_{ip} The supplying cost for product p in depot i

CO_i The opening cost of depot i

cap_p_{ip} Maximum capacity of product p in depot i

cap_s_{ip} Maximum capacity of storage of product p in depot i

Decision Variables

x_{ijpvt} {1, If depot i delivers product p to customer j using vehicle v at time t;0;o.w}

y_{jivt} {1, if customer j delivers vehicle v to depot i at time t;0;o.w}

z_{ipt} {1, if depot i receives product p at time t;0;o.w}

w_{it} {1, if depot i is active at time t;0;o.w}

TH_{ipt} Amount of received product p in depot i at time t

s_{ipt} Amount of stored product p in depot i at the end of time t

f_{ijpvt} Frequency of traveling between depot i and customer j by vehicle v at time t

QP_{ijpt} Quantity of product p can be satisfied by depot i to customer j at time t

NR_{jivt} Number of vehicle v delivered to depot i by customer j at time t

DV_{ivt} Demand of vehicle v for depot i at time t

12.3.1 Objective Function

$$Minimize\ (F) = Min\ (f_1 + f_2 + f_3 + f_4 + f_5)$$

$$f_1 = \sum_{i \in I} \sum_{j \in J} \sum_{p \in P} \sum_{v \in V} \sum_{t \in T} x_{ijpvt} \cdot f_{ijpvt} \cdot d_{ij} \cdot CT_{ijv} \tag{12.1}$$

$$f_2 = \sum_{j \in J} \sum_{i \in I} \sum_{v \in V} \sum_{t \in T} y_{jivt} \cdot NR_{jivt} \cdot d_{ij} \cdot CT_{ijv} \tag{12.2}$$

$$f_3 = \sum_{i \in I} \sum_{p \in P} \sum_{t \in T} S_{ipt} \cdot Ch_{ip} \tag{12.3}$$

$$f_4 = \sum_{i \in I} \sum_{p \in P} \sum_{t \in T} TH_{ipt} \cdot Cs_{ip} \tag{12.4}$$

$$f_5 = \sum_{i \in I} \sum_{t \in T} (w_{it+1} - w_{it}) \cdot CO_i \tag{12.5}$$

12.3.2 Constraints

$$\sum_{p \in P} z_{ipt} \leq M \cdot w_{it}, \quad \forall i \in I, \forall t \in T \tag{12.6}$$

$$\sum_{p \in P} z_{ipt} \geq w_{it}, \quad \forall i \in I, \forall t \in T, \tag{12.7}$$

$$TH_{ipt} \leq M \cdot z_{ipt}, \quad \forall i \in I, \forall p \in P, \forall t \in T, \tag{12.8}$$

$$TH_{ipt} \geq z_{ipt}, \quad \forall i \in I, \forall p \in P, \forall t \in T, \tag{12.9}$$

$$w_{it+1} \geq w_{it}, \quad \forall i \in I, \forall t \in T, \tag{12.10}$$

$$\sum_{i \in I} QP_{ijpt} = D_{jpt}, \quad \forall j \in J, \forall p \in P, \forall t \in T, \tag{12.11}$$

$$\sum_{j \in J} QP_{ijpt} \leq TH_{ipt} + S_{ipt-1}, \quad \forall i \in I, \forall p \in P, \forall t \in T, \tag{12.12}$$

$$S_{ipt-1} + TH_{ipt} - \sum_{j \in J} QP_{ijpt} = S_{ipt}, \quad \forall i \in I, \forall p \in P, \forall t \in T, \tag{12.13}$$

$$TH_{ipt} \leq cap\_p_{ip}, \quad \forall i \in I, \forall p \in P, \forall t \in T, \tag{12.14}$$

$$S_{ipt} \leq cap\_s_{ip}, \quad \forall i \in I, \forall p \in P, \forall t \in T, \tag{12.15}$$

$$QP_{ijpt} \cdot \left(1 - \sum_{v \in V} x_{ijpvt}\right) = 0, \quad \forall i \in I, \forall j \in J, \forall p \in P, \forall t \in T, \tag{12.16}$$

$$QP_{ijpt} \geq \sum_{v \in V} x_{ijpvt}, \quad \forall i \in I, \forall j \in J, \forall p \in P, \forall t \in T, \tag{12.17}$$

$$\left[\left((QP_{ijpt} \div VL_{vp}) \cdot x_{ijpvt}\right) + 0.999\right] = f_{ijpvt}, \quad \forall i \in I, \forall j \in J, \forall p \in P, \forall v \in V, \forall t \in T, \tag{12.18}$$

$$ND_{ivt-1} - \sum_{j \in J} \sum_{p \in P} f_{ijpvt} + \sum_{j \in J} NR_{jivt} = ND_{i,v,t}, \quad \forall i \in I, \forall v \in V, \forall t \in T, \tag{12.19}$$

$$\sum_{j \in J} \sum_{p \in P} f_{ijpvt} \leq ND_{ivt-1}, \quad \forall i \in I, \forall v \in V, \forall t \in T, \tag{12.20}$$

$$DV_{ivt} \leq M \cdot w_{it}, \quad \forall i \in I, \forall v \in V, \forall t \in T, \tag{12.21}$$

$$\sum_{j \in J} NR_{jivt} = DV_{ivt}, \quad \forall i \in I, \forall v \in V, \forall t \in T, \tag{12.22}$$

$$NC_{jvt-1} = \sum_{i \in I} NR_{jivt}, \quad \forall j \in J, \forall v \in V, \forall t \in T, \tag{12.23}$$

$$NR_{jivt} \leq M \cdot y_{jivt}, \quad \forall j \in J, \forall i \in I, \forall v \in V, \forall t \in T, \tag{12.24}$$

$$NR_{jivt} \geq y_{jivt}, \quad \forall j \in J, \forall i \in I, \forall v \in V, \forall t \in T, \tag{12.25}$$

$$NC_{jvt-1} + \sum_{i \in I} \sum_{p \in P} f_{ijpvt} - \sum_{i \in I} NR_{jivt} = NC_{jvt}, \quad \forall j \in J, \forall v \in V, \forall t \in T, \tag{12.26}$$

Integrity and non-negativity constraints:

$$x_{ijpvt} \in \{0,1\}, \quad \forall i \in I, \forall j \in J, \forall p \in P, \forall v \in V, \forall t \in T, \tag{12.27}$$

$$y_{jivt} \in \{0,1\}, \quad \forall j \in J, \forall i \in I, \forall v \in V, \forall t \in T, \tag{12.28}$$

$$z_{ipt} \in \{0,1\}, \quad \forall i \in I, \forall p \in P, \forall t \in T, \tag{12.29}$$

$$w_{it} \in \{0,1\}, \quad \forall i \in I, \forall t \in T, \tag{12.30}$$

$$f_{ijpvt} \geq 0, \quad \forall i \in I, \forall j \in J, \forall p \in P, \forall v \in V, \forall t \in T, \tag{12.31}$$

$$QP_{ijpt}, Integer, \quad \forall i \in I, \forall j \in J, \forall p \in P, \forall t \in T, \tag{12.32}$$

$$NR_{jivt}, Integer, \quad \forall j \in J, \quad \forall i \in I, \quad \forall v \in V, \quad \forall t \in T, \tag{12.33}$$

$$TH_{ipt} \geq 0, \quad i \in I, \forall p \in P, \forall t \in T, \tag{12.34}$$

$$DV_{ivt} \geq 0, \quad \forall i \in I, \forall v \in V, \forall t \in T. \tag{12.35}$$

Equations 12.1 and 12.2 are the objective functions which minimize total cost of both forward and backward distance, respectively. Equation 12.3 is the objective function which minimizes total cost of storage. Equation 12.4 is the objective function which minimizes total cost of supply. Equation 12.5 is the objective function which minimizes cost of opening depot. The constraints 12.6 and 12.7 show that each depot can be supplied when it is activated. The constraints 12.8 and 12.9 ensure that the amount of product each selected depot receives is non-negative. The constraints 12.10 prevent the depots from changing their status more than once. The constraints 12.11 guarantee that all customer demands are met for all products required at all periods. The constraints 12.12 are the flow conservation at depots. The constraints 12.13 show amount of stored product at the end of period. The constraints 12.14 and 12.15 represent capacity restriction. The constraints 12.16 and 12.17 ensure that delivery is accomplished by only one vehicle. The frequency of traveling between depots and customers has been shown in constraints 12.18. The constraints 12.19 represent the number of remaining vehicles at the end of period. The constraints 12.20 require that the frequency of traveled vehicles from depot is lower than or equal to its stationed vehicles. The constraint 12.21 requires that each activated depot can order vehicles. The constraints 12.22 guarantee that all depots' demands of vehicles are met, for all vehicles required and for any period. The constraints 12.23 are the flow balance of stationed vehicles at the end of period.

The constraints 12.24 and 12.25 guarantee that delivery of vehicles from customer to depot is accomplished while the corresponded path was selected. The constraints 12.26 represent the number of remaining vehicles stationed at the corresponded customer at the end of period. The constraints 12.27 through 12.30 require that this variable is binary. The constraints 12.31 through 12.35 restrict all other variables from taking non-negative values.

12.3.3 Linearization

To improve the performance of the proposed mathematical model we act out the following linearization for the nonlinear equations.

As Equation 12.1 is nonlinear, we turn it into the following equations,

$$\text{Equation } 12.1 \rightarrow f_1 = \sum_{i \in I} \sum_{j \in J} \sum_{p \in P} \sum_{v \in V} \sum_{t \in T} f_{ijpvt} \cdot d_{ij} \cdot CT_{ijv} \tag{12.36}$$

$$(QP_{ijpt} \div VL_{vp}) - M \cdot (1 - x_{ijpvt}) \leq f_{ijpvt}, \quad \forall i \in I, \forall j \in J, \forall p \in P, \forall v \in V, \forall t \in T, \tag{12.37}$$

$$f_{ijpvt} \leq M \cdot x_{ijpvt}, \quad \forall i \in I, \forall j \in J, \forall p \in P, \forall v \in V, \forall t \in T, \tag{12.38}$$

As Equation 12.2 is nonlinear, we turn it into the following equations,

$$\text{Equation } 12.2 \rightarrow f_2 = \sum_{j \in J} \sum_{i \in I} \sum_{v \in V} \sum_{t \in T} NR_{jivt} \cdot d_{ij} \cdot CT_{ijv} \tag{12.39}$$

$$NR_{jivt} \leq M \cdot y_{jivt}, \quad \forall j \in J, \forall i \in I, \forall v \in V, \forall t \in T, \tag{12.40}$$

$$NR_{jivt} \geq y_{jivt}, \quad \forall j \in J, \forall i \in I, \forall v \in V, \forall t \in T, \tag{12.41}$$

For constraints 12.16 and 12.17 we use the following equations:

$$QP_{ijpt} \leq M \cdot \sum_{v \in V} x_{ijpvt}, \quad \forall i \in I, \ \forall j \in J, \ \forall p \in P, \ \forall v \in V, \ \forall t \in T, \tag{12.42}$$

$$QP_{ijpt} \geq \sum_{v \in V} x_{ijpvt}, \quad \forall i \in I, \ \forall j \in J, \ \forall p \in P, \ \forall v \in V, \ \forall t \in T, \tag{12.43}$$

$$\sum_{v \in V} x_{ijpvt} \leq 1, \quad \forall i \in I, \ \forall j \in J, \ \forall p \in P, \ \forall t \in T, \tag{12.44}$$

For constraints 12.31 through 12.33 we use the following equations:

$$f_{ijpvt}, integer \quad \forall i \in I, \ \forall j \in J, \ \forall p \in P, \ \forall v \in V, \ \forall t \in T, \tag{12.45}$$

$$QP_{ijpt} \geq 0, \quad \forall i \in I, \ \forall j \in J, \ \forall p \in P, \ \forall t \in T, \tag{12.46}$$

$$NR_{jivt} \geq 0, \quad \forall j \in J, \ \forall i \in I, \ \forall v \in V, \ \forall t \in T, \tag{12.47}$$

12.4 Numerical Illustration

Here, we propose a numerical example to indicate the effectiveness of the proposed mathematical model. The number of customers is three, number of products is three, number of candidate depots is seven, and number of vehicles is two. We consider a five-period supply chain which there is no demand at period five. Other input data are given in Table 12.1.

To facilitate the computations, LINGO 8 package is applied to model the mixed integer code. The output of forward flow for the decision variables is presented in Table 12.2. The amount of received product from supplier in each depot is shown in Table 12.3; meanwhile, the amount of storage of products in all depots for each period is zero. The backward flow for the decision variables is presented in Table 12.4. The number of left vehicles at the end of period, and best objective are presented in Table 12.5.

12.5 Discussions

We proposed a supply network in which one supplier has provided various products for customers in different time periods. The contribution of the proposed model is the flexibility on vehicles and depots and also the location problem of candidate depots. The aim was to minimize not only the total cost and time of the order to delivery process but also the total cost and time of returning vehicles' distances. The effectiveness and validity of the proposed mathematical model can be presented using numerical illustrations. In our further research, we will consider to include qualitative parameters to our proposed problem.

TABLE 12.1

Input Data

Order	Product 1	Product 2	Product 3
First Period			
Customer 1	40	45	60
Customer 2	70	30	50
Customer 3	0	20	30
Second Period			
Customer 1	19	0	18
Customer 2	0	0	13
Customer 3	13	15	17
Third Period			
Customer 1	30	25	17
Customer 2	16	20	18
Customer 3	26	25	20
Fourth Period			
Customer 1	10	15	0
Customer 2	8	16	12
Customer 3	15	0	14

Distance	Customer 1	Customer 2	Customer 3
Depot 1	20	25	10
Depot 2	10	15	17
Depot 3	14	12	13
Depot 4	10	15	12
Depot 5	16	22	24
Depot 6	13	16	20
Depot 7	14	15	16

	Product 1	Product 2	Product 3
Capacity of Vehicle			
Vehicle 1	30	50	20
Vehicle 2	10	15	8
Depot Capacity			
Depot 1	100	85	90
Depot 2	90	80	70
Depot 3	80	75	70
Depot 4	90	100	70
Depot 5	85	65	75
Depot 6	80	70	60
Depot 7	100	70	80

(*Continued*)

TABLE 12.1 (*Continued*)

Input Data

Storage Capacity			
Depot 1	50	50	50
Depot 2	50	50	50
Depot 3	50	50	50
Depot 4	50	50	50
Depot 5	50	50	50
Depot 6	50	50	50
Depot 7	50	50	50

	Vehicle 1	Vehicle 2
Transferring Cost per Unit of Distance	50	30
Number of Vehicles		
Depot 1	14	12
Depot 2	14	12
Depot 3	14	12
Depot 4	14	12
Depot 5	14	12
Depot 6	14	12
Depot 7	14	12

	Product 1	Product 2	Product 3
Supplying Cost			
Depot 1	10	8	11
Depot 2	12	6	10
Depot 3	11	7	15
Depot 4	13	10	9
Depot 5	14	7	14
Depot 6	13	12	14
Depot 7	8	13	9
Holding Cost			
Depot 1	6	7	5
Depot 2	7	8	4
Depot 3	6	5	4
Depot 4	4	7	6
Depot 5	3	5	7
Depot 6	8	7	6
Depot 7	6	6	4
Opening Cost			
Depot 1	2000		
Depot 2	2000		
Depot 3	2000		
Depot 4	2000		
Depot 5	2000		
Depot 6	2000		
Depot 7	2000		

TABLE 12.2

The Forward Path Output

X	Depot	Customer	Product	Vehicle	Period	F	QP
1	2	1	2	1	1	1	45
1	2	1	3	1	1	1	10
1	2	2	1	1	1	3	70
1	2	2	3	1	1	3	50
1	7	1	1	1	1	2	40
1	7	1	3	1	1	3	50
1	7	2	2	1	1	1	30
1	7	3	2	1	1	1	20
1	7	3	3	2	1	4	30
1	1	1	1	1	2	1	19
1	1	3	3	2	2	1	1
1	2	1	3	1	2	1	18
1	2	2	3	1	2	1	13
1	2	3	2	2	2	1	15
1	2	3	3	2	2	2	16
1	7	3	1	2	2	2	13
1	1	2	2	2	3	2	20
1	1	3	3	1	3	1	20
1	2	1	3	1	3	1	17
1	2	2	1	1	3	1	16
1	2	2	3	1	3	1	18
1	2	3	1	1	3	1	26
1	7	1	1	1	3	1	30
1	7	1	2	2	3	2	25
1	7	3	2	1	3	1	25
1	1	2	3	1	4	1	12
1	1	3	3	2	4	1	6
1	2	1	1	2	4	1	10
1	3	1	2	2	4	1	15
1	3	2	1	1	4	1	8
1	3	2	2	2	4	2	16
1	4	3	1	1	4	1	15
1	7	3	3	2	4	1	8

TABLE 12.3

The Amount of Received Product in Each Depot

TH	Product 1	Product 2	Product 3
First Period			
Depot 2	70	45	60
Depot 7	40	50	80
Second Period			
Depot 1	19	0	1
Depot 2	0	15	47
Depot 7	13	0	0
Third Period			
Depot 1	0	20	20
Depot 2	42	0	35
Depot 7	30	50	0
Fourth Period			
Depot 1	0	0	18
Depot 2	10	0	0
Depot 3	8	31	0
Depot 4	15	0	0
Depot 7	0	0	8

TABLE 12.4

The Backward Path Output

Y	Customer	Depot	Vehicle	Period	NR
1	1	2	1	2	7
1	2	7	1	2	7
1	3	7	2	2	4
1	3	1	1	2	1
1	1	2	1	3	2
1	2	7	1	3	1
1	3	1	2	3	6
1	1	4	1	4	2
1	1	4	2	4	2
1	2	3	1	4	2
1	2	3	2	4	2
1	3	1	1	4	3
1	1	4	2	5	2
1	2	3	1	5	2
1	2	3	2	5	2
1	3	1	1	5	1
1	3	1	2	5	2
	BEST OBJECTIVE: 7,0123				

TABLE 12.5

The Number of Left Vehicles at the End of Periods

Number of Left Vehicles	Vehicle 1	Vehicle 2
At the End of Period 1		
Depot 1	14	12
Depot 2	6	12
Depot 3	14	12
Depot 4	14	12
Depot 5	14	12
Depot 6	14	12
Depot 7	7	8
At the End of Period 2		
Depot 1	14	11
Depot 2	11	9
Depot 3	14	12
Depot 4	14	12
Depot 5	14	12
Depot 6	14	12
Depot 7	14	10
At the End of Period 3		
Depot 1	13	15
Depot 2	9	9
Depot 3	14	12
Depot 4	14	12
Depot 5	14	12
Depot 6	14	12
Depot 7	13	8
At the End of Period 4		
Depot 1	15	14
Depot 2	9	8
Depot 3	15	11
Depot 4	15	14
Depot 5	14	12
Depot 6	14	12
Depot 7	13	7
At the End of Period 5		
Depot 1	16	16
Depot 2	9	8
Depot 3	17	13
Depot 4	15	16
Depot 5	14	12
Depot 6	14	12
Depot 7	13	7

References

Aikens CH. Facility location models for distributing planning. *European Journal of Operational Research* 1985;22(3):263–279.

Amiri A. Designing a distribution network in a supply chain system: Formulation and efficient solution procedure. *European Journal of Operational Research* 2004;174:567–576.

Bidhandi HM, Yusuff RM, Hamdan Megat Ahmad MM, Abu Bakar MR. Development of a new approach for deterministic supply chain network design. *European Journal of Operational Research* 2009;198:121–128.

Cárdenas-Barrón LE. Optimizing inventory decisions in a multi-stage multi-customer supply chain: A note. *Transportation Research Part E: Logistics and Transportation Review* 2007;43(5):647–654.

Cordeau JF, Pasin F, Solomon MM. An integrated model for logistics network design. *Annals of Operations Research* 2006;144(1):59–82.

Farahani RZ, Elahipanah M. A genetic algorithm to optimize the total cost and service level for just-in-time distribution in a supply chain. *International Journal of Production Economics* 2008;111:229–243.

Geoffrion AM, Graves GW. Multi-commodity distribution system design by benders decomposition. *Management Science* 1974;20:822–844.

Geoffrion AM, Powers RF. Twenty years of strategic distribution system design: An evolution perspective. *Interfaces* 1995;25:105–128.

Goetschalckx M, Vidal CJ, Dogan K. Modeling and design of global logistics systems: A review of integrated strategic and tactical models and design algorithms. *European Journal of Operational Research* 2002;143:1–18.

Javid P, Parikh TS. Augmenting rural supply chains with a location-enhanced mobile information system. In: *International Conference on Information and Communication Technologies and Development, 2006, ICTD '06*. Berkeley, CA, 2006:110–119.

Kerbache L, Smith JM. Queuing networks and the topological design of supply chain systems. *International Journal of Production Economics* 2004;91:251–272.

Li L, Chen H, Wang J. Research on military product supplier selection architecture based on fuzzy AHP. In: *International Conference on Service Systems and Service Management*, USA, 2004;2:1397–1402.

Meixell MJ, Gargeya VB. Global supply chain design: A literature review and critique. *Transportation Research Part E* 2005;41:531–550.

Min H, Zhou G. Supply chain modeling; past, present and future. *Computers & Industrial Engineering* 2002;43:231–249.

Nagumey A. On the relationship between supply chain and transportation network equilibria: A super network equivalence with computations. *Transportation Research Part E* 2006;42:293–316.

Otto A, Kotzab H. Does supply chain management really pay? Six perspectives to measure the performance of managing a supply chain. *European Journal of Operational Research* 2003;144(2):306–320.

Pan A, Leung SYS, Moon KL, Yeung KW. Optimal reorder decision-making in the agent-based apparel supply chain. *Expert Systems with Applications* 2009;36:8571–8581.

Tayur S, Ganeshan R, Magazine M. (eds.). *Stochastic programming models for managing product variety. Quantitative models for supply chain management*. Kluwer Press, Boston, 1999.

Wang MH, Liu JM, Wang HQ, Cheung WK, Xie XF. On-demand e-supply chain integration: A multi-agent constraint-based approach. *Expert Systems with Application* 2008;34:2683–2692.

Wang, N, Lu JC, Kvam P. Reliability modeling in spatially distributed logistics systems, reliability. *IEEE Transactions* 2006;55(3):525–534.

Wu B, Wang S, Hu J. An analysis of supplier selection in manufacturing supply chain management. In: *International Conference on Service Systems and Service Management*, USA, 2006;2:1439–1444.

Yan H, Yu Z, Cheng TCE. A strategic model for supply chain design with logical constraints: Formulation and solution. *Computers & Operations Research* 2003;30:2135–2155.

Yao MJ, Chu WM. A genetic algorithm for determining optimal replenishment cycles to minimize maximum warehouse space requirements. *Omega* 2008;36:619–631.

Zhao X, Huang X, Sun F. An optimization model for distribution network design with uncertain customer demands and production capacity. In: *International Conference on Services Systems and Services Management, Proceedings of ICSSSM '05*, Chongqing, China, 2005;1:621–625.

13

Food Supply Chain: Two-Stage Model

SUMMARY In this chapter, a three-layer supply network (depots, retailers and customers) is considered. The purposes are determination of proper locations of retailers and a suitable distribution of goods throughout the network minimizing cost of all tours.

13.1 Introduction

Supply chain (SC) is a complex of multi-layers (supply units, production units and customers). Usually there are both upstream and downstream flow in this network. Information is sent by customer to the production units. Producers receive information and forward it to the suppliers where the raw materials are being provided. The raw materials are produced in production units and finally finished products are delivered to the customers. Managing these processes efficiently is called supply chain management (Beamon, 1998). With the increasing population rate, requirements of food products increase, too. Therefore, quality of food products and delivery time are two significant issues in supply chains. A food chain is known as supply chain where the inputs are primary farm productions (Swedish Environmental Protection Agency, 1999). This chain consists of two parts: the first one discusses fresh agricultural products which are delivered to customers with unchangeable intrinsic specifications and the second one corresponds to the processed food products in which agricultural products are used as raw materials to evolve them (van der Vorst, 2000). A new perspective has been suggested in food chain in recent years is an environmental management. Interactions between product and surrounding environment are affected by various activities (Broekmeulen, 2001; Apaiah and Hendrix, 2005). So, food product's quality changes in different stages. Also, preserving product's freshness and delivering it to customer on time requires cost optimization. Obtaining the food quality and on time delivery obliges controlling environmental effects and designing distribution chain properly (Tijskens et al., 2001). Researchers studied on the effects of controlling uncertain factors such as weather variations for biological products on behavior of products, also various process conditions in each node (i.e., depots and retailers) (Van Impe et al., 2001; Peirs et al., 2002; Hertog, 2002; Hertog et al., 2004). Many approaches have been employed to assess uncertainty on the output status. A widely used model was Monte Carlo approach (Nicolaï et al., 1998; Demir et al., 2003; Poschet et al., 2003).

The structure of SCs is similar to network configuration. Many procedures are available in this field. One of these is related to Traveling Salesman Problem (TSP) concept which having N cities, a salesman should start from home city, visit all customers once and come back to the home city finding a minimal route. While there are several salesmen who all start and return to a single home city somehow all customers are visited exactly once is known as multiple Traveling Salesman Problem (mTSP). Now, we suppose that there are multiple depots in a supply network. Any of them has number of salesmen. Multiple Depots, Multiple Traveling Salesmen Problem (MDMTSP) finds tours for all salesmen such that all customers are visited exactly once and the total cost of the tours are minimized. While salesmen depart from depots and arrive to the single destination, is called as multiple departures single destination multiple TSP. This concept has several applications which one of them is for modeling school bus routing. In such problems, buses depart from depots and arrive to the single destination (school). All passengers are serviced exactly once and the total cost of all the tours is minimized (Kara and Bektas, 2006).

Here, we consider a three-layer supply network (depots, retailers and customers). Customers send order lists and wait to deliver them. The purposes are determination of proper locations of retailers and a suitable distribution of goods throughout the network minimizing cost of all tours. We are faced with two distribution chains as follows: distribution chain between retailers and customers and then distribution chain between depots and known retailers. Due to receive products on time by customer, we should design a network efficiently. In this way, knowing locations of retailers, vehicles located at depots move toward retailers. Through this path, vehicles supply several retailers and finally return to the similar or dissimilar depots. Finding proper routes in this phase can help us to deliver products to retailers at right time. Then, the retailers forward products to customers.

13.2 Problem Definition

Here, we propose a network that contains three layers (depots, retailers and customers). Receiving order list from customers, we decide which retailers can supply them. Selection of proper retailers to supply customers depends on satisfying time windows on customers' viewpoints. Also, distribution of food products from depots to retailers plays critical role. MDMTSP approach can be appropriate for this problem. Any salesman located at depot must depart and visit at least two retailers and then go back to the similar or dissimilar depot. In this problem, we suppose that any customer is supplied by at least one retailer. Meanwhile, the total demands are satisfied. A configuration of the proposed problem is shown in Figure 13.1.

13.3 Bi-Stage Model

Here, we present a mathematical model that consists of two stages: the first describes the location-allocation concept between retailers and customers, and the second applies the multiple depot multiple traveling salesman problem (MDMTSP) concept between depots and known retailers. The proposed models are given as follows (see Figures 13.2 and 13.3):

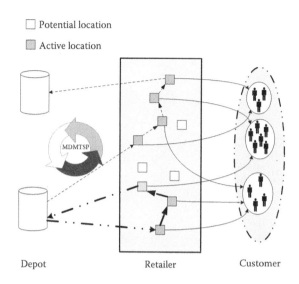

FIGURE 13.1
A configuration of proposed problem.

Notations

Indices

I	set of retailers
Z	set of customers
K	set of depots

Parameters

dis_{iz}	the distance between retailer i and customer z
$time_{iz}$	the transfer time between retailer i and customer z
d_z	demand of customer z
cap_i	capacity of retailer i
c	the establishing cost of retailer
W	maximum time window for customer
M	a large number
dis_{ij}	the distance between retailer i and retailer j
d_{ki}	the distance between depot k and retailer i
m_k	the number of salesman located at depot k
L	maximum number of nodes a salesman may visit
k	minimum number of nodes a salesman must visit

Decision Variables

h_i $\begin{cases} 1 & \text{if retailer } i \text{ is active} \\ 0 & \text{o.w} \end{cases}$

y_{iz} $\begin{cases} 1 & \text{if there is a link between retailer } i \text{ and customer } z \\ 0 & \end{cases}$

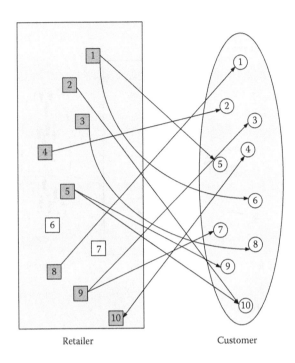

FIGURE 13.2
A configuration of location-allocation section of proposed model.

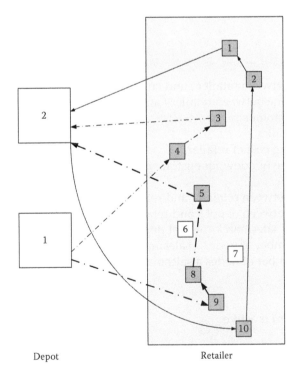

FIGURE 13.3
A configuration of routing process in proposed model.

QP amount of flow between retailer i and customer z

x_{ki} $\begin{cases} 1 & \text{if there is a link between depot } k \text{ and retailer } i \\ 0 \end{cases}$

z_{ik} $\begin{cases} 1 & \text{if there is a link between retailer } i \text{ and depot } k \\ 0 \end{cases}$

y_{ij} $\begin{cases} 1 & \text{if there is a link between retailer } i \text{ and retailer } j \\ 0 \end{cases}$

u_i the number of nodes visited by travelers from depot to node i

13.3.1 Stage 1: Location-Allocation

13.3.1.1 Objective Function

$$\min f = \min(f_1 + f_2),$$

$$f_1 = \sum_{i \in I} h_i \cdot c \tag{13.1}$$

$$f_2 = \sum_{i \in I} \sum_{z \in Z} y_{iz} \cdot dis_{iz} \tag{13.2}$$

13.3.1.2 Constraints

$$\sum_{i \in I} y_{iz} \geq 1, \quad \forall z \in Z, \tag{13.3}$$

$$\sum_{z \in Z} y_{iz} \geq h_i, \quad \forall i \in I, \tag{13.4}$$

$$\sum_{z \in Z} y_{iz} \leq h_i \cdot M, \quad \forall i \in I, \tag{13.5}$$

$$QP_{iz} \leq y_{iz} \cdot M, \quad \forall i \in I, \forall z \in Z, \tag{13.6}$$

$$QP_{iz} \geq y_{iz}, \quad \forall i \in I, \forall z \in Z, \tag{13.7}$$

$$y_{iz} \cdot time_{iz} \leq W, \quad \forall i \in I, \forall z \in Z, \tag{13.8}$$

$$\sum_{i \in I} Qp_{iz} = d_z, \quad \forall z \in Z, \tag{13.9}$$

$$\sum_{z \in Z} Qp_{iz} \leq cap_i, \quad \forall z \in Z, \tag{13.10}$$

$$h_i \in \{0,1\}, \quad \forall i \in I, \tag{13.11}$$

$$y_{iz} \in \{0,1\}, \quad \forall i \in I, \forall z \in Z, \tag{13.12}$$

$$Qp_{iz} \geq 0, \quad \forall i \in I, \forall z \in Z. \tag{13.13}$$

Formulas 13.1 and 13.2 are objective functions minimizing configuration cost of retailers and distribution cost between retailers and customers, respectively. Constraints 13.3 show that any customer can be supplied by at least one retailer. Constraints 13.4 and 13.5 impose that any retailer can supply customers while it is active. Amount of delivered products between retailers and customers is represented by constraints 13.6 and 13.7. Constraints 13.8 satisfy time windows. Constraints 13.9 guarantee that the demand constraint is met. Restriction of capacity is shown by constraints 13.10. Constraints 13.11 and 13.12 imply the binary variables. Non-negativity of the variable is shown by constraints 13.13.

13.3.2 Stage 2: MDMTSP

13.3.2.1 Objective Function

$$\min f = \min (f_1 + f_2),$$

$$f_1 = \sum_{k \in K} \sum_{i \in I} d_{ki} \cdot (x_{ki} + z_{ik}) \tag{13.14}$$

$$f_2 = \sum_{i \in I} \sum_{j \in J} y_{ij} \cdot dis_{ij} \tag{13.15}$$

13.3.2.2 Constraints

$$\sum_{k \in K} x_{ki} + \sum_{k \in K} z_{ik} \leq 1, \quad \forall i \in I, \tag{13.16}$$

$$\sum_{i \in I} x_{ki} = m_k, \quad \forall k \in K, \tag{13.17}$$

$$\sum_{k \in K} x_{kj} + \sum_{i \in I} y_{ij} = 1, \quad \forall j \in J, \tag{13.18}$$

$$\sum_{k \in K} z_{ik} + \sum_{j \in J} y_{ij} = 1, \quad \forall i \in I, \tag{13.19}$$

$$u_i - u_j + (L \cdot y_{ij}) + ((L-2) \cdot y_{ji}) \leq L - 1, \quad \forall i \in I, \ \forall j \in J, \tag{13.20}$$

$$u_i + \left((L-2) \cdot \sum_{k \in K} x_{ki} \right) - \sum_{k \in K} z_{ik} \leq L - 1, \quad \forall i \in I, \tag{13.21}$$

$$u_i + \left((2-K) \cdot \sum_{k \in K} z_{ik} \right) + \sum_{k \in K} x_{ki} \geq 2, \quad \forall i \in I, \tag{13.22}$$

$$x_{ki} \in \{0,1\}, \quad \forall k \in K, \ \forall i \in I, \tag{13.23}$$

$$z_{ik} \in \{0,1\}, \quad \forall k \in K, \ \forall i \in I, \tag{13.24}$$

$$y_{ij} \in \{0,1\}, \quad \forall j \in J, \ \forall i \in I, \tag{13.25}$$

$$u_i \geq 0, \quad \forall i \in I. \tag{13.26}$$

Formulas 13.14 and 13.15 are objective functions minimizing distances between available nodes (depots and retailers). Constraints 13.16 represent that a salesman must visit at least two retailers. Constraints 13.17 show that total available salesmen located at depot must depart toward retailers. Constraints 13.18 require that any retailers be supplied by either depots or other retailers, and either come back to depots or supply other retailers. This concept is represented by constraints 13.19. Constraints 13.20 through 13.22 prevent any sub tour in network. Constraints 13.23 through 13.25 impose variables to be binary. Constraints 13.26 show that variable is non-negative.

13.4 Numerical Illustration

We present numerical example to illustrate effectiveness of our proposed model. To facilitate the computations we apply LINGO package. There are two depots, ten candidate retailers and ten customers in this network. Table 13.1 shows customers' demand. Configuration cost of retailer is 1500 unit of money. Maximum waiting time for customers is set to be 50 unit of time.

Table 13.2 represents capacity of retailers. Distances between retailers and customers are shown in Table 13.3.

Table 13.4 represents transfer times between retailers and customers.

TABLE 13.1

Customers' Demands

Customer	1	2	3	4	5	6	7	8	9	10
Demand	20	25	15	16	20	10	17	25	30	40

TABLE 13.2

Capacity of Retailers

Retailer	1	2	3	4	5	6	7	8	9	10
Capacity	30	20	25	25	50	20	35	20	40	20

TABLE 13.3

Distances between Retailers and Customers

Retailer	Customer									
	1	2	3	4	5	6	7	8	9	10
1	15	16	17	14	8	17	16	12	14	13
2	24	21	18	17	10	17	24	23	19	10
3	13	26	14	25	22	17	8	14	26	25
4	24	17	15	11	15	20	7	13	18	114
5	19	25	28	16	25	17	13	19	10	13
6	24	16	17	14	27	25	23	14	16	18
7	16	14	25	16	11	17	28	15	13	14
8	15	10	14	12	20	15	17	13	16	10
9	21	20	12	14	13	10	15	16	19	14
10	13	15	9	10	14	23	24	15	16	11

TABLE 13.4

Transfer Times between Retailers and Customers

Retailer	Customer									
	1	2	3	4	5	6	7	8	9	10
1	10	16	18	44	10	50	45	18	30	45
2	35	21	28	17	10	47	54	63	49	40
3	33	46	74	85	52	37	48	14	26	25
4	24	17	55	71	15	60	47	33	48	114
5	19	25	80	56	65	67	73	69	30	23
6	44	46	57	64	77	25	33	44	36	48
7	66	74	25	86	90	77	58	55	63	54
8	15	10	84	72	60	45	57	63	56	100
9	21	60	12	64	13	100	15	65	95	14
10	53	55	90	20	34	23	24	15	56	110

The outputs are given as follows: The active retailers to supply customers are shown in Table 13.5. Amount of delivered products to customers by retailers is represented in Table 13.6.

Here, we begin the second stage of our proposed model. There are two depots in the network. The first depot employs two salesmen and the second depot employs one salesman. Maximum and minimum numbers of nodes a traveler may visit are three and one, respectively. Table 13.7 represents distances between depots and retailers. But, our model considers distances that are related to the known retailers specified in the previous stage. Distances among retailers are given in Table 13.8.

As a result, the visited routes are obtained.

TABLE 13.5

The Active Retailers

Retailer	1	2	3	4	5	6	7	8	9	10
	✓	✓	✓	✓	✓	–	–	✓	✓	✓

TABLE 13.6

Amount of Delivery Products to Customers

Retailer	Customer									
	1	2	3	4	5	6	7	8	9	10
1					20	10				
2										20
3								25		
4		25								
5									30	20
8	20									
9			15				17			
10				16						

TABLE 13.7

Distances between Depots and Retailers

Depot	Retailer									
	1	2	3	4	5	6	7	8	9	10
1	10	12	13	8	15	18	12	14	12	15
2	5	8	7	9	10	15	14	12	11	10

TABLE 13.8

Distances between Retailers

Retailer	Retailer									
	1	2	3	4	5	6	7	8	9	10
1	0	12	15	14	∞	20	30	∞	11	20
2	12	0	22	∞	17	14	∞	∞	∞	16
3	15	22	0	11	14	10	20	23	∞	∞
4	14	∞	11	0	∞	22	∞	24	∞	13
5	∞	17	14	∞	0	∞	18	10	15	∞
6	20	14	10	22	∞	0	16	∞	10	20
7	30	∞	20	∞	18	16	0	24	7	13
8	∞	∞	23	24	10	∞	24	0	7	25
9	11	∞	∞	∞	15	10	7	7	0	23
10	20	16	∞	13	∞	20	13	25	23	0

13.5 Discussions

In this work, we presented a supply chain network that contained three layers (depots, retailers and customers). Finding optimal locations of retailers and distribution of food products satisfying time windows were our purposes that are attained in a two-stage mathematical model. This way, we proposed an approach as multiple depot multiple traveling salesman problem (MDMTSP) between depots and retailers.

References

Apaiah KA, Hendrix EMT. Design of a supply chain network for pea-based novel protein foods. *Journal of Food Engineering* 2005;70:383–391.

Beamon BM. Supply chain design and analysis: Models and methods. *International Journal of Production Economics* 1998;55:281–294.

Broekmeulen RACM. Modelling the management of distribution centres. In: Tijskens LMM, Hertog MLATM, Nicolaï BM, (eds.). *Food Process Modelling*, Cambridge: Woodhead Publishing Limited, 2001:432–447.

Demir AD, Baucour P, Cronin K, Abodayeh K. Analysis of temperature variability during the thermal processing of hazelnuts. *Innovative Food Science and Emerging Technologies* 2003:69–84.

Hertog MLATM. The impact of biological variation on postharvest population dynamics. *Postharvest Biology and Technology* 2002;26:253–263.

Hertog MLATM, Lammertyn J, Desmet M, Scheerlinck N, Nicolaï BM. The impact of biological variation on postharvest behaviour of tomato fruit. *Postharvest Biology and Technology* 2004;34:271–284.

Kara I, Bektas T. Integer linear programming formulations of multiple salesman problems and its variations. *European Journal of Operational Research* 2006;174:1449–1458.

Nicolaï BM, Verboven P, Scheerlinck N, De Baerdemaeker J. Numerical analysis of the propagation of random parameter fluctuations in time and space during thermal food processes. *Journal of Food Engineering* 1998;38:259–278.

Peirs A, Scheerlinck N, Berna Perez A, Jancsók P, Nicolaï BM. Uncertainty analysis and modelling of the starch index during apple fruit maturation. *Postharvest Biology and Technology* 2002;26:199–207.

Poschet F, Geerarerd AH, Scheerlinck N, Nicolaï BM, Van Impe JF. Monte Carlo analysis as a tool to incorporate variation on experimental data in predictive microbiology. *Food Micro-Biology* 2003;20:285–295.

Swedish Environmental Protection Agency. A sustainable food supply chain: A swedish case study. Stockholm: Elanders Gotab, 1999.

Tijskens LMM, Koster AC, Jonker JME. Concepts of chain management and chain optimization. In: Tijskens LMM, Hertog MLA TM, Nicolaï BM (eds.). *Food Process Modelling*, 3rd ed. Washington, DC: CRC Press, 2001:448–469.

van der Vorst JGAJ. Effective food supply chains. *Generating, Modeling And Evaluating Supply Chain Scenarios*. Wageningen University, Wageningen, 2000.

Van Impe JF, Bernaerts K, Geeraerd AH, Poschet F, Versyck KJ. Modelling and prediction in an uncertain environment. In: Tijskens LMM, Hertog MLA, Nicolaï BM (eds.). *Food process modelling*, Cambridge: Woodhead Publishing Limited, 2001:156–179.

14

Multi-Period Food Supply Chain: Time-Windows Model

SUMMARY In this chapter, we consider a multiple time-windows supply network of multi-food products with deterministic demands in multi-time periods. A set of vehicles is stationed at each depot and each selected vehicle can transfer various food products based on its capacity.

14.1 Introduction

The agro-food industry is a sector of significant economic and political importance. It is one of the most regulated and protected sectors. As a result of intensive development in the past century, food production has been continuously putting increasing pressure on the environment, and thus increasingly attracting the attention of policy makers.

Supply chain coordination has gained a considerable notice recently from both practitioners and researchers. In monopolistic markets with a single chain or markets with perfect competing retailers, a vertically integrated supply chain maximizes the profit of the chain (e.g., Jeuland and Shugan, 1983; Bernstein and Federgruen, 2005; Cachon and Lariviere, 2005). Therefore, many supply chain contracts try to induce retailers and suppliers to act according to the vertical integration strategy.

Consumers today demand high-quality products in various innovative forms through the entire year at competitive prices. Society imposes constraints on producers in order to economize the use of resources, ensure animal friendly and safe production practices and restrict environmental damage. These demands, together with advancing technology and presence of open markets have changed the production, trade and distribution, namely the supply chain of food products beyond recognition (Trienekens and Omta, 2001).

A supply chain (SC) is an integrated process in which raw materials are acquired, converted into products and then delivered to the consumer (Beamon, 1998). The chain is characterized by a forward flow of goods and a backward flow of information. Food supply chains are made up of organizations that are involved in the production and distribution of plant and animal-based products (Zuurbier et al., 1996). Such SCs can be divided into two main types (van der Vorst, 2000):

- SCs for fresh agricultural products: the intrinsic characteristics of the product remain unchanged, and
- SCs for processed food products: agricultural products are used as raw materials to produce processed products with a higher added value.

The main fact that differentiates food SCs from other chains is that there is a continuous change in quality from the time the raw material leaves the grower to the time the product reaches the consumer (Tijskens et al., 2001). The food SC being considered here consists of six links: primary producers, ingredient preparations, product processing, distributions, retailers and consumers (see Figure 14.1).

Performance measures or goals are considered to design SCs or supply networks by determining the values of the decision variables that yield the desired goals or performance levels (Apaiah et al., 2005; Beamon, 1998). The design of the chain or network changes in accordance with the goal for which the chain is being designed and optimized. Since consumer demands have to be met, it is important to ask the consumer what attributes he/ she desires in the product to be considered in achieving the goals to design the chain; for example, if the goal is quality at any cost, then technologically advanced and consequently expensive equipment can be used to produce the product and it can be transported to the consumer by air for fast delivery. However, if the goal is a low priced product, care has to be taken to minimize production and transportation costs.

A market with two competing supply chains was investigated in the seminal work of McGuire and Staelin (1983). They considered a price (i.e., Bertrand) competition between two suppliers selling through independent retailers. They concluded that for highly substitutable products, a decentralized supply chain Nash equilibrium was preferred by both suppliers. In Coughlan (1985), the author applied this work to the electrical industry and Moorthy (1988) further explained why the decentralized chains could lead to higher profits for the manufacturer and the entire chains. In Bonanno and Vickers (1988), the authors investigated a similar model, and used geometric insights to show that with franchise fees there were some settings in which the manufacturer's optimal strategy was to sell his products using an independent retailer. Thus, in these cases, the manufacturer prefers a decentralized supply chain irrespective of the decision in the other supply chain. Neither of these works considered demand uncertainty.

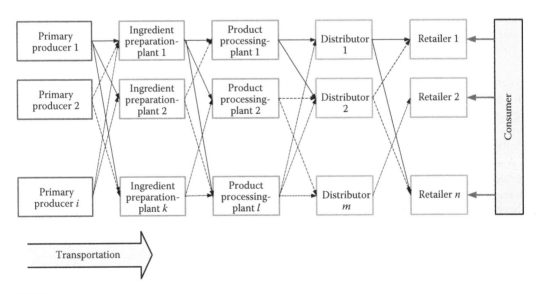

FIGURE 14.1
A food supply chain.

Product specifications have been studied as supplier allocation in supply networks (Xu and Nozick, 2009). Christopher et al. (2009) studied the effects of innovation and knowledge in the supply chain performance. They proposed a knowledge based mechanism for order allocation to suppliers. Also, product allocation in supply network was presented by Francas and Minner (2009). They found out that back order products could be allocated to manufacturing element of a supply network. Order and reorder placements using intelligent agents were studied by Pan et al. (2009). He explored the elements of an intelligent agent for order allocation. Che et al. (2009) proposed a line balancing based model for supplier allocation. They proposed a line balancing technique to assign supplying activities to the supply network elements and order allocation.

The supply chain problem is of an utmost economic importance to businesses due to the elapsing times and costs associated with the provision of a fleet of delivery vehicles for transportation of products to a set of geographically dispersed customers (Pasternack, 1985). Moreover, such problems are also significant in the public sector, where vehicle routes must be determined for bus systems, postal carriers, and other public service vehicles. In each of these instances, the problem typically involves finding a minimal cost of the combined routes for a number of vehicles in order to facilitate delivery from a supply location to a number of customer locations. Since cost is closely associated with distance, a company might attempt to find a minimal distance traveled by the vehicles in order to satisfy customer demands. In doing so, the firm attempts to minimize costs while elevating or at least maintaining an expected level of customer service.

14.2 Problem Definition

The food industry is a complex, global array of diverse businesses that collectively supply much of the food energy consumed by the world population. Only subsistence farmers, those who survive on what they grow, can be considered outside of the scope of the modern food industry. The following aspects are associated with the food industry.

Regulation: Local, regional, national and international rules and regulations for food production and sale, including food quality and food safety, and industry lobbying activities.

Education: Academic, vocational, and consultancy.

Research and development: Food technology.

Manufacturing: Agrichemicals, seed, farm machinery and supplies, agricultural construction, etc.

Agriculture: Raising of crops and livestock, and seafood.

Food processing: Preparation of fresh products for market, and manufacture of prepared food products.

Marketing: Promotion of generic products (e.g., milk board), new products, public opinion, through advertising, packaging, public relations, etc.

Wholesale and distribution: Warehousing, transportation, and logistics.

Retail: Supermarket chains and independent food stores, direct-to-consumer, restaurant, food services.

Vast global transportation network is required by the food industry in order to connect its numerous parts. These include suppliers, manufacturers, warehousing, retailers and the end consumers. There are also companies that, during the food processing process, add vitamins, minerals, and other necessary requirements usually lost during the food preparation. Wholesale markets for fresh food products have tended to decline in importance in some countries as well as in Latin America and some Asian countries as a result of the growth of supermarkets, which procure directly from farmers or through preferred suppliers, rather than going through markets.

The constant and uninterrupted flow of products from distribution centers to store locations is a critical element of food industry operations. Distribution centers run more efficiently, the throughput can be increased, costs can be lowered, and manpower can be better utilized if proper steps are taken in setting up a material handling system in a warehouse. With populations around the world being concentrated in urban areas, purchasing food is increasingly removed from all aspects of food production. This is a relatively recent development, taking place mainly over the past 50 years. Supermarkets are defining retail elements of the food industry, whereas tens of thousands of products are gathered in diverse locations, with continuous, year-round supplies.

Food preparation is another area for which change in recent decades has been dramatic. Today, two food industry sectors are in apparent competition for the retail food values. The grocery industry sells fresh and largely raw products for consumers to be used as ingredients in home cooking. The food service industry offers prepared food, either as finished products, or as partially prepared components for final "assembly."

Sophisticated technologies define modern food production. They include many areas. Agricultural machinery, originally led by tractors, has practically eliminated human labor in many areas of production. Biotechnology is driving much change in areas as diverse as agrochemicals, plant breeding and food processing. Many other aspects of technology are also involved, to the point that hardly any aspect could be found not to have a direct impact on the food industry. Computer technology is also a central force, with computer networks and specialized software providing the support infrastructure to allow for the global movement of the myriad components involved.

The introduction of thousands of new food products each year into retail consumer markets has become a normal expectation of consumers. Food manufacturers have been generating new products and line extensions at an amazing pace in an effort to retain retail shelf space and a share of the consumer's food expenditure. New retail food product introductions expanded annually from around 5,500 in 1985 to 16,900 in 1995, before tapering slightly in 1996 and 1997 (Food Marketing Institute, 1997). Figure 14.2 illustrates categorized food products in a supply network.

Several factors have been considered to be effective as driving forces behind this pace of new introductions. On the demand side, the demand for greater convenience, healthier and safer products, special dietary considerations, product variety, and other product features has been buoyed by greater disposable incomes. On the supply side, retailers have grown their capacities to handle more products, manage categories, and generally become more responsive to even slight changes in consumer preferences through innovations such as electronic data interchange (EDI), efficient consumer response (ECR), category management, and customer loyalty programs (Kahn et al., 1997).

Consumers may consider several alternatives in their shopping experience, almost to the point of being overwhelmed. Couponing, merchandising, and advertisement of new food products have kept pace with the number of new introductions. The introduction of new food products has become a strategic tool applied by manufacturers to achieve or retain

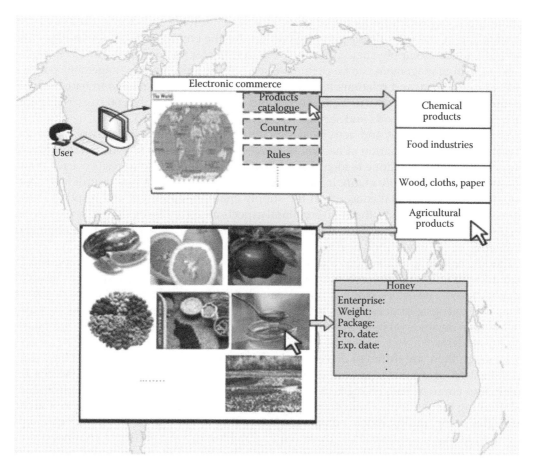

FIGURE 14.2
Categorized food products in a supply network.

prime shelf space. Product life cycles for these new products are considerably short, with industry sources estimating that 96% of these new products are no longer on the shelf after one year of their release (Toops, 1996).

Changes in the retail environment, intensified competition among food manufacturers, and shorter product life cycles have raised the significance of focus on new product development (NPD) efficiency. Increasing or changing development costs associated with a variety of regulatory and internal research activities have similarly heightened interest in NPD. Science and technology have changed the manufacturing capabilities in a way that R&D investment decisions have become very complicated.

The food supply chain can be defined as a complex network of inputs and outputs which starts from primary farm production and everything related to it, goes through different forms and stages of food processing and preservation, very often associated with long distance travel, and reaches the end-of-life phase (Swedish Environmental Protection Agency, 1999). Under the conditions of constantly changing environment, the capability of organizations to restructure and respond is of vital importance. A suitable model for this appears to be supply chains. They can contribute to reducing the uncertainties, providing access to information, and improving the reliability and responsiveness.

Supply chains provide better possibilities for problem solving and sharing the associated benefits and burdens. With the growing concern for the environment, a new perspective has been added to supply chains and their environmental managements. Environmental supply chain management can be defined as a "set of supply chain management policies held, actions taken, and relations formed in response to concerns related to natural environment with regard to the design, acquisition, production, distribution, use, reuse and disposal of the firm's goods and services" (Hagelaar and Vorst, 2002).

As with all industries and activities, agriculture and food industry contribute to the depletion and contamination of natural resources, and thus create serious environmental impacts. A lot has been done to identify and analyze the aspects and impacts of each of the phases of the food supply chain. It has become apparent, however, that in order to incorporate the environmental concerns in the supply chain management and respond to the rising consumer demands, the environmental aspects could not be dealt with separately at each step of the chain. Therefore, a holistic approach is necessary for the identification and evaluation in the whole food chain as such.

Though realizing that supply chain perspective is important, analyzing it appears to be a difficult task. The reason is the great variability of existing goods requiring different supply chains. The actors, being involved, can also differ considerably (in terms of size of companies, geographical locations, types of business in which they operate). The structure of the chain depends on its objectives and it can change over time, additionally complicating the analysis.

Although there are different drives for the developments in the food supply chain (e.g., globalization of markets, greater consumer choice, consumer and media concerns on safety and environment, changes in eating habits, etc.), the incorporation of environmental thinking in the supply chain management is still limited and mainly considered in the strategies of big, multinational companies. According to Hall (2002), environmental supply chain dynamics take place if there is a channel leader with sufficient channel power over the suppliers, technical competencies, and when specific environmental pressure is exerted. Such pressure can be any external factor affecting the company's environmental policy. Two main areas of pressure were identified as regulatory and non-regulatory. Government measures are usually the primary regulatory pressures and are used for correcting market imperfection, but they are not considered sufficient. Non-regulatory pressures are identified as consumer pressure, customer pressure, environmental pressure groups, disclosure requirements, employees and unions, and corporate citizenship (Hall, 2002).

14.3 Time-Windows Model

As supply networks of food products become more dependent on the efficient movement of products among geographically dispersed facilities, there would be more opportunity for disruption. This issue is more significant in food industries.

The development of a robust supply network demands careful attention to both the location of the individual supplier facilities and the opportunities for effective transportation between them. Here, we propose a supply network considering multiple timed-windows. The supplier receives the order and forwards it to depots containing multiple food products. Considering the location of the customers, the depots decide upon sending products in the correct time-windows. The aim is to identify the allocation of orders to depots while satisfying the delivery time and minimizing the transportation cost.

Our proposed model here considers different customers being serviced with one supplier. The supplier provides various products and keeps them in different depots. Each depot uses various types of vehicles to satisfy the orders. All depots are already stationed at the designated locations. Here, we consider a multiple time-windows supply network of multiple food products with deterministic demands in multi-time periods. A set of vehicles is stationed at each depot and each selected vehicle can transfer various food products based on its capacity. A configuration of the proposed problem is presented in Figure 14.3.

The mathematical model follows here.

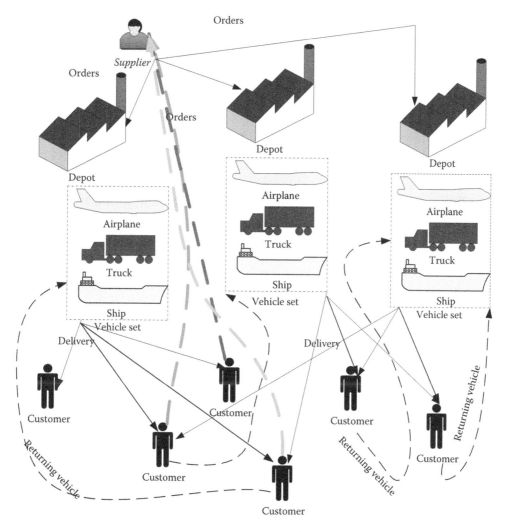

FIGURE 14.3
A configuration of the proposed model.

Notations

P	Set of all food products; $p \in P$
I	Set of all stationed depots; $i \in I$
J	Set of all customers; $j \in J$
T	Set of time periods; $t \in T$
V	Set of all vehicles; $v \in V$
D_{jpt}	Demand for food product p of customer j at period t
TH_{ip}	Maximum throughput of product p at depot i
CA_i	Total capacity of depot i
N_{ivt}	The number of existing vehicles v in depot i at period t
VL_{vp}	Capacity of vehicle v for food product p
d_{ij}	Distance between depot i and customer j
RV_{jvt}	The number of returning vehicles v from customer j at period t
M	A large number
CT_{ijv}	Traveling fixed cost per mile from depot i to customer j using vehicle v
TRT_{ijv}	Traveling time from depot i to customer j using vehicle v
W_{pt}	Time-window of product p at period t
C	Cost for every unit of time

Decision Variables

x_{ijpvt} $\begin{cases} 1 & \text{if depot } i \text{ is selected to deliver food product } p \text{ to customer } j \\ & \text{by vehicle } v \text{ at period } t \\ 0 & \text{otherwise} \end{cases}$

f_{ijpvt} Frequency of traveling for product p between depot i and customer j by vehicle v at period t

QP_{ijpt} Quantity of food product p that can be supplied by depot i to customer j at period t

14.3.1 Objective Function

$$\text{Minimize } F = f_1 + f_2,$$

where,

$$f_1 = \sum_{i \in I} \sum_{j \in J} \sum_{p \in P} \sum_{v \in V} \sum_{t \in T} x_{ijpvt} f_{ijpvt} d_{ij} CT_{ijv}, \tag{14.1}$$

$$f_2 = \sum_{i \in I} \sum_{j \in J} \sum_{p \in P} \sum_{v \in V} \sum_{t \in T} x_{ijpvt} TRT_{ijv} C. \tag{14.2}$$

14.3.2 Constraints

$$\sum_{i \in I} QP_{ijpt} = D_{jpt}, \quad \forall t, j, p, \tag{14.3}$$

$$\sum_{j \in J} QP_{ijpt} \leq TH_{ip}, \quad \forall t, i, p, \tag{14.4}$$

$$QP_{ijpt}\left(1-\sum_{v\in V}x_{ijpvt}\right)=0, \quad \forall t,i,j,p, \tag{14.5}$$

$$QP_{ijpt}\geq\sum_{v\in V}x_{ijpvt}, \quad \forall t,i,j,p, \tag{14.6}$$

$$N_{ivt-1}-\sum_{j\in J}\sum_{p\in P}f_{ijpvt}=N_{ivt}, \quad \forall t,i,v, \tag{14.7}$$

$$\sum_{j\in J}\sum_{p\in P}f_{ijpvt}\leq N_{ivt-1}, \quad \forall t,i,j,v, \tag{14.8}$$

$$RV_{jvt-1}=\sum_{p\in P}\sum_{i\in I}f_{ijpvt}, \quad \forall t,j,v, \tag{14.9}$$

$$RV_{jvt-1}+\sum_{i\in I}N_{ivt}-\sum_{p\in P}\sum_{i\in I}f_{ijpvt}=RV_{jvt}, \quad \forall t,j,v, \tag{14.10}$$

$$x_{ijpvt}TRT_{ijv}\leq W_{pt}, \quad \forall t,i,j,p,v, \tag{14.11}$$

$$\left|\left|\left(\left(\frac{QP_{ijpt}}{VL_{vp}}\right)x_{ijpvt}\right)+0.999\right|\right|\leq f_{ijpvy}, \quad \forall t,i,j,p,v, \tag{14.12}$$

$$x_{ijpvt}\in\{0,1\}, \quad \forall t,i,j,p,v, \tag{14.13}$$

$$f_{ijpvt},QP_{ijpt}\geq 0, \quad Integer, \quad \forall t,i,j,p,v, \tag{14.14}$$

Equations (14.1) and (14.2) are the objective functions corresponding to the total cost and time, respectively. Since time and cost are two separate objectives, we have a multi-objective model. To establish a single objective optimization problem, we consider a cost parameter (C) to turn Equation (14.2) into a cost. This way, we can aggregate the two objectives and thus form a single objective function. The constraints (14.3) guarantee that all customer demands are met, for all food products required and in any period. The constraints (14.4) are the flow conservation at depots; they must receive enough food products from supplier in order to meet all the demands. The constraints (14.5) and (14.6) ensure that delivery is accomplished by only one vehicle. The constraints (14.7) represent the number of remaining vehicles at the end of the period. The constraints (14.8) require that the frequency of traveled vehicles from depot is lower than or equal to its stationed vehicles. The constraints (14.9) present the flow conservation of returning vehicles from customers. The constraints (14.10) update the number of returning vehicles. The constraints (14.11) confine the model to deliver the products to customers using an appropriate vehicle during the time-windows considering the traveling time. The frequencies of traveling between depots and customers are shown by constraints (14.12). In this equation, the number 0.999 is used to have one unit more than the value of the inner parenthesis in order to have an adequate exact value for frequency of traveling. The constraints (14.13) require that the variables be binary. The constraints (14.14) prevent all variables from being negative.

We note that the nonlinear Equation (14.5) can be linearized by using the inequality,

$$QP_{ijpt} \leq M \sum_{v \in V} x_{ijpvt}, \quad \forall t, i, j, p, \tag{14.15}$$

where M is a sufficiently large number; since $\sum_{v \in V} x_{ijpvt}$ is equal to 0 or 1, then (14.15) implies that QP_{ijpt} be 0 (when $\sum_{v \in V} x_{ijpvt} = 0$) or arbitrary (when $\sum_{v \in V} x_{ijpvt} = 1$). The latter is imposed when M is large enough, and the former is guaranteed by the simultaneous incurred inequalities $QP_{ijpt} \leq 0$ and $QP_{ijpt} \geq 0$ (from Equation [14.14]).

14.4 Numerical Example

Here, we propose a numerical example to show the effectiveness of the proposed mathematical model. The number of customers is three, number of products is three, number of depots is two, and number of vehicles is two. We consider two time periods for the supply chain. The demands for various products by different customers in the two periods are given in Table 14.1.

The transferring costs for vehicles 1 and 2 are 20 and 30, respectively. The numbers of vehicles in different depots are equal to 15. Also, the capacity of vehicle one for each product is 30, 50, 20 and the capacity of vehicle two for each product is 10, 15, and 8, respectively. The distances between customers and depots are shown in Table 14.2. The transferring times of vehicles from depots to customers are given in Table 14.3.

The time-windows for each product considering the corresponding time periods are given in Table 14.4.

To facilitate the computations, LINGO software package was used. The output for the decision variables are presented in Tables 14.5 and 14.6. The best objective value in this status is 1030.

TABLE 14.1

The Demands for Food Products

	Product 1	Product 2	Product 3
First Period			
Customer 1	3	5	6
Customer 2	7	3	5
Customer 3	0	2	3
Second Period			
Customer 1	9	0	8
Customer 2	0	0	3
Customer 3	3	5	7

TABLE 14.2

Distances Between Customers and Depots

	Customer 1	Customer 2	Customer 3
Depot 1	20	25	10
Depot 2	10	15	17

TABLE 14.3

Transferring Times of Vehicles

	Depot 1	Depot 2
Vehicle 1		
Customer 1	10	5
Customer 2	12	7
Customer 3	5	9
Vehicle 2		
Customer 1	15	10
Customer 2	17	12
Customer 3	10	14

TABLE 14.4

Time-Windows

	Period 1	Period 2	Period 3
Product 1	15	40	25
Product 2	15	35	25
Product 3	10	30	20

TABLE 14.5

Frequency and Depot Allocations for Food Products Delivery to Customers in Various Time Periods

f	X	Depot	Customer	Product	Vehicle	Time
1	1	1	1	1	1	2
1	1	1	1	3	1	2
1	1	1	2	1	1	2
1	1	1	3	2	1	2
1	1	1	3	3	1	2
1	1	2	1	2	1	2
1	1	2	2	2	1	2
1	1	2	2	3	1	2
1	1	1	1	1	1	3
1	1	1	1	3	1	3
1	1	1	3	1	1	3
1	1	1	3	2	1	3
1	1	1	3	3	1	3
1	1	2	2	3	1	3

TABLE 14.6

Quantity of Food Products to Satisfy Customers by Different
Depots in Various Periods

Q_p	Depot	Customer	Product	Time
3	1	1	1	2
6	1	1	3	2
7	1	2	1	2
2	1	3	2	2
3	1	3	3	2
5	2	1	2	2
3	2	2	2	2
5	2	2	3	2
9	1	1	1	3
8	1	1	3	3
3	1	3	1	3
5	1	3	2	3
7	1	3	3	3
3	2	2	3	3

As seen in Table 14.5, only vehicle one is used to carry the products to customers. This is due to low transferring cost of vehicle one in comparison with that of vehicle two. The reason that vehicle one has adequate duration to deliver all the demands is the length of the time-windows.

14.5 Discussions

We proposed a supply network in which one supplier would provide various food products for customers in different time periods. The contribution of the proposed model is in its flexibility on vehicles and depots. The aim was to minimize the cumulative sum of cost and time of the order to delivery process regarding time-windows.

References

Apaiah KR, Hendrix EMT, Meerdink G, Linnemann AR. Qualitative methodology for efficient food chain design. *Trends in Food Science and Technology* 2005;16(5):204–214.

Beamon BM. Supply chain design and analysis: Models and methods. *International Journal of Production Economics* 1998;55:281–294.

Bernstein F, Federgruen A. Decentralized supply chains with competing retailers under demand uncertainty. *Management Science* 2005;51(1):18–29.

Bonanno G, Vickers J. Vertical separation. *Journal of Industrial Economics* 1988;36:257–265.

Cachon G, Lariviere M. Supply chain coordination with revenue sharing contracts. *Management Science* 2005;51(1):30–44.

Che ZG, Che ZH, Hsu TA. Cooperator selection and industry assignment in supply chain network with line balancing technology. *Expert Systems with Applications* 2009;36:10381–10387.

Christopher W, Craighead G, Hult TM, Ketchen DJ. The effects of innovation–cost strategy, knowledge, and action in the supply chain on firm performance. *Journal of Operations Management* 2009;27:405–421.

Coughlan AT. Competition and cooperation in marketing channel choice: Theory and application. *Marketing Science* 1985;4(Spring):110–129.

Food Marketing Institute. Food Industry Outlook, 1997. [Online]. Available in: http://prestohost23. inmagic.com/Presto/home/home.aspx

Francas D, Minner S. Manufacturing network configuration in supply chains with product recovery. *Omega* 2009;37:757–769.

Hagelaar G, van der Vorst G. Environmental supply chain management: Using LCA to structure supply chains and Wageningen. *International Food and Agribusiness Management Review*, 2002. [Online]. Available in: http://www.sciencedirect.com/

Hall J. Environmental supply chain dynamics, 2002. [Online]. Available in: http://www.sciencedirect.com/

Jeuland AP, Shugan SM. Managing channel profits. *Marketing Science* 1983;2(3):239–272.

Kahn BE, McAlister L. *Grocery revolution: The new focus on the consumer.* Reading, MA: Addison-Wesley Press, 1997.

McGuire TW, Staelin R. An industry equilibrium analysis of downstream vertical integration. *Marketing Science* 1983;2(2):161–191.

Moorthy K. Strategic decentralization in channels. *Marketing Science* 1988;7(4):335–355.

Pan A, Leung SYS, Moon KL, Yeung KW. Optimal reorder decision-making in the agent-based apparel supply chain. *Expert Systems with Applications* 2009;36:8571–8581.

Pasternack BA. Optimal pricing and returns policies for perishable commodities. *Marketing Science* 1985;4(2):166–176.

Swedish Environmental Protection Agency. A sustainable food supply chain: A swedish case study. Stockholm: Elanders Gotab, 1999.

Tijskens LMM, Koster AC, Jonker JME. Concepts of chain management and chain optimization. In: Tijskens LMM, Hertog MLATM, Nicolai BM (eds.), *Food Process Modeling.* Lancaster, Pennsylvania: Woodhead, 2001:145–162.

Toops D. Against all odds. *Food Processing* 1996;Jan:16–17.

Trienekens JH, Omta SWF. Paradoxes in food chains and networks. In: *Proceedings of the Fifth International Conference on Chain and Network Management in Agribusiness and the Food Industry,* the Netherlands, Noordwijk, 2001:452–463.

van der Vorst JGAJ. Effective food supply chains. Generating, modeling and evaluating supply chain scenarios. Wageningen University, Wageningen, 2000:171–190.

Xu N, Nozick L. Modeling supplier selection and the use of option contracts for global supply chain design. *Computers & Operations Research* 2009;36:2786–2800.

Zuurbier PJP, Trienekens JH, Ziggers GW. *Verticale Samenwerking.* Kluwer Bedrijfswetenschappen, Deventer, 1996.

Section II

Reverse Supply Chain Models

15

Return Items in a Multi-Layer Multi-Product Reverse Supply Chain: Clustering Model

SUMMARY In this chapter, the reverse multi-layer multi-product supply chain is considered so that after collecting the returning commodities, clustering is performed via k-mean algorithm. The results are used to perform a sampling process to deliver the commodities to the related layer for rework and repair operations.

15.1 Introduction

Knowledge induction from data becomes a necessity in reverse supply chain to enhance productivity, to understand the process and predict and improve the future system performance. Therefore, especially in recent years, knowledge has received significant attention in reverse supply chain to build a competitive advantage. With the increased environmental concerns and stringent environmental laws, reverse logistics has received growing attention throughout this decade. Reverse logistics can be defined as the logistics activities all the way from used products no longer required by the customer to products again usable in the market.

Although most companies realize that the total processing cost of returned products is higher than the total manufacturing cost, it is found that strategic collections of returned products can lead to repetitive purchases and reduce the risk of fluctuating the material demand and cost. Implementation of reverse logistics especially in product returns would allow not only for savings in inventory carrying cost, transportation cost, and waste disposal cost due to returned products, but also for the improvement of customer loyalty and futures sales. In a broader sense, reverse logistics refers to the distribution activities involved in product returns, source reduction, conservation, recycling, substitution, reuse, disposal, refurbishment, repair and remanufacturing (Stock, 1992).

In recent years data mining has become a very popular technique for extracting information from the database in different areas due to its flexibility of working on any kind of databases and also due to the surprising results (Shahbaz et al., 2010). Data mining is the process in databases to discover and to reveal previously unknown, hidden, meaningful and useful patterns (Fayyad et al., 1996; Baker, 2010). Many approaches, methods and algorithms have been developed in the field of data mining. Data mining techniques are classified as characterization and discrimination, classification, cluster analysis, association analysis, outlier analysis and evolution analysis (Han and Kamber, 2006; Chen et al., 1996). These techniques are briefly described as below. Characterization is used for summarizing the general characteristics of any dataset. However, discrimination is utilized for determining the diversities among different datasets. The products whose sales rates are over 25% for a year in a shopping center are based on the characterization technique. Whereas, comparison of the products whose sales rates increased up to 10% and the products whose sales rates decreased up to 15% is based on the discrimination technique (Dincer, 2006).

Classification is used for determining the class of a new observation utilizing available classes of the observations in training set (Larose, 2005). Grouping the customers as the ones who paid in a three-day period and the ones who paid over a three-day period is based on the classification technique. Decision trees, regression analysis, artificial neural networks, support vector machines, Naïve Bayes algorithm, k-nearest neighbor algorithm and genetic algorithm are among the classification techniques (Liao and Triantaphyllou, 2007).

Cluster analysis is used for clustering similar data structures in any dataset (Tan et al., 2006). Determining the real group of the musical instruments according to their sound signals is based on the clustering technique (Essid et al., 2005). Hierarchical methods, partitioning methods, density-based methods, grid-based methods and heuristic methods are among the clustering techniques.

The association analysis discovers relationships among observations and determines which observations can be realized together (Chen and Weng, 2009). A priori algorithm is one of the techniques used in association analysis. Many data mining techniques detect the exceptions as a noise but the exceptions can contain more information with respect to other observations. For this reason, outlier analysis is used in the stage of analyzing the observations that differ from the data distribution model of available dataset (Hea et al., 2004). As the last technique, the main aim of evolution analysis is to reveal time-varying tendencies of the observations within the dataset (Tan et al., 2009).

Concerning reverse logistics, a lot of researches have been made on various fields and subjects such as reuse, recycling, remanufacturing logistics, etc. Der Laan and Salomon (1997) propose a hybrid manufacturing/remanufacturing system with stocking points for serviceable and remanufacture able products, which will be a part of our framework. Kim et al. (2006) discussed a notion of remanufacturing systems in reverse logistics environment. They proposed a general framework in view of supply planning and developed a mathematical model to optimize the supply planning function. The model determines the quantity of products parts processed in the remanufacturing facilities subcontractors and the amount of parts purchased from the external suppliers while maximizing the total remanufacturing cost saving.

In reuse logistics models, Kroon and Vrijens (1995) reported a case study concerning the design of a logistics system for reusable transportation packages. The authors proposed a MIP (mixed integer programming), closely related to a classical un-capacitated warehouse location model. In recycling models, Barros et al. (1998) proposed a mixed integer program model considered two-echelon location problems with capacity constraints based on a multi-level capacitated warehouse location problem. Krikke et al. (1999) developed a mixed integer program to determine the locations of shredding and melting facilities for the recovery and disposal of used automobiles, while determining the amount of product flows in the reverse logistics network.

Listes (2007) presented a generic stochastic model for the design of networks comprising both supply and return channels, organized in a closed loop system. The author described a decomposition approach to the model, based on the branch-and-cut procedure known as the integer L-shaped method. Salema et al. (2007) studied the design of a reverse distribution network and found that most of the proposed models on the subject are case based and, for that reason, they lack generality. The model contemplates the design of a generic reverse logistics network where capacity limits, multi-product management and uncertainty on product demands and returns are considered. A mixed integer formulation is developed. This formulation allows for any number of products, establishing a network for each product while guaranteeing total capacities for each facility at a minimum cost. But the inventory was not taken into consideration.

The literature presents several studies that examine the implementation of data mining techniques. Gibbons et al. (2000) described a computer component manufacturing scenario which concentrated on the application of data mining techniques to improve information management and process improvement within a manufacturing scenario. Huang and Wu (2005) made an analysis of products quality improvement in ultra-precision manufacturing industry using data mining for developing quality improvement strategies.

Harding et al. (2006) reviewed applications of data mining in manufacturing engineering, in particular production processes, operations, fault detection, maintenance, decision support, and product quality improvement. Liu (2007) developed a data-mining algorithm for designing the conventional cellular manufacturing systems. Hsu (2009) developed a data-mining framework based on two-stage cluster approach to generate useful patterns and rules for standard size charts. The results could provide high tech apparel industries with industrial standards. An empirical study was conducted in an apparel industry to support their manufacturing decision for production management and marketing with various customers' needs. This study proposed data mining techniques in reverse supply chain problem.

15.2 Problem Definition

The reverse supply chain under study is multi-layer, multi-product. In the designed (planned) model, the returned products, after collecting and inspecting, divide into two groups of disassembling and not disassembling products. The products which can be taken parted to the parts will be sent to the disassembling centers and, there, they will convert to the parts. There they divide into reusable and not reusable parts. The not reusable parts will rebut safely and the reusable parts will be sent to the processing center. In the remanufacturing process, according to the production center's demand, the parts which can be used again after processing center will be sent to the remanufacturing center and, after compounding with the other parts, will be changed into new products and can return to the distribution chain. In the recycling process, according to the recycling center's demand the disassembled parts (which can recover again) right after disassembling centers will be sent to the recycling centers for the purpose of producing the secondary materials.

In this research, the aim is to cluster returned items and connect them to the layers of the reverse supply chain by using data mining. Data mining is recognizing the exact, novel, useful, perceptible samples from available inputs in an input station, which cannot be reachable by using the usual process.

In this research, most of returned items are home garbage. After collecting the garbage to take it back to the supply chain by regarding efficiency, we need many experts and time and cost categories. In this study, to ease the process and lessen the human power, we will categorize the returned items in two phases. By using this method we can examine any kind of returned item more tenuously until its efficiency and further use is recognized.

In the first phase, we segregate these objects by the difference in their nature; then, in the second phase, by paying attention to object's quality, we can put it in the correct section and take it back to the appropriate layer of the reverse supply chain. In this research we study four types of returned products which have most usage in recycle and remanufacture categories: (1) plastic, (2) glass, (3) paper, (4) metal.

15.3 Return Items' Clustering Model

Clustering the returned products in the first phase can be performed with a data-mining algorithm called k-means. This algorithm by having the number of clusters, categorizes the inputs and finally specifies the centers which according to them the clusters will be categorized. To perform this algorithm we need to a specified model or scale among the inputs which the amount of that scale should be different in any kind of returned products. By performed studies it is suggested that the suitable standard for exerting this algorithm be the attraction coefficient of any kind products against x-ray glitter. To determine the inputs of this algorithm we should take all of the under control returned products against x-ray glitter and by machines which are equipped by attracted amount calculation system in any returned product we can determine the attraction coefficient of any products. We put these attraction coefficients in k-means algorithm as inputs and by specifying k-cluster we will reach to the first clustering of performing this algorithm, which is shown in Figure 15.1.

For second phase clustering we need to proficiently control and detect the clustered products at different kinds, to determine to which categories each product belongs, described as follow:

- This exact and proficient survey is conducted by experts.
- If the product returned by the customer is reusable or impeccable or has been repaired according to its destruction can be put in distribute system again.
- If the product needs renovation and adding new things will be transmitted to the product assemble phase and if a part has a problem, the part will be replace and remanufactured.

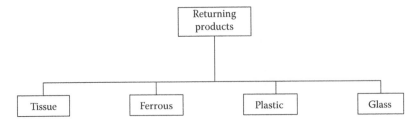

FIGURE 15.1
First phase clustering.

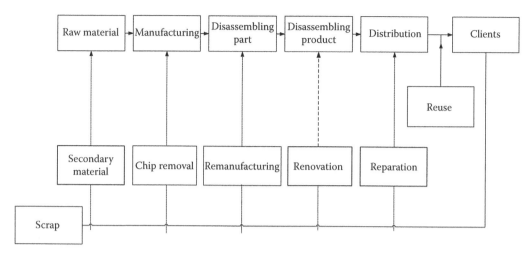

FIGURE 15.2
Phases of reverse supply chain.

- If there is any problem in part, chip removal will be performed, and
- If there is no way to repair, the part would be sent to the first phase of the chain as a secondary material, but,
- If it is impossible to use it as a secondary material, the company has to consider it as a scrap (Figure 15.2)

In this phase of clustering, to avoid wasting time and cost, the suggested method of sampling can be used via an operator and from any specific pallet related to the specific kind. The proficient operator in the basis of special kind will exert sampling.

According to these samples the operator will eventually determine what kind by which percent should be returned to the noteworthy layer.

We introduce the possibilities as follow:

α: The possible percentage that the returnable product will be referred to the supplier chain.

β: The possible percentage that the returnable product will be referred to the manufacturer layer

λ: The possible percentage that the returnable product will be referred to the distributor layer

γ: The possible percentage that the returnable product will be destroyed.

So according to the suggested method clustering we will have in the Figure 15.3.

In the collecting process of returned products from clients and referring them after clustering by nature to any one of supplier, manufacturer, distributor, or destroy sites it may take different operations on returned products in parts or related sub parts, so we can point out one of these operation such as duplication, or output a percent of product from supply chain as waste. We can see some of these operations in reverse supply chain process in Figure 15.4.

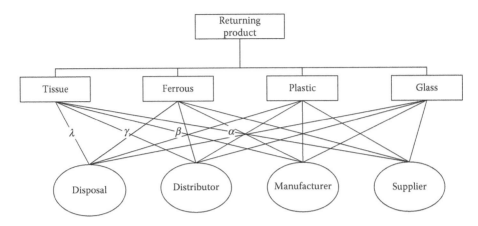

FIGURE 15.3
Second phase clustering.

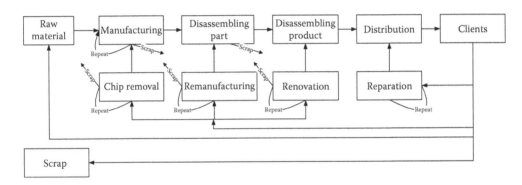

FIGURE 15.4
The recycle items illustrating duplication and output operation as waste in every part.

15.4 Application Study

For example, we radiate the x-ray to the returned products collection with energy photon 1 and will determine the attraction coefficient of all products with the machines which are equipped by the attraction amount calculator of this radiance by the existed products. We use these attraction coefficients as inputs and by attention to the under control kinds (plastic, glass, paper, metal) will determine the amount of clusters as follow:

$$K = 4$$

$$i = 1\,2\,3\,4$$

By performing this algorithm by MATLAB software, the inputs will be clustered and four clusters will be determined. This amount is shown in Table 15.1.

According to the performed researches and diagrams shown in Figures 15.5 through 15.8, there is a distinct attraction against x-ray for different kinds at general standard status.

TABLE 15.1

Each Cluster Center

C_1	0.0307	C_2	0.0320	C_3	0.0332	C_4	0.0343

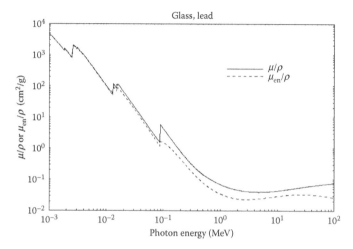

FIGURE 15.5
The glass attraction coefficient against x-ray glitter.

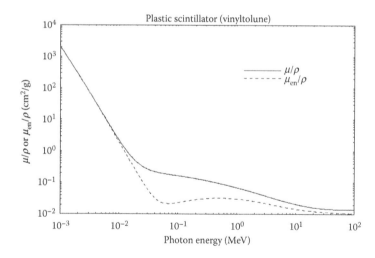

FIGURE 15.6
The plastic attraction coefficient against x-ray glitter.

By examining the given diagrams the attraction coefficient of different stuff against one energy photon of x-ray would be as follow:

0.03027 = the attraction coefficient of plastic

0.03468 = the attraction coefficient of glass

0.03074 = the attraction coefficient of paper

0.03093 = the attraction coefficient of metal

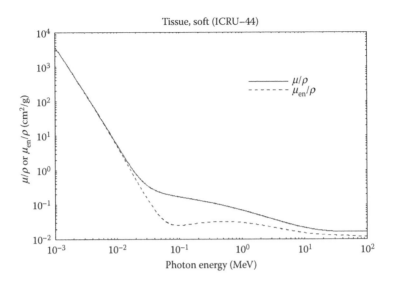

FIGURE 15.7
The soft tissue like paper attraction coefficient against x-ray glitter.

By assuming that the cluster centers that are obtained from performing k-means algorithm are equal to the nearest attraction obtained from remarked diagrams, we can perform clustering the first phase as follow:

$C_1 \sim 0.03074$
$C_2 \sim 0.03093$
$C_3 \sim 0.03027$
$C_4 \sim 0.03468$

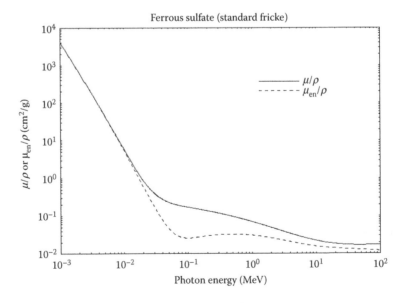

FIGURE 15.8
The metallic stuff attraction coefficient against x-ray glitter.

So, cluster number 1 demonstrates paper stuff, cluster number 2 demonstrates metal stuff, cluster number 3 demonstrates plastic stuff and cluster number 4 demonstrate glassy stuff.

In the second phase of clustering we use experts in any kind of stuff separately. These operators by having enough knowledge about specific stuff pallet and by sampling them will determine that the under control returned product should be returned to which one of supplier (α), manufacturer (β), distribution sites (γ) or with what probability should be destroyed (λ) at all. To determine the probability percentage, first we should find the amount of returned products that will be sent to any of these sites. Thus, we use STRATA method.

For example the operator who is the glass stuff expert will sample from the special pallet of returned glass products, which has 224 stuffs, and determines that 22 of 45 samples should be returned to the supplier site and 17 of 45 samples to the remanufacturing site and 6 of 45 samples to the distribution site. In this sampling we don't see any destructive product. So by using STARTA method we can find the unknown amount (the number of returned glass products which entered to each site) and finally calculate the probabilities α, β, γ, and λ.

If we assume supplier site h, by the given inputs we have:

$$Nh = \frac{22}{45} \cdot 224 = 109.5 \approx 109 \rightarrow \alpha = \frac{109}{224} \cdot 100 = 48.66$$

If we assume manufacturing site h, by the given inputs we have:

$$Nh = \frac{17}{45} \cdot 224 = 84.62 \approx 85 \rightarrow \beta = \frac{85}{224} \cdot 100 = 37.95$$

If we assume distribution site h, by the given inputs we have:

$$Nh = \frac{6}{45} \cdot 224 = 29.87 \approx 30 \rightarrow \gamma = \frac{30}{224} \cdot 100 = 13.39$$

The probability λ will be zero because of not having the destroyable product.

In a distinct operational process, from collecting the products until sending them to the reverse supply chain and also by attention to probabilities one more duplication and waste in every part and sub part which are related to those sites, now by numerical solution we can show that how many returned products will be returned again to the supply chain. In Figure 15.9 we examine this method for returned glass products.

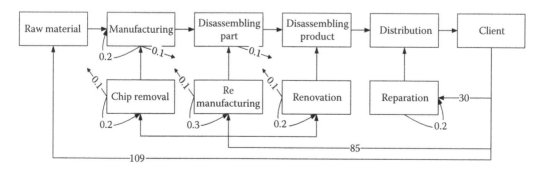

FIGURE 15.9
The recycle options with probabilities of waste and duplication.

The experts according to experiences and backgrounds first of all consider this assumption that from returned products that will be sent to the manufacturing site, 0.3 of them will be sent to the renovation sub part, 0.5 of them will be sent to the remanufacturing sub part and 0.2 of them will be sent to the chip removal sub part. By having these predetermined probabilities the number of products which will be sent to each one of the stated sub parts will be calculated as follow:

The number of products that will be entered to the renovation sub part:

$$0.3 * 85 = 25.5 \approx 26$$

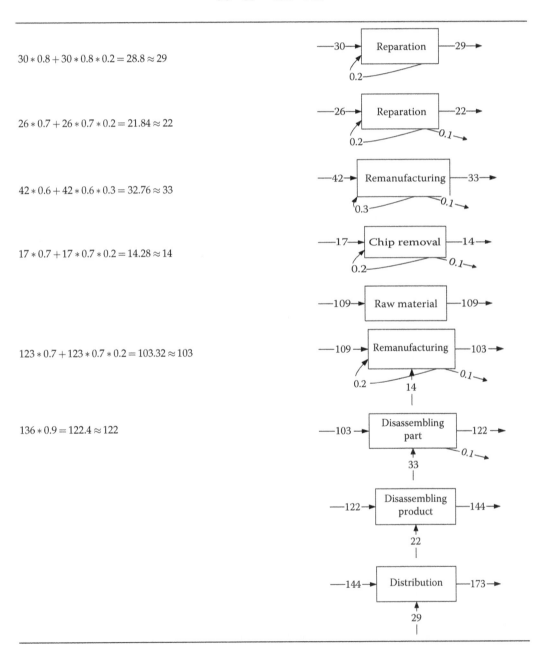

$$30 * 0.8 + 30 * 0.8 * 0.2 = 28.8 \approx 29$$

$$26 * 0.7 + 26 * 0.7 * 0.2 = 21.84 \approx 22$$

$$42 * 0.6 + 42 * 0.6 * 0.3 = 32.76 \approx 33$$

$$17 * 0.7 + 17 * 0.7 * 0.2 = 14.28 \approx 14$$

$$123 * 0.7 + 123 * 0.7 * 0.2 = 103.32 \approx 103$$

$$136 * 0.9 = 122.4 \approx 122$$

The number of products that will be entered to the remanufacturing sub part:

$$0.5 * 85 = 42.5 \approx 42$$

The number of products that will be entered to the chip removal sub part:

$$0.2 * 85 = 17$$

Now we can examine calculation of each part of the performed phases upon diagram to determine the number of recycled products among all of returned glass products.

By assuming different phases in sequence of reverse supply chain and performed calculations, we find out that we can recycle 173 of 224 returned glass products and use it again in the supply chain.

15.5 Discussions

This chapter considered a reverse supply chain including multiple layers and flowing multiple products. After collecting the returned commodities a clustering of the wastes was performed to relate the materials to the corresponding layer. To optimize the clustering and reduce the inspection, material diversification is handled via X-ray. K-mean has been applied to categorize the returned commodities. Probabilities of rework activities were given and the obtained recovered products in each layer and subsequently the total reworked products were determined. Economic analysis emphasizes the effectiveness of the methodology.

References

Baker RSJ. Data mining. *International Encyclopaedia of Education*, 3rd edition, Vol. 7, Oxford, UK: Elsevier, 2010:112–118.

Barros AI, Dekker R, Scholten VA. A two-level network for recycling sand: A case study. *European Journal of Operational Research* 1998;110(2):199–214.

Chen M, Han J, Yu PS. Data mining: An overview from a database perspective. *IEEE Transactions on Knowledge and Data Engineering* 1996;8(6):866–883.

Chen YL, Weng CH. Mining fuzzy association rules from questionnaire data. *Knowledge-Based Systems* 2009;22(1):46–56.

Der Laan E, Salomon M. Production planning and inventory control with remanufacturing and disposal. *European Journal of Operational Research* 1997;102:264–278.

Dincer E. The k-means algorithm in data mining and an application in medicine, MSc thesis, Kocaeli Univesity, Kocaeli, 2005.

Essid S, Richard G, David B. Inferring efficient hierarchical taxonomies for MIR tasks: Application to musical instruments. In: *6th International Conference on Music Information Retrieval*, London, UK, 2005:324–328.

Fayyad, U, Shapiro GP, Smyth P. The KDD process for extracting useful knowledge from volumes of data. *Communications of the ACM* 1996;39(11):27–34.

Gibbons WM, Ranta M, Scott TM. Information management and process improvement using data mining techniques. In: Loganantharaj R, et al. (eds.), *IEA/AIE 2000, LNAI, 1821*, Springer-Verlag, Berlin, 2000:93–98.

Han J, Kamber M. *Data Mining Concepts and Techniques*. San Francisco: Elsevier Inc, 2006:21–27.

Harding JA, Shahbaz M, Srinivas S, Kusiak A. Data mining in manufacturing: A review. *Journal of Manufacturing Science and Engineering* 2006;128:969–976.

Hea Z, Xua X, Huangb JZ, Denga S. Mining class outliers: Concepts, algorithms and applications in CRM. *Expert Systems with Applications* 2004;27(4):681–697.

Hsu C. Data mining to improve industrial standards and enhance production and marketing: An empirical study in apparel industry, *Expert Systems with Applications* 2009;36(3):4185–4191.

Huang H, Wu D. Product quality improvement analysis using data mining: A case study in ultra-precision manufacturing industry. In: Wang L, Jin Y, (eds.) *FSKD 2005, LNAI, 3614*, Berlin: Springer-Verlag, 2005:577–580.

Kim KB, Song IS, Jeong BJ. Supply planning model for remanufacturing system in reverse logistics environment. *Computers & Industrial Engineering* 2006;51(2):279–287.

Krikke HR, Kooi EJ, Schurr PC. Network design in reverse logistics: A quantitative model. In: Stahly P, (ed.) *New trends in distribution logistics*. Berlin: Springer 1999:45–62.

Kroon L, Vrijens G. Returnable containers: An example of reverse logistics. *International Journal of Physical Distribution & Logistics Management* 1995;25(2):56–68.

Larose DT. *Discovering knowledge in data*. New Jersey: Wiley & Sons Inc, 2005:1–25.

Liao TW, Triantaphyllou E. *Recent advances in data mining of enterprise data: Algorithms and applications*. Singapore: World Scientific Publishing 2007:111–145.

Listes O. A generic stochastic model for supply-and-return network design. *Computers & Operations Research* 2007;34(2):417–442.

Liu C. A data mining algorithm for designing the conventional cellular manufacturing systems. In: Orgun MA, Thornton J, (eds.) *AI 2007, LNAI, 4830*, Springer-Verlag, Berlin, 2007:715–720.

Shahbaz M, Athar S, Muhammad M, Ayaz Khan S. Data mining methodology in perspective of manufacturing databases. *The Journal of American Science* 2010;6(11):999–1012.

Salema MIG, Barbosa-Povoa AP, Novais AQ. An optimization model for the design of a capacitated multi-product reverse logistics network with uncertainty. *European Journal of Operational Research* 2007;179:1063–1077.

Stock JK. *Reverse logistics. White Paper*. Oak Brook, Illinois: Council of Logistics Management, 1992.

Tan KC, Teoh EJ, Yua Q, Goh KC. A hybrid evolutionary algorithm for attribute selection in data mining. *Expert Systems with Applications* 2009;36(4):8616–8630.

Tan PN, Steinbach M, Kumar V. *Introduction to data mining*. Boston: Addison-Wesley, 2006:1–37.

16

Sustainable Reverse Supply Chain: Customer Requirement Fulfillment Model

SUMMARY In this chapter, we study the reverse supply chain, considering customers' requirements. The supply chain—including suppliers, producers and customers—is proposed. In the customer layer, we analyze the customer's data to identify and fulfill their needs. The mathematical model is developed for each of the categories.

16.1 Introduction

Reverse Logistics Network, part of the supply chain can be defined as "the accurate and timely transmission of materials used and goods haven't been solved through the supply chain (SC) to the end consumer of the last unit of the good." In other words, reverse logistics is the process of moving and transporting goods and products in the supply chain for return. Design and implementation of reverse logistics network for product returns, inventory, and transportation not only reduce costs but also increase customer loyalty (Lee et al., 2009).

The American Association Reverse Logistics Executive Council, defines reverse logistics as follows:

> Reverse logistics is the process of planning, implementing and controlling the flow of raw materials, semi-manufactured inventories, end products and information to help them in terms of cost, from the point of origin to point of consumption, with the goal of re-creating value or proper disposal (Lu and Bostel, 2007).

Reverse logistics in supply chains start with the pieces that go back—components for recycling or recovery value and are collected for proper disposal. Moving companies turning to the planning, implementation and control of reverse logistics can be divided into three categories:

- Economic factors (direct and indirect)
- Rules and regulations
- Accountability for environmental sensitivities.

Logistics network design is a strategic decision, which usually involves the nature of the facilities, their capacity, number of products, number of classes in the chain, and is associated facilities. All these issues impact the performance of the supply chain. Since the construction or closing of a facility incurs costs and takes time, thus changes in the short run is not possible. Also, investment in strategic network design decisions is less than tactical and operational decisions which provides greater returns on investment (Pishvaee et al., 2010).

Product returns are often regarded as a cost of doing business; reverse supply chain management (RSCM) appears to lack significant benefit or impact worth being studied. Such a mentality is slowly changing. In fact, more researchers and managers start to realize that RSCM can provide competitive advantage instead of posing problems if it is being managed effectively (Price-White, 2002). "Returned goods are a very real part of the supply chain, and focusing on reverse logistics management can account for savings of 1%–2% of sales" (Dave Vehec, vice president of GENCO retail service, in Price-White, 2002). In fact, the significant costs and potential benefits of RSMC call for serious attention. For example, the total processing fees and product value of product returns cost suppliers in the US more than $100 billion per year (Stock et al., 2002). Thus, ignoring the issue of product returns can be very costly. Hewlett Packard product returns used to be treated as a low-level divisional problem until thorough analysis revealed that the total cost of product returns was equivalent to 2% of total outbound sales (Guide et al., 2003a). The idea of utilizing used products is indeed not groundbreaking as it has been around for decades. Recovering used products for remanufacturing and recycling emerged in the 1960s and has been implemented by manufacturers specializing in remanufacturing/recycling or those that manufacture and remanufacture/recycle their own products (Guide et al., 2003b). Recovering values of products traveling upstream is also a familiar research topic, being referred to by terminologies such as product recovery, recycling, and closed-loop manufacturing. The newest terminologies for this similar concept emerging in the last few years include closed-loop supply chain, reverse logistics, reverse supply chain, green supply chains, and environmentally responsible manufacturing. Yet, most of this research is highly operations-oriented (Dowlashahi, 2004) and mainly based on the physical structures of upstream product movements and providing quantitative models to manage reverse movement of products. There are hardly any studies that investigate the benefits of RSCM and develop a model/framework to thoroughly examine RSCM practices and its effects (Guide et al., 2003a,b). In fact, there is little consensus on the definition of RSCM. The terms listed above such as reverse logistics, reverse supply chain, etc., bear various meanings from study to study. Most importantly, it appears that most research views managing returned products as a problem of product manufacturers. Thus, while forward SCM emphasizes the chain perspective in their research and practice, RSCM illustrates a clear lack of the chain perspective. Meanwhile, RSCM becomes increasingly important for a number of reasons. The most noticeable is the change in legislation in regard to managing industrial waste. Many industries are specifically required to take back their own products once being disposed, particularly if they contain hazardous materials. Given the rising number of environmentally conscious countries and consumers, legislation will only become stricter. Second, today customers are more educated and demanding, and tend to have less tolerance for imperfect products (Krikke et al., 2004; Avittathur and Shah, 2004). They are also better informed of their rights to have the option to return products as they wish. As a result, return policies are becoming much more lenient (Guide et al., 2003a,b). Furthermore, wholesalers and retailers also push for extended return warranties to reduce their own risks. Given the intense global competition, that means manufacturers often have to provide take-back warranties for any unsold, damaged, end-of-season, or obsolete products to sell their products. Third, shorter product life cycles, frequent new product introductions, product leasing and upgrade options are also among the reasons to increase the volume of reverse flows. It is estimated that 6% of all retail purchases in the US is returned by customers. This is equal to around $52 billion worth of returned goods annually (Tibben-Lembke, 2004). Fourth, companies that incorporate remanufacturing/recycling in their business model, such as Xerox and Kodak, and invest in technologies that allow them to do so effectively, find it crucial to be able to recover and sort their used products efficiently. Finally, new business models such as catalogue and mail order—and especially e-business—affect product flows forward and backward significantly (DiMaggio, 2000; Krikke et al., 2004). Wehkam, a large mail order company in the Netherlands has 28% of product returns, or about 10,000 items per day (Van Nunen and Zuidwijk, 2004). The Home Shopping Network ships around 32 million packages each year, of which 6.4 million are returned, making it a real challenge to keep track of both forward and reverse flows of products and payments. On average, 20% of products sold by e-retailers are returned and the figure could be as high as 35% for certain products such as clothing (Trebilcock, 2002).

Due to such changes in the operational environment, companies find themselves having to deal with an increasing number of products that are unwanted, used, end-of-life, defective, or obsolete. Yet managing this problem is not easy given the novelty of problem recognition and its unpredictability in nature (Blackburn et al., 2004). Product return flows are characterized as uncertain and unpredictable in terms of quantity, timing, type, and status, making it very challenging to handle (Minner, 2001; Van Nunen and Zuidwijk, 2004). The RSCM problem so far has been viewed from a silo approach and thought as a time consuming and costly problem, while yielding few benefits (Guide et al., 2003a,b). A new way of looking at the whole product return process is needed to better understand and handle the issue. This chapter intends to provide a thorough literature review on the topic of RSCM and develop a framework to research the practices and benefits of RSCM in the future. The literature review aims at reflecting the progress as well as shortfalls of research on RSCM. The framework is developed from a chain perspective taking into consideration the interactions between forward and reverse chains as well as those among the chain members. This study also discusses the meanings and focus of various definitions of RSCM, leading to developing a definition of RSCM that reflects its overall concept. Notice that both research journals and practitioner-related journals are included due to limited academic research on this topic up-to-date. The literature review also examines the potential benefits and challenges of researching and implementing RSMC. This study also briefly discusses the continuing research direction and a plan based on this chapter to validate and improve the proposed framework.

16.2 Problem Definition

The supply chain under study here is composed of supplier, manufacturer, and customers. The backward information flow in this SC is from customers to other layers. First, we collect a list of customers' views. In this case, we assume that we analyze the customers' views in three areas: transport, production and quality. We use the coding system for getting the customers opinions. Then, by using the k-means algorithm, which is one of the data analyzing algorithms, we cluster the data so that similar data go to the same cluster. By collecting customers' opinions and mining their requirements, we are aiming to satisfy them. The customers' opinions are in three parts: transportation, production and quality. Using a k-means clustering technique, the opinions are classified and then, using a mathematical program, we present an approach to fulfill customers' needs.

The definitions of indices, parameters and decision variables employed for mathematical formulation are as follows:

Indices

i	Index for factors in the third category (customer relationship costs)
j	Index for factors in the second category (competitive price)
k	Index for factors in the first category (product variety)
l	Customers' opinions collection method (manually distributing questionnaires, SMS, web, fax)

Parameters

C_k	Process capability for k factor
h_k	Cost of needed items to run for k factor
B	Total cost available to the required items
Co_j	Operating costs for j factor
Q_j	The number of sales with respect to the j factor
θ	Profit margins
W_j	The maximum price that the customer is willing to pay for each factor j
S_{il}	Operating labor cost i getting through of l method
F_{il}	Factor IT costs i getting through of l method
V_{il}	Speed of the poll i factor getting through of l method
n_{il}	Number of acquired factor i getting through of l method
t_{il}	When using the comments we have l method for i factor
usl_k	Upper control limit for the k factor
lsl_k	Lower control limit for the k factor

Decision Variables

Z_{il}	If the i'th factor from the poll cost category that taken from the l'th poll method is chosen, 1, otherwise zero.
y_j	if the j'th factor from the competitive cost category is chosen, 1, otherwise zero.
x_k	if the k'th factor from the product diversification category is chosen, 1, otherwise zero.
pr_j	if the k'th factor from the product diversification category is chosen, 1, otherwise zero. The product cost is chosen according to j'th factor.

16.3 Customers' Requirement Model

The objective function of product diversification:

$$MaxZ = \sum_k c_k . x_k \tag{16.1}$$

The aim of the objective function of product diversification is to maximize the total capability of processes of the chosen factors in the category of product diversification.

$$\sum_k x_k \geq 1 \tag{16.2}$$

Constraint 16.2 indicates that at least one of the factors of this category (product diversification) should get the value of 1.

$$\sum_k h_k \cdot x_k \leq B \tag{16.3}$$

Constraint 16.3 indicates that the total cost of the necessary items to run on the k'th factor, should be less than the total available cost for the required items.

$$x_k \in \{0, 1\} \tag{16.4}$$

Constraint 16.4 indicates that if the k'th factor from the product diversification category is chosen, 1, otherwise zero.

$$c_k = \frac{usl_k - lsl_k}{6\delta_k} \tag{16.5}$$

Process capability index, is calculated from the above formula:

- If $c_p > 1$, the process can observe the limitation of acceptable characteristics.
- If $c_p = 1$, the process nearly can observe the limitation of acceptable characteristics.
- If $c_p < 1$, the process can't observe the limitation of acceptable characteristics.

The objective function of competitive price:

$$Max \sum_j y_j (\mathrm{Pr}_j . Q_j) - co_j \tag{16.6}$$

The aim of the objective function of competitive price is to maximize the profit that comes from multiplying the product selling price according to j'th factor to the number of product sales according the j'th factor minus factor costs.

$$\theta \sum_j co_j \leq pr_j \tag{16.7}$$

Constraint 16.7 indicates the maximum benefit that manufacturer can take from the customer, in costs; the profit margin should be less than the product price.

$$\sum_j y_j . w_j \geq pr_j \tag{16.8}$$

Constraint 16.8 indicates that the product price must be less than the maximum price that the customer tends to pay for each factor of j.

$$y_j \in \{0,1\} \tag{16.9}$$

Constraint 16.9 indicates that if the j'th factor from the competitive price category is chosen, 1, otherwise zero.

$$\sum_j y_j \geq 1 \tag{16.10}$$

$$pr_j \geq 0 \tag{16.11}$$

Constraint 16.11 indicates that the product price is positive.

$$pr_j \leq M \cdot y_j \quad \forall j, \tag{16.12}$$

Objective function of poll cost:

$$Min \sum_i \sum_l Z_{il} \cdot (S_{il} + F_{il}) \tag{16.13}$$

The aim of objective function of poll cost is minimize the total cost (total cost of technology and human resources)

$$Max \sum_i \sum_l V_{il} \cdot Z_{il} \tag{16.14}$$

The aim of objective function of poll cost is maximize the poll speed that achieves from different methods:

$$V_{il} = \sum_i \frac{n_{il}}{t_{il}} \quad \forall l \tag{16.15}$$

Poll speed that resulting from dividing the number of comments on time.

$$Z_{il} \in \{0,1\} \tag{16.16}$$

$$\sum_i \sum_l Z_{il} \geq 1 \tag{16.17}$$

Constraint 16.16 indicates that if i'th factor from the poll cost category that taken from the l'th poll method is chosen, 1, otherwise zero.

We note that the nonlinear Equation 16.6 can be linearized by using the inequality

$$pr_j \leq M \cdot y_j, \quad \forall j, \tag{16.18}$$

where M is a sufficiently large number; since y_j is equal to 0 or 1, then (16.18) implies that Pr_j be 0 (when $y_j = 0$) or arbitrary (when $y_j = 1$). The latter is imposed when M is large enough, and the former is guaranteed by the simultaneous incurred inequalities $pr_j \leq 0$ and $pr_j \geq 0$ (from Equation 16.11).

Linearization of the objective function:

$$Max \sum_j y_j (Pr_j \cdot Q_j) - co_j \tag{16.19}$$

$$\theta \sum_j co_j \leq pr_j \tag{16.20}$$

$$\sum_j y_j \cdot w_j \geq pr_j \tag{16.21}$$

$$y_j \in \{0,1\} \tag{16.22}$$

$$\sum_j y_j \geq 1 \tag{16.23}$$

$$pr_j \geq 0 \tag{16.24}$$

$$pr_j \leq M \cdot y_j \quad \forall j, \tag{16.25}$$

16.3.1 Sampling by Using STRATA Method

By using STRATA method, we consider 1000 customers for getting decision. The sampling is done by STRATA method that explained below:

Sample size for method l n_l
Population size for method l N_l
The total population size N
The total sample size n

$$n_1 = \frac{N_1}{N} \times n$$

STRATA sampling for this problem:

$$N = 1000 \; n = 100 \; N_l$$

$$n_1 = \frac{150}{1000} \times 100 = 15$$

$$n_2 = \frac{450}{1000} \times 100 = 45$$

$$n_3 = \frac{350}{1000} \times 100 = 35$$

$$n_4 = \frac{100}{1000} \times 100 = 10$$

16.3.2 Average Factors Weight

After STRATA sampling we calculate the average weight of every 16 factors separately:
Average weight for every 16 factors respectively from left to right:
3.33, 3.52, 3.84, 3.49, 3.17, 3.69, 3.38, 3.15, 4.02, 3.99, 3.05, 3.72, 3.70, 3.95, 3.65, 3.25

16.3.3 *k*-Means Algorithm

k-Means is one of the simplest unsupervised learning algorithms that solve the well-known clustering problem. The procedure follows a simple and easy way to classify a given data set through a certain number of clusters (assume k clusters) fixed *a priori*. The main idea is to define k centroids, one for each cluster. These centroids should be placed in a cunning way because of different location causes different result. So, the better choice is to place them as far away from each other as possible. The next step is to take each point belonging to a given data set and associate it to the nearest centroid. When no point is pending, the first step is completed and an early group age is done. At this point, we need to re-calculate k new centroids as bar centers of the clusters resulting from the previous step. After we have these k new centroids, a new binding has to be done between the same data set points and the nearest new centroid. A loop has been generated. As a result of this loop, we may notice that the k centroids change their location step by step until no more changes are done. In other words, centroids do not move any more. Finally, this algorithm aims at minimizing an objective function—in this case a squared error function. The objective function

$$J = \sum_{j-1}^{k} \sum_{i-1}^{n} \left\| x_i^{(j)} - c_j \right\|^2 ,$$

where $\left\| x_i^{(j)} - c_j \right\|^2$ is a chosen distance measure between a data point $x_i^{(j)}$ and the cluster centre c_j, is an indicator of the distance of the n data points from their respective cluster centers.
The algorithm is composed of the following steps:

1. Place k points into the space represented by the objects that are being clustered. These points represent initial group centroids.

2. Assign each object to the group that has the closest centroid.

3. When all objects have been assigned, recalculate the positions of the k centroids.

4. Repeat Steps 2 and 3 until the centroids no longer move. This produces a separation of the objects into groups from which the metric to be minimized can be calculated.

Although it can be proved that the procedure will always terminate, the *k*-means algorithm does not necessarily find the most optimal configuration, corresponding to the global objective function minimum. The algorithm is also significantly sensitive to the initial randomly selected cluster centers. The *k*-means algorithm can be run multiple times to reduce this effect.

16.4 Application Study

Here, we are going to cluster 16 factors that we took the customers comments from them, in three categories including customers' demands. Three random points that we picked up for three clusters respectively are

3.28 = first cluster (product variation)

3.42 = second cluster (competitive price)

3.61 = third cluster (comments following)

We encode *k*-means algorithm in MATLAB software. Here are the results:

16.4.1 Results of *k*-means Algorithm

The centers of obtained clusters:

The first cluster center is: 3.1550

The second cluster center is: 3.4300

The third cluster center is: 3.8200

Clustering is performed for the first category (product variation):

The First Cluster is: 3.17

The First Cluster is: 3.15

The First Cluster is: 3.05

The First Cluster is: 3.25

Clustering is performed for the second category (competitive price):

The Second Cluster is: 3.33

The Second Cluster is: 3.52

The Second Cluster is: 3.49

The Second Cluster is: 3.38

Clustering is performed for the third category (after sale service):

The Third Cluster is: 3.84

The Third Cluster is: 3.69

The Third Cluster is: 4.02

The Third Cluster is: 3.99

The Third Cluster is: 3.72

The Third Cluster is: 3.70

The Third Cluster is: 3.95

The Third Cluster is: 3.65

Here are the factors that placed in the first cluster (product variation):

1. Warranty
2. Effectiveness
3. Damages resulting from delay
4. Deterioration of handling

Here are the factors that placed in the second cluster (competitive price):

1. Proper packaging
2. Quality compared to competitors
3. Apparent quality of the product
4. Inspection

Here are the factors that placed in the third cluster (after sale services):

1. Services
2. Desired performance
3. Efficiency
4. Labor productivity
5. Being fresh
6. Provide services in all areas
7. Order online
8. Availability

We encode and run every of our three objective functions separately in LINGO software.

16.4.2 Calculation and Determining the Mathematical Model Parameters

- Process capability values

$$c_1 = 1.2,\ c_2 = 1.6,\ c_3 = 1,\ c_4 = 1.4$$

- The values of costs required for running on k'th factor

$$h_1 = 5,\ h_2 = 6,\ h_3 = 8,\ h_4 = 7$$

- The total available cost for required stuff
$$B = 17$$
- Total product sale according to j'th factor

$$Q_1 = 50, \; Q_2 = 40, \; Q_3 = 38, \; Q_4 = 44$$

- Costs of j'th factor

$$CO_1 = 6, \; CO_2 = 7, \; CO_3 = 10, \; CO_4 = 5$$

- The maximum price that the customer is willing to pay according to each of the factor j'th

$$W_1 = 8, \; W_2 = 10, \; W_3 = 13, \; W_4 = 9$$

- The profit margin

$$\theta = 1.15$$

- Labor cost matrix for each i and l:

$$s = \begin{bmatrix} 5 & 6 & 5 & 7 \\ 5 & 6 & 7 & 6 \\ 6 & 9 & 6 & 7 \\ 5 & 7 & 7 & 5 \\ 7 & 6 & 5 & 8 \\ 5 & 9 & 7 & 5 \\ 5 & 6 & 8 & 5 \\ 9 & 6 & 7 & 8 \end{bmatrix}$$

- IT cost matrix for each i and l:

$$f = \begin{bmatrix} 5 & 7 & 6 & 9 \\ 5 & 7 & 7 & 9 \\ 5 & 6 & 9 & 5 \\ 6 & 7 & 5 & 9 \\ 9 & 8 & 7 & 9 \\ 5 & 7 & 5 & 9 \\ 5 & 6 & 9 & 6 \\ 5 & 6 & 9 & 8 \end{bmatrix}$$

- Poll speed matrix for each i and l:

$$v = \begin{vmatrix} 15.0 & 34.2 & 19.8 & 19.8 \\ 15.0 & 30.0 & 17.4 & 15.0 \\ 15.0 & 28.2 & 24.0 & 30.0 \\ 17.4 & 26.4 & 18.6 & 15.0 \\ 18.0 & 40.2 & 21.6 & 6.0 \\ 19.8 & 32.4 & 19.8 & 15.0 \\ 12.0 & 36.0 & 30.0 & 30.0 \\ 19.8 & 24.0 & 18.0 & 19.8 \end{vmatrix}$$

16.4.3 Using AHP for Multi-Objective Optimization

According to the doubly objective function of the above problem, for its optimization we use the weighting method to objective function by means of analytic hierarchy process. We compare two objective functions in terms of three criteria that are competitive advantage, economical aspect and strategic perspective. The results will be in the following tables; Matrix of Binary comparison of objective-criteria is based in Table 16.1.

Criteria binary comparison-criteria is based in Table 16.2:

$$\psi_1 = \text{Total weight for objective } 1 = 0.557 \times 0.503 + 0.639 \times 0.348 + 0.488 \times 0.149 = 0.575$$

$$\psi_2 = \text{Total weight for objective } 2 = 0.443 \times 0.503 + 0.361 \times 0.348 + 0.512 \times 0.149 = 0.425$$

New objective function for the comments following category:

$$MAX\left[\psi_1\left(\sum_i \sum_l v_{il} \cdot z_{il}\right) - \psi_2\left(\sum_i \sum_l z_{il} \cdot (S_{il} + F_{il})\right)\right]$$

16.4.4 Mathematical Model Results

After solving the model by LINGO software:

Objective function of product variation:

- The value of objective function of product variation became 3.
- The deterioration and effectiveness due to displacement factors, got the value of 1, this means that in the category of product variation, the two factors of deterioration and effectiveness due to displacement are effective.

$$x_2 = 1, \ x_4 = 1$$

TABLE 16.1

The Objective-Criteria Matrix

	Strategic Perspective	Economical Aspect	Competitive Advantage
First Objective Function	0.557	0.639	0.488
Second Objective Function	0.443	0.361	0.512

TABLE 16.2

Criteria-Criteria Matrix

	Strategic Perspective	Economical Aspect	Competitive Advantage	W
Strategic Perspective	1	3	2	0.503
Economical Aspect	1/3	1	5	0.348
Competitive Advantage	1/2	1/5	1	0.149

The objective function of competitive price:

- The value of the objective function of product variation became 1662.
- The value of product price according to j'th factor:

$$pr_1 = 8, \ pr_2 = 10, \ pr_3 = 13, \ pr_4 = 9$$

The objective function of after sale services:

- After solving by LINGO software the value of objective function became 39.15.
- Here is the factor that got the value of 1 in this category that means that they are effective:

$$z_{12} = 1, \ z_{13} = 1, \ z_{22} = 1, \ z_{32} = 1, \ z_{34} = 1, \ z_{42} = 1, \ z_{52} = 1, \ z_{53} = 1, \ z_{61} = 1,$$
$$z_{62} = 1, \ z_{72} = 1, \ z_{73} = 1, \ z_{74} = 1, \ z_{82} = 1$$

16.5 Discussions

In this chapter we examined the issue of reverse supply chain, considering customer requirements. So, we considered supply chain with three layers of supplier, manufacturer and customer. In customers' layer, data mining the customers' comments were proposed to identify and fulfill their requests. First, by preparing a list of customers' comments, we collected their opinions in cases that we need, and we assumed to evaluate the customers' comments in three categories of transportation, production and quality. Then by use of k-means algorithm, that is one of the algorithms that uses for clustering, we did the clustering to categorize the data. By doing this, we can identify the customers' requests and eventually fulfill them. Then, by considering the factors in each category, we developed a mathematical model for each category.

The proposal for future works:

- In customers' satisfactory issues, the operational field of geographical servicing by using positioning and allocation models can be considered.
- The considered data in this model are deterministic, while the uncertain data with fuzzy and probabilistic approach can be considered in the modeling.
- In this chapter we used k-means algorithm, while for more efficiency, in large volume data, the algorithms based on artificial intelligence can be used.
- In one of the models we used the multi-objective weighting algorithm, which can use methods such as ideal planning or adoption planning.

References

Avittathur B, Shah J. Tapping product returns through efficient reverse supply chains: Opportunities and issues. *IIMB Management Review* 2004;16(4):84–93.

Blackburn J, Guide D Jr, Souza G, Van Wassenhove L. Reverse supply chains for commercial returns. *California Management Review* 2004;46(2):6–22.

DiMaggio J. Reverse psychology. *Warehousing Management* 2000;7(4):30–32.

Dowlathshahi S. Developing a theory of reverse logistics. *Interfaces* 2004;30(3):143–155.

Guide D Jr, Jayaraman V, Linton J. Building contingency planning for closed-loop supply chains with product recovery. *Journal of Operations Management* 2003a;21:259–279.

Guide D Jr, Tuenter R, Van Wassenhove L. Matching demand and supply to maximize from remanufacturing. *Manufacturing & Service Operations Management* 2003b;5(4):303–331.

Guide D Jr, Van Wassenhove L. Closed-loop supply chains: Practice and potential. *Interfaces* 2003a;33(6):1–32.

Guide D Jr, Harrison T, Van Wassenhove L. The challenge of closed-loop supply chains. *Interfaces* 2003b;33(6):3–6.

Krikke H, Blanc I, van de Velde S. Product modularity and the design of closed-loop supply chains. *California Management Review* 2004;46(2):23–39.

Lee J, Gen M, Rhee K. Network model and optimization of reverse logistics by hybrid genetic algorithm. *Computers & Industrial Engineering* 2009;56:951–964.

Lu Z, Bostel N. A facility location model for logistics systems including reverse flows: The case of remanufacturing activities. *Computers & Operations Research* 2007;34:299–323.

Minner S. Strategic safety stock in reverse logistics supply chains. *International Journal of Production Economics* 2001;71:417–428.

Pishvaee MS, Zanjirani Farahani R, Dullaert W. A mimetic algorithm for bi-objective integrated forward/reverse logistics network design. *Computers and Operations Research* 2010;37(6):1100–1112.

Price-White C. Check it out and back in again. *Frontline Solutions* 2002;June:24–29.

Stock J, Speh T, Shear H. Many happy (product) returns. *Harvard Business Review* 2002;80(7):16–17.

Trebilcock B. Return to sender. *Warehousing Management* 2002;9(4):24–27.

Tibben-Lembke S. Strategic use of secondary market for retail consumer goods, California returns. *California Management Review* 2004;46(2):6–22.

Van Nunen J, Zuidwijk R. E-enabled closed-loop supply chains. *California Management Review* 2004;42(2):40–54.

17

Multi-Layer Multi-Product Reverse Supply Chain: Defects and Pricing Model

SUMMARY   In this chapter, a multi-layer reverse supply chain is considered and modeled with respect to the defects of the returned items. During the process of producing a product and getting it to the end consumer, there will be some costs, and one of these costs is the cost of rejecting the product. We considered three kinds of defects that might cause a product to be rejected, including defects due to improper transportation, defects due to improper production, and defects due to improper packing. Then, an interactive pricing is implemented for the returning products.

17.1 Introduction

The forward supply chain (FSC) is composed of a series of activities in the process of converting raw materials to finished goods. The manager's objective of investing in the forward supply chain is to improve performance in areas such as procurement, demand management and order fulfillment, among others (Cooper et al., 1997). Improvement initiatives can take several forms, including supplier development programs and customer relationship management. In contrast, the reverse supply chain (RSC) refers to the series of activities necessary to retrieve a product from a customer and either dispose of it or recover value (Guide and van Wassenhove, 2002; Prahinski and Kocabasoglu, 2005).

Reverse supply chains deal with products at the end of their lifecycle. Reverse supply chain management aims at product value recovery at least cost possible. Not all reverse supply chains are identical; however, they are all designed to carry out five main processes: product acquisition, reverse logistics, inspection and disposition, remanufacturing or refurbishing, and marketing (Blackburn et al., 2004). To most companies, product returns have been viewed as a nuisance; as a result, their legacy today is a reverse supply chain designed to minimize cost (Guide Jr et al., 2003a).

American Association Reverse Logistics Executive defined reverse logistics as the process of planning, implementing and controlling the flow of raw materials, semi-manufactured inventories, end products and information to help them in terms of cost, from the point of origin to point of consumption, with the goal of re-creating value or proper disposal (Lu and Bostel, 2007). Reverse logistics in supply chains start with the pieces returned; components for recycling or recovery value are collected for proper disposal. Movers turning to the planning, implementation and control of reverse logistics can be divided into three categories:

- Economic factors (direct and indirect)
- Rules and regulations
- Accountability for environmental sensitivities

Logistics network design is a strategic decision, which usually involves the nature of the facilities, their capacity, number of products, number of classes in the chain, and associated facilities. Since the construction of the facility or closing costs, and they take a long time, change in the short run, it is not possible. Also, investing in strategic network design decisions, tactical and operational decisions are greater return on investment (Pishvaee et al., 2010). One way of minimizing the environmental impact of waste is to use reverse supply chains to increase the amount of product materials recovered from the waste stream. Reverse supply chain is a process by which a manufacturer systematically accepts previously shipped products or parts from the point of consumption for possible reuse, remanufacturing, recycling, or disposal. Thus reverse logistics has important environmental dimensions (Álvarez-Gil et al., 2007; Linton et al., 2007; Ciliberti et al., 2008; Zhu et al., 2008) as well as dimensions relating to value reclamation (Logozar et al., 2006; Alshamrani et al., 2007; Kumar and Putnam, 2008; Mutha and Pokharel, 2009; Pokharel and Mutha, 2009; Ilgin and Gupta, 2010).

Research on reverse supply chain has been growing since the Sixties and research on strategies and models on reverse logistics (RL) can be seen in the publications in and after the 1980s. However, efforts to synthesize the research in an integrated broad-based body of knowledge have been limited. Most research focuses only on a small area of RL systems, such as network design, production planning or environmental issues. Fleischmann et al. (1997) studied RL from the perspectives of distribution planning, inventory control and production planning. Carter and Ellram (1998) focused on the transportation and packaging, purchasing and environmental aspects in their review of RL literature. Krikkey et al. (2003) studied the interactions between sustainability and supply chains by considering environmental issues regarding product design, product life extension and product recovery at end-of-life.

Most of academic researchers have been focusing on reverse supply chain structure design. As identified by Guide et al. (2006), time value affects commercial product return value recovery greatly, as a large proportion of the product value usually erodes away due to long processing time. To address that, they proposed that reverse supply chain structure should follow two fundamental structures: efficient (centralized) and responsive (decentralized), similar to a forward supply chain. Jayaraman et al. (1999) examined the closed-loop logistics structure using a 0-1 model, solving the location of remanufacturing/distribution facilities, transportation, production and stocking of the optimal quantities.

After industry practices and further researches, people start to realize that remanufacturing could be a profit generating process, depending on the quantity and quality of product returns and on the demand for remanufactured products. Guide Jr et al. (2003b) used cellular telephone industry as an example and examined how acquisition prices and selling prices affect profitability of a remanufacturing process. This research is a big step in the field of reverse supply chain management since it firstly takes acquisition management into consideration for profit generation.

The effective implementation of reverse logistics does not preclude achieving one goal at the expense of the other. Considering this, many world class companies have realized that reverse logistics practices, combined with source reduction processes, can be used to gain competitive advantage and at the same time can achieve sustainable development (Maslennikova and Foley, 2000; Neto et al., 2008; Seuring and Muller, 2008; Hu and Bidanda, 2009; Lee et al., 2010). Firms engaged in reverse supply chains are in the process of investment recovery and certainly would receive direct (input materials, cost reduction, value added recovery) and indirect benefits (impeding legislation, market protection, green image and improvement in customer/supplier relations). Guide Jr. and Van Wassenhove (2009) revealed that US $700 million of perfectly operating product that could be recovered were destroyed. They found that a US firm, ReCellular, has gained economic advantage through refurbishing cell phones. Manufacturer HP showcased that returns of its products could cost around 2% of total outbound sales and only half of them were being recovered (Guide et al., 2006).

Customer demand is considered as one of the major driving forces for reverse supply chain practices. Research suggests that there is an increasing customer demand for green products and for organizations to engage in environmental supply chain practices (New et al., 2000). The stakeholders of large firms have also become more concerned about customer attitudes and are more conscious of environmental issues and want to be socially responsible themselves (Bowen, 2000). The impact of customer demand is felt equally by manufacturing and retail businesses. For example, nowadays vehicle manufacturers are not competing on cost alone, but also on environmentally responsible features. In turn, manufacturers are forcing their strategic suppliers to obtain environmental accreditation, such as that of the Eco-Management and Audit Scheme (Lamming and Hampson, 1996). Similarly, big retailers are pressuring their suppliers to be more environmentally responsible (Hall, 2001).

17.2 Problem Definition

Nowadays, reverse supply chain and air pollution and environmental considerations result in better configuration and control of supply chain. In supply chain for different products, providing appropriate quality and services for customers is of importance. During processing the product several defects may occur leading to customer dissatisfaction and the product is returned. In this research, we try to examine the return of products in a multi-layer chain that contains the supplier, producer, distributor, retailer and finally customer by proposing a mathematical model with the goal of decreasing the costs of returned product and increasing the transaction from produced products that satisfy the customers. We examine three kinds of return product in different layers of a five-layer supply chain. The return causes are

1. Return product for problem in manufacturing
2. Return product for problem in packing
3. Return product for problem in transfer

A configuration of the problem is depicted in Figure 17.1.

For computing the cost of these three kinds of return product, we consider one probability of occurring for the maximum occurrence, one measure for frequency and the number of occurring times of each of return products and one coefficient of single cost for penalty of each of these products in different layers. By using these three measures we obtain the costs of each return product.

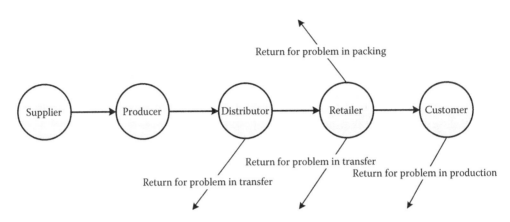

FIGURE 17.1
A configuration of the problem.

In this research, we use the following model for calculation of maximum benefit from produced and selling products.

$$\text{Benefit} = (\text{price} * \text{produced number}) - (\text{price} * \text{demand}) - \text{cost}$$

According to this, we choose two ways to price—one is pricing of the product by producer. The producer can develop a pricing procedure so that the benefit is obtained and the specified selling price is fixed (Fazlollahtabar et al., 2012). The second group of price is pricing of product from customer opinions; this price is by consumers and according to their desire to purchase various goods. For this type of pricing we make use of the model developed by Rezaie et al. (2012) based on willingness to pay.

Let us investigate the concept of willingness to pay (WTP). Any potential customer has a maximum WTP, which is called reservation price interchangeably. Therefore, a customer would buy a product only if its price is lower than his maximum WTP. We can compute the WTP of different customers for a given price interval $[p_1, p_2]$ as follows,

$$WTP = \int_{p_1}^{p_2} w(x)dx \tag{17.1}$$

where $w(x)$ is the WTP function. As stated, the maximum demand (D) is obtained to be $D = d(p)$, that is, the maximum demand is when the price is zero. Here, we can obtain the price-response function using WTP,

$$d(p) = \int_{p}^{\infty} w(x)dx \tag{17.2}$$

In our proposed model, assuming WTP is a uniform probability distribution function, the demand function is considered to be linear using the proposed integral. As a result, the following linear demand function is obtained, $d(p) = D - mP$, where $D = d(0)$ is the maximum demand, m is the gradient of the demand curve and P is the price.

17.3 Defect Model

Here, we develop the mathematical model. The indices, parameters, decision variables, and the complete mathematical model are given below:

Indices

i Index for supplier ($i = 1, ..., I$)
j Index for producer ($j = 1, ..., J$)
k Index for distributor ($k = 1, ..., K$)
l Index for retailer ($l = 1, ..., L$)
m Index for customer ($m = 1, ..., M$)
n Index for product ($n = 1, ..., N$)

Parameters

α_{njk} Probability of return product n produced by producer j by distributor k for problem in transfer

β_{njl} Probability of return product n produced by producer j by retailer l for problem in transfer

γ_{njl} Probability of return product n produced by producer j by retailer l for problem in packing

ρ_{njm} Probability of return product n produced by producer j by customer m for problem in manufacture

δ_{njk} Coefficient of return cost for product n produced by producer j from distributor layer for problem in transfer

μ_{njl} Coefficient return cost for product n produced by producer j from retailer layer for problem in transfer

θ_{njl} Coefficient return cost for product n produced by producer j from retailer layer for problem in packing

λ_{njm} Coefficient return cost for product n produced by producer j from customer layer for problem in manufacture

$X_{nj} \cdot \alpha_{njk} \cdot \delta_{njk}$ Return cost for product n produced by producer j from distributor layer for problem in transfer

$X_{nj} \cdot \beta_{njl} \cdot \mu_{njl}$ Return cost for product n produced by producer j from retailer layer for problem in transfer

$X_{nj} \cdot \gamma_{njl} \cdot \theta_{njl}$ Return cost for product n produced by producer j from retailer layer for problem in packing

$X_{nj} \cdot \rho_{njm} \cdot \lambda_{njm}$ Return cost for product n produced by producer j from customer layer for problem in manufacture

CT_{nj} Maximum return cost for product n produced by producer j for problem in transfer from distributor layer

CD_{nj} Maximum return cost for product n produced by producer j for problem in transfer from retailer layer

CB_{nj} Maximum return cost for product n produced by producer j for problem in packing

CP_{nj} Maximum return cost for product n produced by producer j for problem in manufacture

C_{nj}	Return cost for product n produced by producer j
d_{nj}	Demand for product n produced by producer j
Z_1	Objective function one
Z_2	Objective function two
Y_{mnj}	Maximum fee that customer m is willing to pay for product n produced by producer j

Decision Variables

P_{nj}	Price of product n produced by producer j
P'_{mnj}	Price of product n produced by producer j from customer m viewpoint
X_{nj}	The number of product n produced by producer j

17.3.1 The Mathematical Model

$$\text{Max } Z_1 = \sum_j \sum_n (p_{nj} * X_{nj}) - \sum_j \sum_n c_{nj} \tag{17.3}$$

$$\text{Max} Z_2 = \sum_m \sum_j \sum_n (p'_{mnj} * X_{nj}) - \sum_j \sum_n c_{nj} \tag{17.4}$$

s.t.

$$\sum_k X_{nj} \cdot \alpha_{njk} \cdot \delta_{njk} \leq CT_{nj}, \quad \forall j, n \tag{17.5}$$

$$\sum_l X_{nj} \cdot \beta_{njl} \cdot \mu_{njl} \leq CD_{nj}, \quad \forall j, n \tag{17.6}$$

$$\sum_l X_{nj} \cdot \gamma_{njl} \cdot \theta_{njl} \leq CB_{nj}, \quad \forall j, n \tag{17.7}$$

$$\sum_m X_{nj} \cdot \rho_{njm} \cdot \lambda_{njm} \leq Cp_{nj}, \quad \forall j, n \tag{17.8}$$

$$\sum_k X_{nj} \cdot \alpha_{njk} \cdot \delta_{njk} + \sum_l X_{nj} \cdot \beta_{njl} \cdot \mu_{njl}$$
$$+ \sum_l X_{nj} \cdot \gamma_{njl} \cdot \theta_{njl} + \sum_m X_{nj} \cdot \rho_{njm} \cdot \lambda_{njm} = C_{nj}, \quad \forall j, n \tag{17.9}$$

$$p'_{mnj} = \int_{c_{nj}/x_{nj}}^{Y_{mnj}} w(x)dx, \quad \forall m, n, j \tag{17.10}$$

$$c_{nj} \leq P_{nj} \cdot X_{nj}, \quad \forall n, j \tag{17.11}$$

$$P_{nj} = \varepsilon_{nj} - \sigma_{nj} \cdot d_{nj}, \quad \forall n, j \tag{17.12}$$

$$\sigma_{nj} = \frac{-\sum_n \sum_j (d_{nj} - \bar{d})(p_{nj} - \bar{p})}{\sum_n \sum_j (p_{nj} - \bar{p})^2}, \quad \forall n, j \tag{17.13}$$

$$\varepsilon_{nj} = \bar{d} + \sigma_{nj} \cdot \bar{p}, \quad \forall n, j \tag{17.14}$$

$$\bar{P} = \frac{\sum_n \sum_j p_{nj}}{R} \tag{17.15}$$

$$\bar{d} = \frac{\sum_n \sum_j d_{nj}}{R} \tag{17.16}$$

$$R = n \times j, \quad \forall n, j \tag{17.17}$$

$$x, p, p' \geq 0, \quad \forall m, n, j \tag{17.18}$$

In the proposed mathematical model Equations 17.3 and 17.4 are the objective functions maximize the total benefit. Equations 17.5 through 17.8 limit the maximum return cost in a different layer. Equation 17.9 is to calculate the cost of the returned that maybe occurred throughout the supply chain. Equation 17.10 is to limit the price of consumer to an upper bound. Equation 17.11 is to limit the maximum return cost. Equations 17.12 through 17.17 show the pricing from consumer viewpoint. Relation 17.18 shows the sign and the kind of decision variables.

17.3.2 The Willingness to Pay Function

The equation below is the cost model from the consumer's view, and $w(x)$ is the consumer's tendency function to pay.

$$p'_{mnj} = \int_{c_{nj}/x_{nj}}^{Y_{mnj}} w(x)dx, \tag{17.19}$$

$w(x)$ follows the continuously distributed function and for its determination, we polled consumers. For this purpose, the groups of customers were asked to tell their opinions about productions in the form of giving a score in a certain range.

$$w(x) = \frac{1}{b-a} \tag{17.20}$$

In the above equation, the a and b are determined by polling customers.

b: The maximum score given by consumers to the productions.

a: The minimum score given by consumers to the productions.

17.4 Computational Results

Here, a hypothetical numerical example is illustrated to show the applicability and effectiveness of the proposed model. We consider three producers, three distributors, three retailers, three customers and also three products in the supply chain. The numerical values for probability of return product n produced by producer j by distributor k for problem in transfer are given in Table 17.1.

The numerical values for probability of return product n produced by producer j by retailer l for problem in transfer are given in Table 17.2.

Also the numerical values for probability of return product n produced by producer j by retailer l for problem in packing are given in Table 17.3.

And the numerical values for probability of return product n produced by producer j by customer m for problem in manufacture are given in Table 17.4.

The numerical values for coefficient return cost for product n produced by producer j from distributor layer for problem in transfer are given in Table 17.5.

Also the numerical values for coefficient return cost for product n produced by producer j from retailer layer for problem in transfer are given in Table 17.6.

And the numerical values for coefficient return cost for product n produced by producer j from retailer layer for problem in packing are given in Table 17.7.

And the numerical values for Coefficient return cost for product n produced by producer j from customer layer for problem in manufacture are given in Table 17.8.

TABLE 17.1

Probability of Return for the Problem in Transfer by Distributor

	$n = 1$			$n = 2$			$n = 3$		
α_{njk}	1	2	3	1	2	3	1	2	3
1	0.01	0.02	0.05	0.04	0.09	0.05	0.03	0.04	0.07
2	0.03	0.05	0.04	0.06	0.08	0.05	0.04	0.06	0.07
3	0.08	0.09	0.06	0.04	0.05	0.02	0.06	0.08	0.04

TABLE 17.2

Probability of Return for the Problem in Transfer by Retailer

	$n = 1$			$n = 2$			$n = 3$		
β_{njl}	1	2	3	1	2	3	1	2	3
1	0.02	0.04	0.03	0.07	0.08	0.04	0.05	0.02	0.01
2	0.05	0.06	0.05	0.05	0.05	0.06	0.07	0.05	0.04
3	0.07	0.04	0.07	0.03	0.05	0.03	0.09	0.07	0.05

TABLE 17.3

Probability of Return for the Problem in Packing

γ_{njl}	$n=1$			$n=2$			$n=3$		
	1	2	3	1	2	3	1	2	3
1	0.03	0.05	0.03	0.04	0.06	0.05	0.04	0.03	0.03
2	0.03	0.04	0.03	0.05	0.05	0.06	0.07	0.06	0.04
3	0.05	0.01	0.02	0.09	0.02	0.08	0.08	0.07	0.06

TABLE 17.4

Probability of Return for the Problem in Manufacture

ρ_{njm}	$n=1$			$n=2$			$n=3$		
	1	2	3	1	2	3	1	2	3
1	0.03	0.01	0.03	0.03	0.04	0.05	0.02	0.04	0.03
2	0.02	0.03	0.04	0.04	0.02	0.01	0.03	0.04	0.05
3	0.02	0.04	0.05	0.05	0.02	0.02	0.01	0.03	0.05

TABLE 17.5

Coefficient of Return Cost for Transfer from Distributor

δ_{njk}	$n=1$			$n=2$			$n=3$		
	1	2	3	1	2	3	1	2	3
1	19	18	15	15	19	11	14	18	10
2	19	18	10	18	10	16	17	19	16
3	16	17	19	19	17	18	18	15	19

TABLE 17.6

Coefficient of Return Cost for Transfer from Retailer

μ_{njl}	$n=1$			$n=2$			$n=3$		
	1	2	3	1	2	3	1	2	3
1	15	18	16	17	16	18	19	18	15
2	15	10	18	15	16	18	19	18	15
3	18	19	15	18	10	17	15	17	19

TABLE 17.7

Coefficient of Return Cost for Packing

θ_{njl}	$n=1$			$n=2$			$n=3$		
	1	2	3	1	2	3	1	2	3
1	16	18	17	18	16	17	18	15	16
2	17	18	15	19	15	17	18	16	15
3	18	17	15	15	17	16	16	15	17

TABLE 17.8

Coefficient of Return Cost for Manufacturer

λ_{njm}	$n = 1$			$n = 2$			$n = 3$		
	1	2	3	1	2	3	1	2	3
1	15	17	16	15	19	10	17	15	19
2	19	16	10	16	17	18	17	10	19
3	18	16	15	19	17	16	19	17	18

TABLE 17.9

Maximum Willingness to Pay

Y_{mnj}	$m = 1$			$m = 2$			$m = 3$		
	1	2	3	1	2	3	1	2	3
1	340	345	360	345	330	385	360	355	350
2	340	350	350	360	395	345	350	345	355
3	360	350	360	325	340	350	345	305	340

Table of maximum fee that customer m willingness to pay for product n produced by producer j are given in Table 17.9.

The maximum return cost matrix for product n produced by producer j for problem in transfer from distributor layer is given below:

$$CT_{nj} = \begin{bmatrix} 270 & 240 & 240 \\ 255 & 275 & 275 \\ 260 & 290 & 255 \end{bmatrix}$$

The maximum return cost matrix for product n produced by producer j for problem in transfer from retailer layer is given below:

$$CD_{nj} = \begin{bmatrix} 240 & 215 & 290 \\ 245 & 245 & 252 \\ 250 & 240 & 260 \end{bmatrix}$$

Also the maximum return cost matrix for product n produced by producer j for problem in packing is given below:

$$CB_{nj} = \begin{bmatrix} 270 & 240 & 235 \\ 280 & 250 & 290 \\ 240 & 255 & 255 \end{bmatrix}$$

Maximum return cost matrix for product n produced by producer j for problem in manufacture is given below:

$$CP_{nj} = \begin{bmatrix} 240 & 250 & 265 \\ 240 & 150 & 245 \\ 235 & 195 & 240 \end{bmatrix}$$

The matrix of demand for product n produced by producer j is given below:

$$d_{nj} = \begin{bmatrix} 40 & 60 & 45 \\ 10 & 90 & 65 \\ 25 & 75 & 40 \end{bmatrix}$$

In the proposed problem, if the scoring range is rated between zero and 10, and if the minimum score given by consumers is 3 and the maximum score is 10, according to Equation 17.20, the value of $w(x)$ is calculated as follows:

$$w(x) = \frac{1}{8-3} = 0.2.$$

And the Equation 17.19 becomes the following equation:

$$p'_{mnj} = 0.2 \times \left[Y_{mnj} - \left(\frac{C_{nj}}{X_{nj}} \right) \right]. \tag{17.21}$$

17.4.1 The Model Solutions

After solving the problem's model, the following solutions were obtained:
  Here is the value of objective function that was obtained:
  MaxZ = 332603.4

17.4.2 The Decision Variables

1. After solving the model, we can see the values for a number of the products generated by each one of the manufacturers, in the matrix below:

$$X_{nj} = \begin{bmatrix} 11.63 & 2.3 & 316.9 \\ 78 & 44.75 & 139.5 \\ 16 & 186.8 & 17 \end{bmatrix}$$

The values of X_{nj} matrix show the optimized number of productions by each one of manufacturers. For example the number 11.63 shows that the optimized number of productions for the product number 1 by the manufacturer number 1 is 11.63.

2. The price value from the manufacturer's view was obtained as follows. The P_{nj} values show the price that each one of the manufacturers, for every generated product according to their costs, considered,

$$P_{nj} = \begin{bmatrix} 280 & 260 & 275 \\ 310 & 230 & 255 \\ 295 & 245 & 280 \end{bmatrix}$$

TABLE 17.10

The Obtained Prices for Customers

p'_{mnj}	m = 1			m = 2			m = 3		
	1	2	3	1	2	3	1	2	3
1	188.5	55.2	232.7	191.6	45.5	249	201.4	61.8	226.2
2	220.5	202	225	227	244.7	222	220.5	212.3	228.5
3	213	224.5	200.9	190.3	217.9	194.4	203.3	195.2	187.9

Also, the price values from the consumer's view were calculated according to the Table 17.10. The application of the decision variable P'_{mnj}, is determination of product price in view of the end consumer, according to their tendency for buying a particular product, which might be different according to product type and product traction from the customer's view and their need to that product. As we know, in the real world, the consumers have a tendency to buy cheaper products; hence, almost all of the obtained values for the decision variable, are less than the price from the manufacturer's view.

17.5 Discussions

In this research, we investigated the Reverse Supply Chain problem in five layers including supplier, manufacturer, distributor, retailer and end customer by presenting a mathematical model, and with the objective of profit maximization. During the process of producing a product and getting it to the end consumer, there will be some costs and one of these costs is the cost of rejecting the product. We considered three kinds of defects that might cause a product to be rejected, including defects due to improper transportation, defects due to improper production, and defects due to improper packing. We considered two different viewpoints for pricing the produced products, the first is pricing from the manufacturer's view, and the second is pricing from the end consumer's view. For this kind of pricing, we considered consumer's tendencies in buying a product. The objective model of the problem is profit maximization of selling products, and according to the two kinds of prices that we considered, the problem will have two objective functions. We conclude that the manufacturers should try to increase the production of the products of which the consumers tend to buy more, and also the product distribution network should attend the consumer's tendency in different regions and the product traction from the consumer's view, and finally decisions in the product distribution process should be optimized.

The proposal for future works:

- The considered data in this model are deterministic, while the uncertain data with fuzzy and probabilistic approach can be considered in the modeling.
- Develop factors affecting the interests of customers and their entry into the modeling process.
- Attention on decision risk for the consideration of the dynamics of the real world.
- Extraction costs with activity-based costing method in the supply chain.

- In one of the models we used the multi-objective weighting algorithm, which one can use with the methods such as ideal planning or adoption planning.
- Market studies and entry into a mathematical model to classify customers.

References

Alshamrani A, Mathur K, Ballou RH. Reverse logistics: Simultaneous design of delivery routes and returns strategies. *Computers and Operations Research* 2007;34:595–619.

Álvarez-Gil MJ, Berrone P, Husillos FJ, Lado N. 2007. Reverse logistics, stakeholders' influence, organizational slack, and managers' posture. *Journal of Business Research* 2007;60:463–473.

Blackburn JD, Guide Jr. VDR, Souza GC, Wassenhove LNV. Reverse supply chains for commercial returns. *California Management Review* 2004;46(2):6–22.

Bowen F. Environmental visibility: A trigger for organizational response? *Business Strategy and Environment* 2000;9:92–107.

Carter CR, Ellram LM. Reverse logistics: A review of the literature and framework for future investigation. *Journal of Business Logistics* 1998;19(1):85–102.

Ciliberti F, Pontrandolfo P, Scozzi B. Logistics social responsibility: Standard adoption and practices in Italian companies. *International Journal of Production Economics* 2008;113:88–106.

Cooper MC, Douglas ML, Pagh JD. Supply chain management: More than a new name for logistics. *The International Journal of Logistics Management* 1997;8(1):19.

Fazlollahtabar H, Akbari F, Mahdavi I. Optimizing e-Shopping sales efficiency using virtual intelligent agent and pricing concept. *Asian Journal of Information and Communications* 2012;4(1):28–36.

Fleischmann M, Bloemhof-Ruwaard JM, Dekker R, van der Laan E, van Nunen JAEE, van Wassenhove LN. Quantitative models for reverse logistics: A review. *European Journal of Operational Research* 1997;103(1):1–17.

Guide Jr. D, Teunter R, Van Wassenhove L. Matching demand and supply to maximize profits from remanufacturing. *Manufacturing & Service Operations Management* 2003a;5(4):303–316.

Guide Jr. VDR, Jayaraman V, Linton JD. Building contingency planning for closed-loop supply chains with product recovery. *Journal of Operations Management* 2003b;21(2003):259–279.

Guide Jr. VDR, Souza G, Van Wassenhove LN, Blackburn JD. Time value of commercial product returns. *Management Science* 2006;52(8):1200–1214.

Guide Jr. VDR, Van Wassenhove LN. The reverse supply chain. *Harvard Business Review* 2002;80(2):25–26.

Guide Jr. VDR, Van Wassenhove LN. 2009. The evolution of closed-loop supply chain research. *Operations Research* 2009;57(1):10–18.

Hall J. Environmental supply chain innovation. *Greener Management International* 2001;35:105–119.

Hu G, Bidanda N. Modeling sustainable product lifecycle decision support systems. *International Journal of Production Economics* 2009;122:366–375.

Ilgin MA, Gupta SM. Environmentally conscious manufacturing and product recovery (ECMPRO): A review of the state of the art. *Journal of Environmental Management* 2010;91:563–591.

Jayaraman V, Guide Jr. VDR, Srivastava R. A closed-loop logistics model from remanufacturing. *The Journal of the Operational Research Society* 1999;50(5):497–508.

Krikkey H, Bloemhof-Ruwaard J, Van Wassenhove L. Concurrent product and closed-loop supply chain design with an application to refrigerators. *International Journal of Production Research* 2003;41(16):3689–3719.

Kumar S, Putnam V. Cradle to cradle: Reverse logistics strategies and opportunities across three industry sectors. *International Journal of Production Economics* 2008;115:305–315.

Lamming R, Hampson J. The environment as a supply chain management issue. *British Journal of Management* 1996;7(Special issue):S45–S62.

Lee DH, Dong M, Bian W. The design of sustainable logistics network under uncertainty. *International Journal of Production Economics* 2010;128(1):159–166. doi: 10.1016/j.ijpe.2010.06.009.

Lintoln N. Concurrent product and closed-loop supply chain design with an application to refrigerators, *International Journal of Production Research* 2003;41(16):3689–3719.

Linton JD, Klassen R, Jayaraman V. Sustainable supply chains: An introduction. *Journal of Operations Management* 2007;25:1075–1082.

Logozar K, Radonjic G, Bastic M. Incorporation of reverse logistics model into in-plant recycling process: A case of aluminum industry. *Resources, Conservation and Recycling* 2006;49:49–67.

Lu Z, Bostel N. A facility location model for logistics systems including reverse flows: The case of remanufacturing activities. *Computers & Operations Research* 2007;34:299–323.

Maslennikova I, Foley D. Xerox's approach to sustainability. *Interfaces* 2000;30(3):226–233.

Mutha A, Pokharel S. Strategic network design for reverse logistics and remanufacturing using new and old product modules. *Computers and Industrial Engineering* 2009;56:334–346.

Neto JQF, Bloemhof-Ruwaard JM, Van Nunen JAEE, Van Heck E. Designing and evaluating sustainable logistics networks. *International Journal of Production Economics* 2008;111:195–208.

New S, Green K, Morton B. Buying the environment: the multiple meanings of green supply. In: Fineman S (ed.), *The Business of Greening*, London: Routledge, 2000:33–53.

Pishvaee MS, Farahani RZ, Dullaert W. A memetic algorithm for bi-objective integrated forward/reverse logistics network design. *Computers & Operations Research* 2010;37(6):1100–1112.

Pokharel S, Mutha A. Perspectives in reverse logistics: A review. *Resources, Conservation and Recycling* 2009;53:175–182.

Prahinski C, Kocabasoglu C. Empirical research opportunities in reverse supply chains. *Omega* 2005;34(6):519–532.

Rezaie B, Esmaeili F, Fazlollahtabar H. Developing the concept of pricing in a deterministic homogenous vehicle routing problem with comprehensive sensitivity analysis. *International Journal of Services and Operations Management* 2012;12(1):20–34.

Seuring S, Muller M. From a literature review to a conceptual framework for sustainable supply chain management. *Journal of Cleaner Production* 2008;16(15):1699–1710.

Zhu Q, Sarkis J, Lai K-H. Green supply chain management implications for "closing the loop." *Transportation Research Part E* 2008;44:1–18.

18

Reverse Supply Chain Vehicle Routing Problem: Similarity Pattern Model

SUMMARY In this chapter, a mathematical model is proposed in order to reduce the cost of waste collection and transporting for the recycling process. A mixed-integer nonlinear programming model is provided, including a waste collection routing problem and the processes following after garbage unloading. There is a balance between the distance among trashcans and the similarity of the trashcans in terms of the types of waste in order to select the optimal route for each garbage transport vehicles. For this purpose, a similarity pattern is designed. By following the similarity pattern for route selection, recovery rate of waste will be increased being shown in the model.

18.1 Introduction

In general, a waste collection system involves the collection and transportation of solid waste to disposal facilities. This essential service is receiving increasing attention from many researchers due to its impact on the public concern for the environment and population growth, especially in urban areas. Because this service involves a very high operational cost, researchers are trying to reduce the cost by improving the routing of waste collection vehicles, finding the most suitable location of disposal facilities and the location of collection waste bins, as well as minimizing the number of vehicles used.

We address a waste collection VRP with consideration of similarity of the trashcans in terms of the types of waste, which is in the trashcans and the processes following after garbage unloading. The waste collection problem consists of routing vehicles to collect customers' waste while minimizing travel cost. This problem is known as the Waste Collection Vehicle Routing Problem (WCVRP). WCVRP differs from the traditional VRP in that the waste collecting vehicles must empty their loads at disposal sites. The vehicles must be empty when returning to the depot. The problem is illustrated in Figure 18.1 for one disposal site with a set of vehicles.

Weigel and Cao (1999) present a case study of application of VRPTW algorithms for Sears' home delivery problem and technician dispatching problem. They follow a cluster-first-route-second method and discuss three main routines: origin-destination (OD) matrix construction, route assignment, and route improvement routines. They apply a shortest-path algorithm to a geographic information system (GIS) to obtain OD matrix—that is, travel time between any two stops. For the route assignment routine (clustering), an algorithm called multiple-insertion, which is similar to the parallel insertion algorithm of Potvin and Rousseau (1993), is developed. As an objective function, the weighted combination of travel time, wait time, and time window violation is used. They propose an

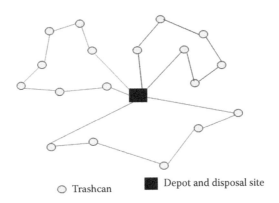

○ Trashcan   ■ Depot and disposal site

FIGURE 18.1
A route sequence of some vehicle considering disposal operations with single disposal sites.

intra-route improvement algorithm and a neighborhood inter-route improvement algorithm that improves the solution quality by transferring and exchanging stops between two routes. In order to enhance the improvement performance, tabu search is applied to the improvement algorithms.

Chang and Wang (1997) used a fuzzy goal MIP model for vehicle routing and scheduling in a solid waste collection system. Shih and Lin (1999) reported an approach to resolve the collection, vehicle scheduling, and routing problems for infectious waste management by using dynamic programming (DP) and integer programming (IP) methods for the periodic vehicle routing problem in cost only.

Tung and Pinnoi (2000) modify Solomon's insertion algorithm and apply it to a waste collection problem in Hanoi, Vietnam. In addition to the considerations of the standard VRPTW, they consider a landfill operation that is the dumping of the collected garbage at the landfill, and inter-arrival time constraints between two consecutive visits at a stop. They incorporate the landfill operation by assuming that a vehicle starts a new route from the depot after landfill. Or-opt and 2-opt algorithms are adopted to improve the solution quality.

Angelelli and Speranza (2002a) address the periodic vehicle routing problem with intermediate facilities (PVRP–IF). When a vehicle visits an intermediate facility, its capacity will be renewed. They propose a tabu search algorithm with four move operators: move a customer in the same day, change the visiting schedule, redistribution of customers, and simplification of intersection. Initial solutions are built by assigning a visiting schedule randomly to each customer and constructing vehicle routes on each day using iterative insertion procedure. Angelelli and Speranza (2002b) applied their algorithm for estimating the operating costs of different waste-collection systems: traditional system with three-man crew, side-loader system, and side-loader system with demountable body. The differentiator between their problem and ours is the time windows of the stops and the facilities. Our problem requires explicit consideration of time windows.

Eisenstein and Iyer (1997) use a Markov decision process to model the residential waste collection problem in the city of Chicago. They model the weight and time required to collect waste from a city block as normally distributed random variables. Action in their Markov decision process is the choice of route that visits the dumpsite once or twice. Teixeira et al. (2004) apply a heuristic approach for a PVRP for the separate collection of three types of waste: glass, paper, and plastic/metal. The approach has three phases: define a zone for each vehicle, define waste type to collect on each day, and select the sites to visit and sequence them. Mourao and Almeida (2000) model the residential garbage collection problem in a quarter of Lisbon, Portugal as a capacitated arc routing problem, and propose two lower-bounding methods and a route-first, cluster-second heuristic method. Chang et al. (1997) discuss how combining GIS functions with analytical models can help analyze alternative solid waste collection strategies for a metropolitan city in Taiwan.

A real life waste collection vehicle routing problem with time windows assuming multiple disposal trips and drivers' lunch breaks was addressed by Kim et al. (2006). They assumed a weekly predetermined schedule and presented a route construction algorithm that was an extension of Solomon's insertion algorithm (Solomon, 1987) and address a real life waste collection VRPTW with consideration of multiple disposal trips and drivers' lunch breaks. Ombuki-Berman et al. (2007) address the same problem by using a multi-objective genetic algorithm on a set of benchmark data from real-world problems obtained by Kim et al. (2006).

Benjamin and Beasley (2010) improve the results when minimizing travel distance using a tabu search and variable neighborhood search and a combination of these. A very similar problem, with only one disposal site, is addressed by Tung and Pinnoi (2000), where they modify Solomon's insertion algorithm and apply it to a waste collection problem in Hanoi, Vietnam. Nuortio et al. (2006) present a guided variable neighborhood thresholding meta-heuristic for the problem of optimizing the vehicle routes and schedules for collecting municipal solid waste in Eastern Finland. Solid waste collection is furthermore considered by Li et al. (2008) for the city of Porto Alegre, Brazil. Their problem consists of designing daily truck schedules over a set of previously defined collection trips, on which the trucks collect solid waste in fixed routes and empty loads in one of several operational recycling facilities in the system. They use a heuristic approach to solve the problem. Buhrkal et al. (2012) study the Waste Collection Vehicle Routing Problem with Time Window, which is concerned with finding cost optimal routes for garbage trucks such that all garbage bins are emptied and the waste is driven to disposal sites while respecting customer time windows and ensuring that drivers are given the breaks that the law requires. They propose an adaptive large neighborhood search algorithm for solving the problem and illustrate the usefulness of the algorithm by showing that the algorithm can improve the objective of a set of instances from the literature as well as for instances provided by a Danish garbage collection company.

18.2 Problem Definition

In this section, a mixed-integer nonlinear programming model is provided including a waste collection vehicle routing problem (WCVRP) and the following processes after garbage unloading, so that, there is a balance between the distance between trashcans, and the similarity of the trashcans in terms of the types of waste which is in the trashcans, in order to select the optimal route for each garbage transport vehicles. For this purpose, a similarity pattern is designed. Figure 18.2 presents a general diagram of our model.

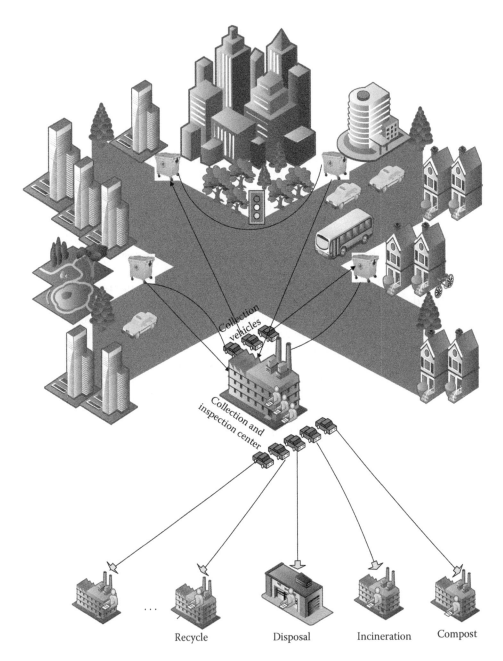

FIGURE 18.2
General diagram of the problem.

The problem is defined on a graph where the set of nodes $N = \{1, ..., n\}$ consists of a depot and a disposal site which are considered as one node $\{1\} \in N$, $n-1$ customers $\{2, ..., n\} \in N$ and the set of arcs is $G = \{(i, j)|i, j \in N, i \neq j\}$. Let M be the set of vehicles in VRP network and let K be the set of types of waste and the types of centers that we will transport the classified waste after unloading them. It is assumed that all vehicles in VRP network can have different capacity O_m. The objective of the WCVRP is to find a set of routes for the vehicles, minimizing total travel cost and satisfying vehicle capacity, such that all customers are visited exactly once, and route of each vehicle is chosen so that the similar trashcan be placed in one route. For this purpose, similar pattern was designed based on the probability of presence of the type of waste in trashcan P_{ki}, and dissimilarity is shown as a penalty in the objective function.

18.3 Similarity Pattern Model

The problem can be modelled using three types of variables: r_{ijm} is one if and only if vehicle $m \in M$ uses arc $(i, j) \in G$, VC_k represents the total volume of transmitted kth type of waste from collection site to kth center, U_k represents the number of trips required to transport VC_k and x_k represents the number of vehicles required to satisfy the x_k. A mathematical model for the present WCVRP is

$$Min$$

$$C(x) = \sum_{(i,j) \in G} \left(C_{ij} + \sum_{k=1}^{K} |P_{ki} - P_{kj}| \right) \cdot \sum_{m=1}^{M} r_{ijm} + \sum_{k=1}^{K} ((x_k \cdot CP_k) + (U_k \cdot CT_k))$$

$$+ \left(\sum_{m=1}^{M} E_m + \sum_{m=1}^{M} \left(\sum_{j=2}^{N} r_{1jm} \cdot LS_m \right) \right)$$

$$St:$$

$$\sum_{j=2}^{N} r_{1jm} = 1 \quad \forall \ m \in M \tag{18.1}$$

$$\sum_{i=2}^{N} r_{i1m} = 1 \quad \forall \ m \in M \tag{18.2}$$

$$\sum_{i=1}^{N} \sum_{m=1}^{M} r_{ijm} = 1 \quad \forall j \in N \backslash \{1\}, \quad i \neq j \tag{18.3}$$

$$\sum_{i=2}^{N} r_{ijm} = \sum_{k=2}^{N} r_{jkm} \quad \forall j \in N \backslash \{1\}, \quad m \in M \tag{18.4}$$

$$\sum_{(i,j) \in G} r_{ijm} \cdot ld_i \leq O_m \quad \forall \ m \in M \tag{18.5}$$

$$\sum_{(i,j) \in G} r_{ijm} \cdot ld_i \geq 0 \quad \forall \ m \in M \tag{18.6}$$

$$r_{iim} = 0 \quad \forall i \in N, \quad m \in M \tag{18.7}$$

$$r_{ijm} + r_{jim} \leq 1 \quad \forall i, j \in N\{1\}, \quad m \in M \tag{18.8}$$

$$z_{ij} \begin{cases} 1 & \text{if } \sum_{k=1}^{K} |P_{ki} - P_{kj}| \leq \alpha \\ 0 & \text{o.w.} \end{cases} \tag{18.9}$$

$$z_{1j} = 1 \quad \forall j \in N \tag{18.10}$$

$$z_{i1} = 1 \quad \forall i \in N \tag{18.11}$$

$$d_{ij} \begin{cases} 1 & \text{if } L_{ij} \leq \beta \\ 0 & \text{o.w.} \end{cases} \tag{18.12}$$

$$r_{ijm} \leq d_{ij} + z_{ij} \quad \forall m \in M \tag{18.13}$$

$$VC_k = \left(\sum_{i=2}^{N} ld_i \cdot P_{ki} \right) \cdot \frac{\sum_{m=1}^{M} \sum_{(i,j) \in G} r_{ijm} \cdot z_{ij}}{\sum_{m=1}^{M} \sum_{(i,j) \in G} r_{ijm}} \quad \forall k \in K \backslash k = 1 \tag{18.14}$$

$$VC_1 = \sum_{i=2}^{N} ld_i - \sum_{k=2}^{K} VC_k \tag{18.15}$$

$$U_k = \begin{cases} \left\lfloor \dfrac{VC_k}{C\max_k} \right\rfloor + 1 & \text{if } \dfrac{VC_k}{C\max_k} - \left\lfloor \dfrac{VC_k}{C\max_k} \right\rfloor \geq \dfrac{2}{3} \quad \forall k \in K \\ \left\lfloor \dfrac{VC_k}{C\max_k} \right\rfloor & \text{o.w.} \quad \forall k \in K \end{cases} \tag{18.16}$$

$$x_k \geq \left\lceil \frac{U_k}{TR} \right\rceil \quad \forall k \in K \tag{18.17}$$

$$\sum_{k=1}^{K} P_{ki} = 1 \quad \forall i \in N \tag{18.18}$$

$$x_k \geq 0, \quad \forall k \in K; \quad r_{ijm} \in \{0,1\}, \quad \forall (i,j) \in G, \quad m \in M \tag{18.19}$$

The objective function minimizes the travel cost under the restriction of the following constraints. All m vehicles must leave (18.1) and return (18.2) to the depot. Constraint (18.3) ensures that all customers are serviced exactly once. Inflow and outflow must be equal except for the depot nodes (18.4). Vehicle capacity is given by (18.5) and (18.6). Constraints (18.7) and (18.8) are for the amount of binary variables. How to calculate the parameters of the similarity and proximity of trashcans (z_{ij} and d_{ij}) is shown in the Constraints (18.9 through 18.12). Constraint (18.13) shows the balance between similarity and proximity of trashcans for routing. Calculation of VC_k is shown in constraints (18.14) and (18.15). These two constraints show that in contrast of disposal volume (VC_1), the rate of recovery has a direct relationship with following the similarity pattern. Constraints (18.16) and (18.17) calculate U_k and x_k, respectively. Constraint (18.18) shows that P_{ki} is a possibility. Finally Constraint (18.19) imposes non-negativity and binary variables.

18.4 Computational Results

In this section, we solved the model by using LINGO software and analyzed the output results of the model. Tables 18.1 and 18.2 show the results of solved model for large size in which the value of indices are: $M = 7$, $K = 7$ and $n = 11$. Objective function value is equal to 915086, which was obtained after 37 minutes.

In this example, we assume that there are 10 trashcans and one depot in an area. The presence of seven types of wastes in the trashcan is clear. Also there are seven different vehicles for collecting wastes. These vehicles should service the trashcans, where possible, similar trashcans should be serviced by a same vehicle so that the same wastes are collected by a specific vehicle. As a result, rate of recovery will increase. According to output results in Table 18.1, let us consider the nonzero decision variables corresponding to the third vehicle: R(1, 3, 3), R(3, 6, 3), R(6, 8, 3), R(8, 7, 3) and R(7, 1, 3). These variables indicate

TABLE 18.1

Results of the Model Obtained from LINGO

Variable	Value
R(1, 2, 2)	1.000000
R(1, 3, 3)	1.000000
R(1, 4, 1)	1.000000
R(1, 5, 7)	1.000000
R(1, 9, 5)	1.000000
R(1, 10, 6)	1.000000
R(1, 11, 4)	1.000000
R(2, 1, 2)	1.000000
R(3, 6, 3)	1.000000
R(4, 1, 1)	1.000000
R(5, 1, 7)	1.000000
R(6, 8, 3)	1.000000
R(7, 1, 3)	1.000000
R(8, 7, 3)	1.000000
R(9, 1, 5)	1.000000
R(10, 1, 6)	1.000000
R(11, 1, 4)	1.000000

TABLE 18.2

Obtained Decision Variables

k	VC_k	U_k	x_k
1	14.2	7	7
2	11.3	5	5
3	3.666667	4	4
4	0.8666667	0	0
5	1.266667	0	0
6	0.4	0	0
7	0.8	1	1

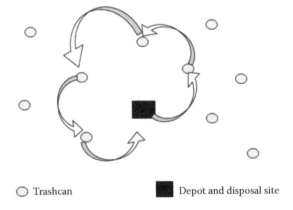

○ Trashcan   ■ Depot and disposal site

FIGURE 18.3
Route of 3rd vehicle in VRP network.

that the third vehicle began to move from node 1 (depot) to node 3 in order to service it. And then move to node 6, node 8, node 7, respectively and finally, it returns to node 1 (depot) after giving service to those nodes. Now, look at Table 18.2, total volume of recyclable waste that should be transferred to recycle center is equal to 3.66 ($VC_3 = 3.66$) and for transferring this volume we need four trips ($U_3 = 4$) and four vehicles ($x_3 = 4$). Finally we have the minimal cost for maximal volume of recycle. We can see the route of third vehicle in these VRP network in Figure 18.3.

18.5 Discussions

In this chapter, a mixed-integer nonlinear programming model has been provided including a waste collection vehicle routing problem (WCVRP) and the following processes after garbage unloading. A mathematical modeling formulation is given for the general WCVRP with consideration of similarity of the trashcans in terms of the types of waste which is in the trashcans in order to maintain the quality of wastes and to increase the recovery rate and to decrease the disposal rate. These vehicles should service the trashcans, where possible, similar trashcans should be serviced by a same vehicle so that the same wastes are collected by a specific vehicle.

References

Angelelli E, Speranza MG. The application of a vehicle routing model to a waste-collection problem: Two case studies. *Journal of the Operational Research Society* 2002a;53:944–52.

Angelelli E, Speranza MG. The periodic vehicle routing problem with intermediate facilities. *European Journal of Operational Research* 2002b;137:233–47.

Benjamin AM, Beasley JE. Metaheuristics for the waste collection vehicle routing problem with time windows, driver rest period and multiple disposal facilities. *Computers & Operations Research* 2010;37:2270–2280.

Buhrkal K, Larsen A, Ropke S. The waste collection vehicle routing problem with time windows in a city logistics context. *Procedia-Social and Behavioral Sciences* 2012;39:241–254.

Chang N-B, Lu HY, Wei YL. GIS technology for vehicle routing and scheduling in solid waste collection systems. *Journal of Environmental Engineering* 1997;123:901–933.

Chang NB, Wang SF. A fuzzy goal programming approach for the optimal planning of metropolitan solid waste management systems. *European Journal of Operational Research* 1997;99:287–303.

Eisenstein DD, Iyer AV. Garbage collection in Chicago: A dynamic scheduling model. *Management Science* 1997;43(7):922–933.

Kim BI, Kim S, Sahoo S. Waste collection vehicle routing problem with time windows. *Computers & Operations Research* 2006;33:3624–3642.

Li J-Q, Borenstein D, Mirchandani PB. Truck scheduling for solid waste collection in the city of Porto Alegre, Brazil. *Omega* 2008;36:1133–1149.

Mourao MC, Almeida MT. Lower-bounding and heuristic methods for a refuse collection vehicle routing problem. *European Journal of Operational Research* 2000;121:420–434.

Nuortio T, Kytöjoki J, Niska H, Bräysy O. Improved route planning and scheduling of waste collection and transport. *Expert Systems with Applications* 2006:30:223–232.

Ombuki-Berman BM, Runka A, Hanshar FT. Waste collection vehicle routing problem with time windows using multiobjective genetic algorithms. *Brock University Technical Report # CS-07-04*, 2007:91–97.

Potvin JY, Rousseau JM. A parallel route building algorithm for the vehicle routing and scheduling problem with time windows. *European Journal of Operations Research* 1993;66:331–40.

Shih L, Lin Y. Optimal routing for infectious waste collection. *Journal Environmental Engineering* 1999;125: 479–484.

Solomon MM. Algorithms for the vehicle routing and scheduling problems with time window constraints. *Operations Research* 1987;35:254–265.

Teixeira J, Antunes AP, Sousa JP. Recyclable waste collection planning—a case study. *European Journal of Operational Research* 2004;158:543–554.

Tung DV, Pinnoi A. Vehicle routing-scheduling for waste collection in Hanoi. *European Journal of Operational Research* 2000;125:449–468.

Weigel D, Cao B. Applying GIS and OR techniques to solve Sears' technician-dispatching and home-delivery problems. *Interfaces* 1999;29(1):112–130.

19

Reverse Supply Chain: Waste Pricing Model

SUMMARY In this chapter, residuals, after transferring by costumer, send to production station, sorting station and different manufacturing processes (melting, forging, clamping, painting, etc.) for reproduction. After completion of several production processes, reproduced products are resent to costumers. Considering different cost factors and also pricing concept and reproduced parts, the mathematical model of optimizing manufacturing cost is developed. The model is a useful tool in strategic decision-making for municipalities.

19.1 Introduction

An effective supply chain is a competitive advantage for firms helping them to be capable with environmental turbulences. A supply chain is a network of supplier, production, distribution centers and channels between them configured to acquire raw materials, convert them to finished products, and distribute final products to customers. Supply chain network design is one of the most important strategic decisions in supply chain management. In general, network structure decisions contain setting the numbers, locations and capacities of facilities and the quantity of flow between them (Amiri 2006).

Recently, many companies such as Kodak, Xerox and HP have concentrated on remanufacturing processes and obtained significant achievements in this area (Uster et al., 2007). Meade et al. (2007) classify driving forces that led to increased interest and investment in reverse supply chain into two groups: environmental factors and business factors. The first group explores environmental impacts of used products, environmental legislations and growing environmental consciousness of customers. The design and establishment of the supply chain network is a very important decision to be effective for several years, during which the parameters of the business environment (e.g., demand of customers) may change (Meepetchdee and Shah, 2007).

In Korea, the extended product responsibility is in force system from 2003 that the obligation is given as a producer as it recycles more than a constant amount of the waste that can be recycled (Biehl et al., 2007; Ko and Evans, 2007; Lieckens and Vandaele, 2007). Reverse logistics is defined by the European working group REVLOG as "the process of planning, implementing and controlling flows of raw materials, in process inventory, and finished goods, from the point of use back to point recovery or point of proper disposal." In a broader sense, reverse logistics refers to the distribution activities involved in product returns, source reduction, conservation, recycling, substitution, reuse, disposal, refurbishment, repair and remanufacturing (Stock, 1992).

Concerning reverse logistics, many works have been studied in different areas and operations included such as reuse, recycling, remanufacturing logistics, etc. In reuse logistics models, Kroon and Vrijens (1995) conducted a case study focusing the design of a logistics system for reusable transportation packages. The authors developed an MIP (mixed-integer programming), closely related to a classical un-capacitated warehouse location model.

In recycling models, Barros et al. (1998) developed a mixed-integer program model by considering two-echelon location problems with capacity constraints based on a multi-level capacitated warehouse location problem. Pati et al. (2008) developed a model based on a mixed-integer goal programming model (mIGP) to solve the problem. The model studied the inter-relationship between multiple objectives of a recycled paper distribution network.

In remanufacturing models, Kim et al. (2006) discussed a notion of remanufacturing systems in reverse logistics environment. Jayaraman et al. (1999) presented a mixed-integer program to determine the optimal number and locations of remanufacturing facilities for the electronic equipment. Lee et al. (2007) proposed the reverse logistics network problem (rLNP) minimizing total reverse logistics various shipping costs. This research offers an efficient MILP model for multi-stage reverse logistics network design that could support recovery and disposal activities.

While the body of literature for reverse supply chain network design implied, mixed-integer programming (MIP) models were the models used commonly. These models include simple incapacitated facility location models to complex capacitated multi-stage or multi-commodity models. The usual objective of the models was to determine the least cost system design, that usually involves making tradeoffs among fixed opening costs of facilities and transportation costs. Melo et al. (2009) and Klibi et al. (2010) presented comprehensive reviews on supply chain network design problems to support variety of future research directions.

19.1.1 Reverse Supply Network

Fleischmann et al. (1997) presented a comprehensive review on the application of mathematical modeling in reverse logistics management. As one of the focal works in reverse supply chain network design, Barros et al. (1998) proposed a MILP model for a sand recycling network. A heuristic algorithm is also used to solve the problem. Jayaraman et al. (1999) developed a MILP model for reverse logistics network design under a pull system based on customer demands for recovered products. The objective of the proposed model was to minimize the total costs. Also, Krikke et al. (1999) designed a MILP model for a two-stage reverse supply chain network for a copier manufacturer. In this model, both the processing costs of returned products and inventory costs were considered in the objective function to minimize the total cost. Jayaraman et al. (2003) extended their prior work to solve the single product two-level hierarchical location problem involving the reverse supply chain operations of hazardous products. They also developed a heuristic to handle relatively large-sized problems. Min et al. (2006) proposed a mixed-integer nonlinear programming (MINLP) model and a genetic algorithm that could solve a multi-period reverse logistics network design problem involving both spatial and temporal consolidation of returned products. Aras et al. (2008) developed a MINLP model for determining the locations of collection centers in a simple reverse supply chain network. The important point about this work was the capability of presented model for determining the optimal buying price of used products with the objective of maximizing profit. They developed a heuristic based on tabu search to solve the model. Pati et al. (2008) proposed a mixed-integer goal programming (MIGP) for paper recycling logistics network. The considered goals included: (1) minimizing the positive deviation from the planned budget allocated for reverse logistics activities, (2) minimizing the positive deviation from the maximum limit of non-relevant wastepaper and (3) minimizing the negative deviation from the minimum desired waste collection.

Demand uncertainty and uncertainty in the type and quantity of returned products are the important elements being considered in the design of reverse and closed-loop supply chain networks. According to this fact, Listes and Dekker (2005) proposed a stochastic mixed-integer programming (SMIP) model for a sand recycling network design to maximize the total profit. This research was an extension of the work done by Barros et al. (1998). Lieckens and Vandaele (2007) combined the traditional MILP models with queuing models to cope with high degree of uncertainty and some dynamic aspects in a reverse logistics network design problem. Because this extension introduced nonlinear relationships, the problem was defined as a MINLP model. A genetic algorithm was developed to solve the proposed model.

19.1.2 Framework for Remanufacturing

Various types of remanufacturing systems exist based on the working industry. In most industries—for example, computer, mobile phone, copy machine, and automotive industries—the remanufacturing process varies from each other in terms of specific "process" itself. However, there also exist common types of remanufacturing processes being categorized as process characteristics such as collection, disassembly, refurbishment, and assembly. In this sense, the following remanufacturing system is considered without loss of generality. Remanufacturing system begins with returned products including end-of-life product from customers. Then, they are collected to the collection centers. Since a product includes several parts, the returned products are disassembled to remanufacturing and the rest, beyond the remanufacturing capacity, are sent to the remanufacturing subcontractor centers. The furnished products from the collection site are disassembled in the disassembly site. Disassembled parts are classified into the reusable parts and non-reusable parts. Finally parts in inventory are supplied to the manufacturing shops according to the company's own production plan. For example, in a mobile phone industry, manufacturers collect and test the used phones. If they are working, they go to the secondary market for rental or resale. Otherwise, they go into the remanufacturing system to reuse parts or segments, that is, PCB, display, speaker, and microphone. Defective phones are going to disassembly, cleaning, and reassembly with new parts or modules if necessary (Hajji et al., 2009).

De Brito et al. (2003) discussed network structures and report cases pertaining to the design of remanufacturing networks by the original equipment manufacturers or independent manufacturers, the location of remanufacturing facilities for copiers, especially Canon copiers and other equipment and the location of IBM facilities for remanufacturing in Europe. They also presented case studies on inventory management for remanufacturing networks of engine and automotive parts for Volkswagen and on Air Force depot buffers for disassembly, remanufacturing and reassembly. Finally, they presented case studies on the planning and control of reverse logistics activities, and in particular inventory management cases for remanufacturing at a Pratt & Whitney aircraft facility, yielding decisions of lot sizing and scheduling.

Bostel et al. (2005) proposed a review of problems and models based on the hierarchical planning horizon and degree of correlation between forward and reverse flows. Strategic planning models are focused on network design problems, while tactical and operational models address a number of specific problems. In this context they discussed a number of inventory management models with reverse flows, including periodic and continuous review deterministic and stochastic inventory models. The special issue published by Verter and Boyaci (2007) contains three papers on optimization models for facility location and capacity planning for remanufacturing and a paper on assessing the benefits of remanufacturing options. Guide and Van Wassenhove (2006) discuss assumptions of models for reverse supply chain activities, and in particular operational issues for remanufacturing and remanufactured product market development. Several papers deal with optimal policies for remanufacturing activities, pertaining to acquisition, pricing, order quantities, and lot sizing for products over a finite life cycle.

In recent years, there has been considerable interest in inventory control for joint manufacturing and remanufacturing systems in forward–reverse logistics networks. As mentioned by El-Sayed et al. (2010), a forward–reverse logistic network establishes a relationship between the market that releases used products and the market for new products. When the two markets coincide, and the manufacturing and remanufacturing activities are strongly connected, the system is called a closed-loop network; otherwise it is called an open-loop network (Salema et al., 2007).

Dobos (2003) found optimal inventory policies in a reverse logistics system with special structure while assuming that the demand is a known function in a given planning horizon and the return rate of used items is a given function. Dobos (2003) minimized the sum of the holding costs in the stores and costs of the manufacturing, remanufacturing and disposal. The necessary and sufficient conditions for optimality were derived from the application of the maximum principle of Pontryagin (Seierstad and Sydsaeter, 1987). Their results were constrained to deterministic demand and return process with no consideration on the dynamics of production facilities. Taking a closer look at the dynamic characteristic of the production planning problems, one can notice that the stochastic optimal control theory, such as in Akella and Kumar (1986), Dehayem et al. (2009), Hajji et al. (2009) and references therein, is not yet used in reverse logistics.

Kibum et al. (2006) discussed the remanufacturing process of reusable parts in reverse logistics, where the manufacturing has two alternatives for supplying parts: either ordering the required parts to external suppliers, or overhauling returned products and bringing them back to "as new" condition. The study presented in Chung et al. (2008) analyzed a closed-loop supply chain inventory system by examining used products returned to a reconditioning facility where they are stored, remanufactured, and then shipped back to retailers for retail sale. The findings of the study presented in Chung et al. (2008) demonstrated that the proposed integrated centralized decision-making approach can substantially improve efficiency. The majority of the previous works are based on mathematical programming. An example of such models can be found in El-Sayed et al. (2010), where a multi-period multi-echelon forward–reverse logistics network model is developed. The control of the manufacturing and remanufacturing production facilities, based on their time dynamics, is very limited in the literature.

19.2 Problem Definition

The reverse logistics network discussed in this research is a multi-stage logistics network including customer, collection, disassembly, refurbish and disposal centers.

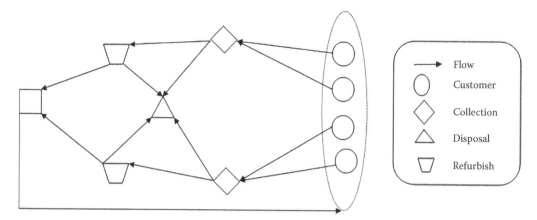

FIGURE 19.1
Structure of a reverse supply network.

As illustrated in Figure 19.1, in the reverse flow, returned products are collected in collection centers and, after inspection, the recoverable products are shipped to disassembly facilities, and scrapped products are shipped to disposal centers. With this strategy, excessive transportation of returned products (especially scrapped products) is prevented and the returned products can be shipped directly to the appropriate facilities. The disassembled parts from products in disassembly facilities are shipped to refurbish and disposal centers through a push system. After the refurbish process, the refurbished parts are delivered to customers as new parts.

A predefined percentage of demand of each customer zone is assumed to result in return products and a predefined value is determined as an average disposal rate. The average disposal rate is associated with the quality of returned products; because high quality returns have a capability for recovery process (remanufacturing and de-manufacturing) and low quality returns should be entered to a safe disposal process.

In the above situations, the remanufacturing company is interested in minimizing total remanufacturing cost so that eventually it can maximize total profit. To achieve the goal, while meeting part demands from manufacturing centers, the company determines how many returned products should be thrown into the remanufacturing process such as refurbishing and disassembling for "as new" condition. The other issues to be addressed by this study are to choose the location and determine the number of collection, disassemble, refurbish and disposal centers and to determine the quantity of flow between network facilities.

19.3 Waste Collection Model

The following notation is used in the formulation of proposed model.

Indices

i Index of collection/inspection center $i = 1, …, I$
j Index of disassembly center $j = 1,…, J$
k Index of refurbish center $k = 1, …, K$

m	Index of disposal center $m = 1, ..., M$
n	Index of customer $n = 1,..., N$
p	Index of product $p = 1, ..., P$
l	Index of part $l = 1, ..., L$

Parameters

d_{np}	Demand of customer n for refurbished products p
r_{np}	Returns of used products p from customer n
s_l	Average disposal fraction part l
re_{ijp}	Exit of returned product p from collection center i to disassembly center j
cc_{ip}	Capacity of handling returned products p at collection/inspection i
cs_{jp}	Capacity of handling recoverable products p at disassembly center j
cr_{kl}	Capacity of handling refurbished parts k at refurbish center k
cd_{ml}	Capacity of handling scrapped parts l at disposal center m
π_{lp}	The number of disassembled parts l from products p
a_{ip}	set-up cost of collection/inspection center i for returned product p
b_{jp}	set-up cost of disassembly center j for recoverable product p
c_{kl}	set-up cost of refurbish center k for part l
o_{ml}	set-up cost of disposal center m for part l
e_{nip}	Shipping cost per unit of returned products p from customer n to collection/inspection center i
q_{ijp}	Shipping cost per unit of recoverable products p from collection/inspection center i to disassembly center j
t_{jkl}	Shipping cost per unit of parts l from disassembly center j to refurbish center k
v_{jml}	Shipping cost per unit of parts l from disassembly center j to disposal center m
α_{ip}	The idle cost of collection/inspection center i for product p
β_{jp}	The idle cost of disassembly center j for product p
γ_{kl}	The idle cost of refurbish center k for part l
λ_{ml}	The idle cost of disposal center m for part l
ci_{ip}	The inspection cost of returned products p in collection/inspection center i
ca_{jp}	The disassembly cost of recoverable products p in disassembly center j
cp_{kl}	The refurbish cost of disassembled parts l in refurbish center k
ch_{ml}	The disposal cost of disassembled parts l in disposal center m

Decision Variables

QE_{nip}	Quantity of returned products p shipped from customer n to collection/inspection center i
QQ_{ijp}	Quantity of recoverable products p shipped from collection/inspection center i to disassembly center j
QT_{jkl}	Quantity of parts l shipped from disassembly center j to refurbish center k
QV_{jml}	Quantity of parts l shipped from disassembly center j to disposal center m

$$X_{ip} = \begin{cases} 1 & \text{if a collection center } i \text{ is set up} \\ 0 & \text{otherwise} \end{cases} \quad \forall i, p$$

$$Y_{jp} = \begin{cases} 1 & \text{if a disassembly center } j \text{ is set up} \\ 0 & \text{otherwise} \end{cases} \quad \forall j, p$$

$$G_{kl} = \begin{cases} 1 & \text{if a refurnish center } k \text{ is set up} \\ 0 & \text{otherwise} \end{cases} \qquad \forall k,l$$

$$\delta_{ml} = \begin{cases} 1 & \text{if a disposal center } m \text{ is set up} \\ 0 & \text{otherwise} \end{cases} \qquad \forall m,l$$

Using the above indices and parameters, the mathematical formulation for this problem can be stated as follows:

$$
\begin{aligned}
\min Z = \Bigg\{ &\sum_{i=1}^{I}\sum_{p=1}^{P}(a_{ip}*X_{ip})\sum_{j=1}^{J}\sum_{p=1}^{P}(b_{jp}*Y_{jp}) + \sum_{k=1}^{K}\sum_{l=1}^{L}(c_{kl}*G_{kl}) + \sum_{m=1}^{M}\sum_{l=1}^{L}(o_{ml}*\delta_{ml}) \\
&+ \sum_{n=1}^{N}\sum_{i=1}^{I}\sum_{p=1}^{P}(ci_{ip}+e_{nip})*(QE_{nip})\sum_{i=1}^{I}\sum_{j=1}^{J}\sum_{p=1}^{P}(ca_{jp}+q_{ijp})*(QQ_{ijp}) \\
&+ \sum_{j=1}^{J}\sum_{k=1}^{K}\sum_{l=1}^{L}(cp_{kl}+t_{jkl})*(QT_{jkl}) + \sum_{j=1}^{J}\sum_{m=1}^{M}\sum_{l=1}^{L}(ch_{ml}+v_{jml})*(QV_{jml}) \\
&+ \sum_{i=1}^{I}\sum_{p=1}^{P}\alpha_{ip}*\left(X_{ip}*cc_{ip}-\sum_{n=1}^{N}QE_{nip}\right) + \sum_{j=1}^{J}\sum_{p=1}^{P}\beta_{jp}*\left(Y_{jp}*cs_{jp}-\sum_{i=1}^{I}QQ_{ijp}\right) \\
&+ \sum_{k=1}^{K}\sum_{l=1}^{L}\gamma_{kl}*\left(G_{kl}*cr_{kl}-\sum_{j=1}^{J}QT_{jkl}\right) + \sum_{m=1}^{M}\sum_{l=1}^{L}\lambda_{ml}*\left(\delta_{ml}*cd_{ml}-\sum_{j=1}^{J}QV_{jml}\right)
\end{aligned}
\tag{19.1}
$$

subject to

$$\sum_{i=1}^{I}QE_{nip} \geq r_{np}*d_{np} \quad \forall p \in P, n \in N \tag{19.2}$$

$$re_{ijp}*\sum_{n=1}^{N}QE_{nip} = \sum_{j=1}^{J}QQ_{ijp} \quad \forall i \in I, p \in P \tag{19.3}$$

$$QT_{jkl} = (1-s_1)\sum_{p=1}^{P}\sum_{i=1}^{I}\pi_{lp}*QQ_{ijp} \quad \forall l \in L, k \in K, j \in J \tag{19.4}$$

$$QV_{jml} = s_1\sum_{p=1}^{P}\sum_{i=1}^{I}\pi_{lp}*QQ_{ijp} \quad \forall l \in L, m \in M, j \in J \tag{19.5}$$

$$\sum_{n=1}^{N}QE_{nip} \leq cc_{ip}*X_{ip} \quad \forall i \in I, p \in P \tag{19.6}$$

$$\sum_{i=1}^{I} QQ_{ijp} \le cs_{jp} * Y_{jp} \quad \forall j \in J, p \in P \tag{19.7}$$

$$\sum_{j=1}^{J} QT_{jkl} \le cr_{kl} * G_{kl} \quad \forall k \in K, l \in L \tag{19.8}$$

$$\sum_{j=1}^{J} QV_{jml} \le cd_{ml} * \delta_{ml} \quad \forall m \in M, l \in L \tag{19.9}$$

$$QE_{nip}, QQ_{ijp}, QT_{jkl}, QV_{jml}, \ge 0 \tag{19.10}$$

$$X_{ip} \in \{0,1\} \tag{19.11}$$

$$Y_{jp} \in \{0,1\} \tag{19.12}$$

$$G_{kl} \in \{0,1\} \tag{19.13}$$

$$\delta_{ml} \in \{0,1\} \tag{19.14}$$

Objective function 19.1 minimizes the total cost, which includes set-up costs, transportation costs, operation costs and idle costs of facilities. This means that our model tries to minimize both the costs from remanufacturing process and the utilization of remanufacturing facilities at the same time. Constraint 19.2 ensures that the demands of all customers are satisfied and returned products from all customers are collected. Constraint 19.3 represents the balance equation for the products that are entered to disassembly center and are exited from collection center. Constraints 19.4 and 19.5 assure the flow balance at disassembly, refurbish and disposal centers. Equations 19.6 through 19.9 are capacity constraints on facilities. Constraint 19.10 checks for the non-negativity of decision variables and the last four Constraints check for binary variables.

19.3.1 Pricing

The purpose of this study is minimization of the costs to maximize the profit. However, the organization's profit for refurbishing parts is increased in a way to consider different factors of cost and various techniques of pricing. First the total costs of refurbishing each of parts according to represented function in mathematical model are computed, and then the costs of each parts are got, using an equation. The equation considers a 20% profit overhead.

$$1.2 \, (price_1 * \text{quantity}_1) = \text{total cost}_1$$

$price_1$: the price of each of part l

quantity$_1$: the total quantities of each of refurbished parts l

total cost$_1$: the total costs of refurbishing each of parts l

19.4 Illustrative Example

Using a numerical example, the applicability of the model in the proposed framework is illustrated and some insights into the proposed model are gained. A small set of stochastic data is prepared. It is assumed that there are three types of products and five types of parts from those products, too; three collection/inspection sites, five disassembly sites, four refurbish sites, three disposal sites and five customers. Rate of return of used products p from customer n, rate of exit returned product p from collection center i to disassemble center j and average disposal fraction parts are considered to be 0.3, 0.8 and 0.5, respectively. The capacity of collection/inspection and disassembly sites is set to be in range of [500, 1000] units of product and the capacity refurbish and disposal sites are in range of [5000, 10,000] units of part. The set-up cost of the facilities is set to be in range of [100, 200] units, transportation cost of products and parts between facilities are set to be in range of [1, 10] units, idle cost of the facilities are set to be in range of [1, 20] units and operation costs are set to be in range of [5, 20] units. The LINGO software is used for solving our mixed-integer programming model. Some input data are listed in Tables 19.1 through 19.3.

The results are shown in Tables 19.4 through 19.11. Table 19.4 shows the number of collected products from customers at collection/inspection site. Table 19.5 shows the number of recoverable products at disassembly site. Table 19.6 shows the number of refurbished parts at refurbished site. Table 19.7 shows the number of scrapped parts at disposal site. X_{ip}, Y_{jp}, G_{kl}, δ_{ml} decision variables are determined the number of optimal quantity sites.

TABLE 19.1

Demand of Customer n for Refurbished Products p

	d_{np}		
		p	
n	1	2	3
1	100	140	100
2	160	100	200
3	700	90	160
4	90	170	200
5	130	140	130

TABLE 19.2

The Number of Disassembled Parts l from Products p

	π_{lp}		
		p	
l	1	2	3
1	16	10	13
2	12	14	12
3	8	13	19
4	11	6	19
5	5	14	5

TABLE 19.3

Shipping Cost Per Unit of Returned Products p from Customer n to Collection/Inspection Center i

	e_{nip}								
	1			**2**			**3**		
				i					
n	**1**	**2**	**3**	**1**	**2**	**3**	**1**	**2**	**3**
1	9	3	4	2	4	6	9	8	8
2	8	10	7	4	7	6	6	2	10
3	9	5	3	8	9	8	2	9	6
4	7	9	6	4	10	6	2	5	5
5	5	10	7	3	5	4	4	2	4

TABLE 19.4

Quantity of Returned Products p Shipped from Customer n to Collection/Inspection Center i

	QE_{nip}								
	1			**2**			**3**		
				i					
n	**1**	**2**	**3**	**1**	**2**	**3**	**1**	**2**	**3**
1	0	0	30	0	0	42	0	30	0
2	0	0	48	0	0	30	0	60	0
3	0	0	211	0	0	27	0	48	0
4	0	0	27	0	0	51	0	60	0
5	0	0	39	0	0	45	0	42	0

TABLE 19.5

Quantity of Recoverable Products p Shipped from Collection/Inspection Center i to Disassembly Center j

	QQ_{ijp}														
	1					**2**					**3**				
						j									
i	**1**	**2**	**3**	**4**	**5**	**1**	**2**	**3**	**4**	**5**	**1**	**2**	**3**	**4**	**5**
1	0	0	0	0	0	0	0	0	0	0	0	0	0	0	0
2	0	0	0	0	0	0	0	0	0	0	0	0	192	0	0
3	0	0	284	0	0	0	0	156	0	0	0	0	0	0	0

TABLE 19.6

Quantity of Parts l Shipped from Disassembly Center j to Refurbish Center k

QT_{jkl}

		l																		
	1				2				3				4				5			
	k																			
j	1	2	3	4	1	2	3	4	1	2	3	4	1	2	3	4	1	2	3	4
1	0	0	0	0	0	0	0	0	0	0	0	0	0	0	0	0	0	0	0	0
2	0	0	0	0	0	0	0	0	0	0	0	0	0	0	0	0	0	0	0	0
3	4300	4300	4300	4300	3948	3948	3948	3948	3974	3974	3974	3974	3854	3854	3854	3854	2282	2282	2282	2282
4	0	0	0	0	0	0	0	0	0	0	0	0	0	0	0	0	0	0	0	0
5	0	0	0	0	0	0	0	0	0	0	0	0	0	0	0	0	0	0	0	0

TABLE 19.7

Quantity of Parts l Shipped from Disassembly Center j to Disposal Center m

QV_{jml}

	l														
	1			2			3			4			5		
	m														
j	1	2	3	1	2	3	1	2	3	1	2	3	1	2	3
1	0	0	0	0	0	0	0	0	0	0	0	0	0	0	0
2	0	0	0	0	0	0	0	0	0	0	0	0	0	0	0
3	4300	4300	4300	3948	3948	3948	3974	3974	3974	3854	3854	3854	2282	2282	2282
4	0	0	0	0	0	0	0	0	0	0	0	0	0	0	0
5	0	0	0	0	0	0	0	0	0	0	0	0	0	0	0

TABLE 19.8

The Number of Collection/Inspection Sites

	X_{lp}		
		p	
l	1	2	3
1	0	0	0
2	0	0	1
3	1	1	0

TABLE 19.9

The Number of Disassembly Sites

	Y_{jp}		
		p	
j	1	2	3
1	0	0	0
2	0	0	0
3	1	1	1
4	0	0	0
5	0	0	0

TABLE 19.10

The Number of Refurbish Sites

	G_{kl}				
			l		
k	1	2	3	4	5
1	1	1	1	1	1
2	1	1	1	1	1
3	1	1	1	1	1
4	1	1	1	1	1

TABLE 19.11

The Number of Disposal Sites

	δ_{ml}				
			l		
m	1	2	3	4	5
1	1	1	1	1	1
2	1	1	1	1	1
3	1	1	1	1	1

Now using the obtained costs and quantity, we compute the prices as follows:

total cost$_1$ = 634,724
quantity$_1$ = 12,900
1.2 ($price_1$ * 12,900) = 634,724
$price_1$ = 41

total cost$_2$ = 462,211
quantity$_2$ = 15,792
1.2 ($price_2$ * 15,792) = 462,211
$price_2$ = 24

total cost$_3$ = 348,543
quantity$_3$ = 15,896
1.2 ($price_3$ * 15,896) = 348,543
$price_3$ = 18

total cost$_4$ = 408,435
quantity$_4$ = 11,562
1.2 ($price_4$ * 11,562) = 408,435
$price_4$ = 29

total cost$_5$ = 326,016
quantity$_5$ = 6846
1.2 ($price_5$ * 6846) = 326,016
$price_5$ = 40

19.5 Discussions

In this research, reverse logistics network problem is addressed for treating a remanufacturing problem that is one of the most important problems in the environmental situation for the recovery of used products and materials. Based on this system, a general framework was proposed in view of supply planning and developed a mathematical model to optimize the supply planning function.

The model determines the quantity of products/parts processed in the remanufacturing facilities while minimizing the total remanufacturing cost. Our research results can be guidelines on the relevant research. The proposed remanufacturing framework and model can be a useful tool to the various industries after customizing for specific industries. However, as the proposed model is introduced as a general framework, many future works can be conducted. Above all, the proposed framework with remanufacturing can be effectively enhanced by adopting more industry practices and so the mathematical model does. Since the proposed model is formulated as mixed-integer programming, the computational burden for optimal solution increases exponentially as the size of problem rises. Thus, an efficient heuristic algorithm needs to be developed in order to solve the large-scale problems. Many possible future research directions can be defined in the area of logistics network design under uncertainty. Time complexity is not addressed in this research, however, since the computational time increases significantly when the size of problem and the number of scenarios increase, therefore developing efficient exact or heuristic solution methods are also a critical need in this area.

References

Akella R, Kumar PR. Optimal control of production rate in a failure prone manufacturing system. *IEEE Transactions on Automatic Control AC* 1986;31:116–126.

Amiri A. Designing a distribution network in a supply chain system: Formulation and efficient solution procedure. *European Journal of Operational Research* 2006;171:567–576.

Aras A, Aksen D, Tanugur AG. Locating collection centers for incentive-dependent returns under a pick-up policy with capacitated vehicles. *European Journal of Operational Research* 2008;191:1223–1240.

Barros AI, Dekker R, Scholten VA. A two-level network for recycling sand: A case study. *European Journal of Operational Research* 1998;110:199–214.

Biehl M, Prater E, Realff MJ. Assessing performance and uncertainty in developing carpet reverse logistics systems. *Computers & Operations Research* 2007;34:443–463.

Bostel N, Dejax P, Lu Z. The design, planning and optimization of reverse logistic networks. *Logistics Systems: Design and Optimization: Springer* 2005:171–212.

Chung SL, Wee HM, Po-Chung Y. Optimal policy for a closed-loop supply chain inventory system with remanufacturing. *Mathematical and Computer Modelling* 2008;6:867–881.

De Brito MP, Dekker R, Flapper SDP. Reverse logistics a review of case studies. ERIM report series reference No.ERS-2003-012-LIS2003, 2003.

Dehayem NFI, Kenne JP, Gharbi A. Hierarchical decision making in production and repair/replacement planning with imperfect repairs under uncertainties. *European Journal of Operational Research* 2009;198:173–189.

Dobos I. Optimal production–inventory strategies for a HMMS-type reverse logistics system. *International Journal of Production Economics* 2003;81–82:351–360.

El-Sayed M, Afia N, El-Kharbotly A. 2010. A stochastic model for forward–reverse logistics network design under risk. *Computers & Industrial Engineering* 2010;58:423–431.

Fleischmann M, Bloemhof RJ, Dekker R, Van der Laan E, Van Nunen J, Van Wassenhove L. Quantitative models for reverse logistics: A review. *European Journal of Operational Research* 1997;103:1–17.

Guide VDR, Van Wassenhove LN. Closed-loop supply chains, feature issue (Part 1). *Production and Operations Management* 2006;15(3):345–350.

Hajji A, Gharbi A, Kenne JP. Joint replenishment and manufacturing activities control in two stages unreliable supply chain. *International Journal of Production Research* 2009;47:3231–3251.

Jayaraman V, Guide VDRJ, Srivastava RAJ. A closed loop logistics model for remanufacturing. *Journal of the Operational Research Society* 1999;50:497–508.

Jayaraman V, Patterson RA, Rolland E. The design of reverse distribution networks: Models and solution procedures. *European Journal of Operational Research* 2003;150:128–149.

Kibum K, Iksoo S, Juyong K, Bongju J. Supply planning model for remanufacturing system in reverse logistics environment. *Computers & Industrial Engineering* 2006;51:279–287.

Kim KB, Song IS, Jeong BJ. Supply planning model for remanufacturing system in reverse logistics environment. *Computers & Industrial Engineering* 2006;51:279–287.

Klibi W, Martel A, Guitouni A. The design of robust value-creating supply chain networks: A critical review. *European Journal of Operational Research* 2010;203:283–293.

Ko HJ, Evans GW. A genetic algorithm-based heuristic for the dynamic integrated forward/reverse logistics network for 3PLs. *Computers & Operations Research* 2007;34:346–366.

Krikke HR, Van Harten A, Schuur PC. Reverse logistic network re-design for copiers. *Operations Research Spektrum* 1999;21: 381–409.

Kroon L, Vrijens G. Returnable containers: An example of reverse logistics. *International Journal of Physical Distribution & Logistics Management* 1995;25:56–68.

Lee JE, Rhee KG, Gen M. Designing a reverse logistics network by priority-based genetic algorithm. *Presented at the Proc. of International Conference on Intelligent Manufacturing Logistics Systems,* Kitakyushu, Japan, 2007.

Lieckens K, Vandaele N. Reverse logistics network design with stochastic lead times. *Computers & Operations Research* 2007;34:395–416.

Listes O, Dekker R. A stochastic approach to a case study for product recovery network design. *European Journal of Operational Research* 2005;160:268–287.

Meade L, Sarkis J, Presley A. 2007. The theory and practice of reverse logistics. *International Journal of Logistics Systems and Management* 2007;3:56–84.

Melo MT, Nickel S, Saldanhada GF. Facility location and supply chain management a review. *European Journal of Operational Research* 2009;196:401–412.

Meepetchdee Y, Shah N. Logistical network design with robustness and complexity considerations. *International Journal of Physical Distribution & Logistics Management* 2007;37:201–222.

Min H, Ko CS, Ko HJ. The spatial and temporal consolidation of returned products in a closed-loop supply chain network. *Computers & Industrial Engineering* 2006;51:309–320.

Pati RK, Vrat P, Kumar P. A goal programming model for paper recycling system. *Omega* 2008;36:405–417.

Salema MIG, Barbosa-Povoa AP, Novais AQ. An optimization model for the design of a capacitated multi-product reverse logistics network with uncertainty. *European Journal of Operational Research* 2007;179:1063–1077.

Seierstad A, Sydsaeter K. Optimal control theory with economic applications. Amsterdam: North-Holland, 1987.

Stock JK. Reverse logistics. White paper, Oak Brook, IL: Council of Logistics Management, 1992.

Uster H, Easwaran G, Akcali E, Cetinkaya S. Benders decomposition with alternative multiple cuts for a multi-product closed-loop supply chain network design model. *Naval Research Logistics (NRL)* 2007;54:890–907.

Verter V, Boyaci T. Special issue on reverse logistics. *Computers and Operations Research* 2007;34:295–298.

20

Multiple Item Reverse Supply Chain: Comprehensive Mathematical Model

SUMMARY   This chapter presents a comprehensive mathematical programming model with the objective of minimizing the total costs of reverse supply chains, including transportation, fixed opening, operation, maintenance, and remanufacturing costs of centers. The proposed model considers the design of a multi-layer, multi-product reverse supply chain that consists of returning, disassembly, processing, recycling, remanufacturing, materials, and distribution centers.

20.1 Introduction

With the increased environmental concerns and stringent environmental laws, companies focus on setting up a reverse supply chain either because of environmental regulations or to reduce their operating costs by reusing products or components. According to the American Reverse Logistics Executive Council, reverse logistics is defined as: "The process of planning, implementing, and controlling the efficient, cost effective flow of raw materials, in-process inventory, finished goods and related information from the point of consumption to the point of origin for the purpose of recapturing value or proper disposal (Rogers and Tibben-Lembke, 1999)."

Implementation of reverse logistics would allow not only for cost savings in inventory carrying, transportation, and waste disposal, but also for the improvement of customer loyalty and future sales (Kannan, 2009; Lee et al., 2009). A group of companies has gone further and achieved economic gains from the adoption of environment-friendly logistic networks. For instance, Nike, the shoe manufacturer, encourages consumers to bring their used shoes to the store where they had purchased them. These shoes are then shipped back to Nike's plants and made into basketball courts and running tracks. By donating the material to the basketball courts and donating funds for building and maintaining these courts, Nike has enhanced the value of its brand.

In a broader sense, reverse logistics refers to the distribution activities involved in product returns, source reduction, conservation, recycling, substitution, reuse, disposal, refurbishment, repair and remanufacturing (Stock, 1992). Reusable parts can be removed from the product and returned to a manufacturer where they can be reconditioned and assembled into new products (Liu et al., 2006). Recycling (with or without disassembly) includes the treatment, recovery, and reprocessing of materials contained in the used products or components in order to replace the virgin materials in the production of new goods (He et al., 2006). Remanufacturing is the process of removing specific parts of the waste product for further reuse in new products. Disposal is the processes of incineration or landfill.

For the last decade, increasing concerns over environmental degradation and increased opportunities for cost savings or revenues from returned products prompted some researchers to formulate more effective reverse logistics strategies. In remanufacturing models, Kim et al. (2006) discussed a notion of remanufacturing systems in reverse logistics environment. They proposed a general framework in view of supply planning and developed a mathematical model to optimize the supply planning function. The model determines the quantity of product parts processed in the remanufacturing facilities subcontractors and the amount of parts purchased from the external suppliers while maximizing the total remanufacturing cost saving. Aras et al. (2008) develop a non-linear model and tabu search solution approach for determining the locations of collection centers and the optimal purchase price of used products in a simple profit maximizing reverse logistics network. Teunter et al. (2008) dealt with the question of when companies should use shared resources for production and remanufacturing and when they should use specialized resources. In their study, Zuidwijk and Krikke (2008) considered two strategic questions in the context of closed-loop supply chains to establish how much a company should invest in product design and how much in the production processes to process their returned products. They formulated the problem as both an integer linear programming and a rule of thumb-based problem.

Du and Evans (2008) minimize tardiness and total costs for location and capacity decisions in a closed-loop logistics network operated by third party logistics (3PL) providers. To solve the bi-objective MILP model, a hybrid scatter search method is developed. Kannan et al. (2010) developed a mathematical model for a case of battery recycling. However, they did not consider uncertainty of parameters. Amin and Zhang (2012) designed a network based on product life cycle. They utilized mixed-integer linear programming to configure the network. Du and Evans (2008) developed a bi-objective model for a reverse logistics network by considering minimization of the overall costs, and the total tardiness of cycle time.

Jayaraman et al. (2003) proposed a general mixed-integer programming model and solution procedure for a reverse distribution problem focused on the strategic level. The model decides whether each remanufacturing facility is open considering the product return flow.

Ko and Evans (2007) consider a network operated by a 3PL service provider and they present an mixed integer nonlinear programming (MINLP) model for the simultaneous design of the forward and return network. They develop a genetic algorithm-based heuristic to solve the complex developed model.

Pati et al. (2008), they developed an approach based on a mixed-integer goal programming model (MIGP) to solve the problem. The model studies the inter-relationship between multiple objectives of a recycled paper distribution network. The objectives considered are reduction in reverse logistics cost.

Salema et al. (2007) have proposed an MILP model to analyze the problem of closed-loop supply chains. They consider multi-product returns with uncertain behavior but limit their consideration of demand for returned products to factories and not to secondary markets or spare markets. Thus, a supplier network that may be required to remanufacture a new product to meet the market demand is not considered. Also, this model is not suitable for modular products.

Sheu et al. (2005) formulated a linear multi-objective programming model to optimize the operations of both integrated logistics and corresponding used-product reverse logistics in a given green-supply chain. Factors such as the used-product return ratio and corresponding subsidies from governmental organization for reverse logistics were considered in the model formulation. The authors also proposed a real world case study for a Taiwan based notebook computer manufacturer.

Fleischmann et al. (2001) extended a forward logistics model to a reverse logistics system and discussed the differences. They utilized mixed-integer linear programming model. Kannan et al. (2009) proposed a model using genetic algorithm and particle swarm techniques. They applied the model by considering two cases including a tire manufacturer and a plastic goods manufacturer. Shi et al. (2010) proposed a mathematical model to maximize the profit of a remanufacturing system by developing a solution approach based on the Lagrangian relaxation method.

Schultmann et al. (2003) developed a hybrid method to establish a closed-loop supply chain for spent batteries. The model included a two-stage (collection point-sorting – recycling or disposal) facility location optimization problem. The authors found the optimal sorting centers to open serve the recycling facilities through a mixed-integer linear programming model, which minimizes the total cost, and implemented the model in GAMS (General Algebraic Modeling System) and solved it using a branch-and-bound algorithm. As a hybrid method, it also approached to a simulation under different scenarios for a steel-making process. Listes (2007) presented a generic stochastic model for the design of networks comprising both supply and return channels, organized in a closed-loop system. The author described a decomposition approach to the model, based on the branch-and-cut procedure known as the integer L-shaped method. Wang and Hsu (2010) proposed an interval programming model where the uncertainty has been expressed by fuzzy numbers. Gupta and Evans (2009) proposed a non-preemptive goal programming approach to model a closed-loop supply chain network.

Pishvaee et al. (2010) considered minimization of the total costs, and maximization of the responsiveness of a logistics network. Min et al. (2006) proposed a mixed-integer non-linear programming model to minimize the total reverse logistics costs for the reverse logistics problem involving both spatial and temporal consolidation of returned products. Fuente et al. (2008) proposed an integrated model for supply chain management (IMSCM) in which the operation of the reverse chain had been built based on the existing processes of the forward chain.

Finally, Lee and Dong (2008) developed an MILP model for integrated logistics network design for end-of-lease computer products. They consider a simple network with a single production center and a given number of hybrid distribution-collection facilities to be opened which they solve using tabu search. However, all of researches are found for some cost in reverse logistics. Our study focuses on a general framework and states total cost in reverse supply chain.

This chapter proposes a multi-layers, multi-product reverse supply chain problem which consists of returning center, disassembly center, processing center, manufacturing center, recycling center, material center and distribution center and minimizes the total costs in the reverse supply chain for returned products.

20.2 Problem Definition

In forward logistics, suppliers offer raw materials to manufacturers. These manufacturers deliver finished products to distributors who finally distribute them to customers. In reverse logistics, collectors and recyclers play important roles for reuse, recycle, remanufacturing and disposal.

The reverse supply chain under study is multi-layer, multi-product. In the designed (planned) model, the returned products after collecting and inspecting divides into two groups of disassembling and not disassembling products. The products which can be taken apart to the parts will be sent to the disassembling centers and there, they will convert to the parts. There they divide into reusable and not reusable parts. The not reusable parts will rebut safely and the reusable parts will be sent to the processing center. Some of the products that don't need to be disassembled according to their variety will be transmitted to the processing center right after collecting centers; then considering to the variety of product and the request of manufacturing centers, will be sent to them. In the remanufacturing process, according to the production center's demand, the parts that can be used again, after processing center will be sent to the remanufacturing center and after compounding with the other parts will be changed into new products and can return to the distribution chain. In the recycling process according to the recycling center's demand the disassembled parts (which can recover again) right after disassembling centers will be sent to the recycling centers for the purpose of producing the secondary materials.

20.3 Proposed Comprehensive Model

In this chapter the reverse supply chain model has been considered for returned products with the purpose of minimizing the reverse supply chain costs.

Assumptions

- The quantity of return, disassembly, processing, manufacturing, recycling, material and distribution centers are determined.
- Some products will transport straight from return centers to the processing centers.
- Some parts will transport straight from disassembly centers to the recycling centers.

Indices, parameters, and decision variables

Indices

i index of returning centers
j index of disassembly centers
k index of processing center
f index of manufacturing center
r index of recycling center
w index of material
p index of products
m index of parts
l index of distribution centers
c index of clients

Parameters

a_{ip}	the capacity of returning center i for product p
b_{jm}	The capacity of disassembly center j for parts m
u_{km}	The capacity of processing center k for part m
d_{rm}	The capacity of recycling center r for part m
h_{fm}	The capacity of manufacturing center f for parts m
E_{lm}	The capacity of distribution center l for part m
DM_{fm}	the manufacturing center's demand f for part m
$DRCP_{rp}$	the recycling center's demand r for product p
$DRCM_{rm}$	the recycling center's demand r for part m
DD_{lm}	the distribution center's demand l for part m
DC_{cm}	the client's demand c for part m
DMA_{wm}	the material center's demand w for part m
n_{mp}	the produced part's amount m from disassembling one product p
$CSRD_{ijp}$	unit cost of transportation from returning center i to disassembly center j for product p
$CSRP_{ikp}$	unit cost of transportation from returning center i into the processing center k for product p
$CSDP_{jkm}$	unit cost of transportation from disassembly center j into processing center k for part m
$CSDRC_{jrm}$	unit cost of transportation from disassembly center j into the recycling center r for part m
$CSPM_{kfm}$	unit cost of transportation from processing center k into the manufacturing center f for part m
$CSPRC_{krm}$	unit cost of transportation from processing center into the recycling center r for part m
$CSRCM_{rwm}$	unit cost of transportation from recycling center r into the material center w for part m
$CSPDC_{flm}$	unit cost of transportation from manufacturing center f into the distribution center l for part m
$CSDC_{lcm}$	unit cost of transportation from distribution center l into the clients c for part m
$FOCD_{jm}$	the fixed opening cost for disassembly center j for part m
$FOCP_{km}$	the fixed opening cost for processing centers k for part m
$FOCR_{ip}$	the fixed opening cost for returning centers i for product p
$FOCRC_{rm}$	the fixed opening cost for recycling centers r for part m
RMC_{fm}	unit cost of remanufacturing in manufacturing center f for part m
IC_{ip}	unit cost of maintaining in returning center i for product p
OCD_{jm}	unit cost of operations in disassembly center j for part m
OCP_{km}	unit cost of operations in processing center k part m
$OCRC_{rm}$	unit cost of operations in recycling center r part m
NRS_{min}	the minimum amount of returning center for opening and operations
NRS_{max}	the maximum amount of returning centers for operations and opening
NDS_{min}	the minimum amount of disassembling centers for opening and operations
NDS_{max}	the maximum quantity of disassembling centers for opening and operations
NPS_{min}	the minimum amount of processing centers for opening and operations
NPS_{max}	the maximum amount of processing centers for opening and operations
$NRCS_{min}$	the minimum amount of recycling centers for opening and operations
$NRCS_{max}$	the maximum amount of recycling centers for opening and operations

Decision Variables

φ_{ijp}	amount shipped from returning center i to disassembling center j for product p
δ_{ikp}	amount shipped from returning center i into the processing center k for product p
G_{jkm}	amount shipped from disassembly center j into the processing center k for part m
O_{jrm}	amount shipped from disassembly center j into the recycling center r for part m
Q_{kfm}	amount shipped from processing center k into the manufacturing center f for part m
S_{krm}	amount shipped from processing center k into the recycling center r for part m
ρ_{rwm}	amount shipped from recycling center r into the material center w for part m
T_{flm}	amount shipped from manufacturing center f into the distribution center l for part m
V_{lcm}	amount shipped from distribution center l into the clients c for part m
α_{jm}	if the disassembly center j is open for part m, 1 or otherwise 0
β_{km}	if processing center k is open for part m, 1 or otherwise 0
γ_{ip}	if the returning center i is open for product p, 1 or otherwise 0
λ_{rm}	if recycling center r is open for part m, 1 or otherwise 0
μ_{fm}	the part's flow amount m in manufacturing center f
X_{ip}	the product's flow amount p in returning center i
Y_{jm}	the part's flow amount m in disassembly center j
θ_{km}	the part's flow amount m in processing center k
τ_{rm}	the part's flow amount m in recycling center r

20.3.1 Mathematical Formulation

The formulation of the mathematical model is given below:

$$
\begin{aligned}
Min\, Z = {} & \sum_{i=1}^{I}\sum_{j=1}^{J}\sum_{p=1}^{P} csrd_{ijp}\Phi_{ijp} \\[6pt]
& + \sum_{i=1}^{I}\sum_{k=1}^{K}\sum_{p=1}^{P} csrp_{ikp}\delta_{ikp} + \sum_{j=1}^{J}\sum_{k=1}^{K}\sum_{m=1}^{M} csdp_{jkm}G_{jkm} + \sum_{j=1}^{J}\sum_{r=1}^{R}\sum_{m=1}^{M} csdrc_{jrm}O_{jrm} \\[6pt]
& + \sum_{k=1}^{K}\sum_{f=1}^{F}\sum_{m=1}^{M} csrm_{kfm}Q_{kfm} + \sum_{k=1}^{K}\sum_{r=1}^{R}\sum_{m=1}^{M} csprc_{krm}S_{krm} \\[6pt]
& + \sum_{r=1}^{R}\sum_{w=1}^{W}\sum_{m=1}^{M} csrcm_{rwm}\rho_{rwm} + \sum_{f=1}^{F}\sum_{l=1}^{L}\sum_{m=1}^{M} cspdc_{flm}T_{flm} \\[6pt]
& + \sum_{l=1}^{L}\sum_{c=1}^{C}\sum_{m=1}^{M} csdc_{lcm}V_{lcm} + \sum_{j=1}^{J}\sum_{m=1}^{M} focd_{jm}\alpha_{jm} \\[6pt]
& + \sum_{k=1}^{K}\sum_{m=1}^{M} focp_{km}\beta_{km} + \sum_{i=1}^{I}\sum_{p=1}^{P} focr_{ip}\gamma_{ip} + \sum_{r=1}^{R}\sum_{m=1}^{M} focrc_{rm}\lambda_{rm} \\[6pt]
& + \sum_{f=1}^{F}\sum_{m=1}^{M} rmc_{fm}\mu_{fm} + \sum_{i=1}^{I}\sum_{p=1}^{P} ic_{ip}X_{ip} + \sum_{j=1}^{J}\sum_{m=1}^{M} ocd_{jm}Y_{jm} + \sum_{k=1}^{K}\sum_{m=1}^{M} ocp_{km}\theta_{km} \\[6pt]
& + \sum_{r=1}^{R}\sum_{m=1}^{M} ocrc_{rm}\tau_{rm}
\end{aligned}
\tag{20.1}
$$

s.t.

$$\sum_{j=1}^{J} \Phi_{ijp} \leq a_{ip}\gamma_{ip} \quad \forall i, p \tag{20.2}$$

$$\sum_{k=1}^{K} \delta_{ikp} \leq a_{ip}\gamma_{ip} \quad \forall i, p \tag{20.3}$$

$$X_{ip} \leq a_{ip}\gamma_{ip} \quad \forall i, p \tag{20.4}$$

$$\sum_{k=1}^{K} G_{jkm} \leq b_{jm}\alpha_{jm} \quad \forall j, m \tag{20.5}$$

$$\sum_{r=1}^{R} O_{jrm} \leq b_{jm}\alpha_{jm} \quad \forall j, m \tag{20.6}$$

$$Y_{jm} \leq b_{jm}\alpha_{jm} \quad \forall j, m \tag{20.7}$$

$$\sum_{f=1}^{F} Q_{kfm} \leq u_{km}\beta_{km} \quad \forall k, m \tag{20.8}$$

$$\sum_{r=1}^{R} S_{krm} \leq u_{km}\beta_{km} \quad \forall k, m \tag{20.9}$$

$$\theta_{km} \leq u_{km}\beta_{km} \quad \forall k, m \tag{20.10}$$

$$\sum_{w=1}^{W} \rho_{rwm} \leq d_{rm}\lambda_{rm} \quad \forall r, m \tag{20.11}$$

$$\sum_{l=1}^{L} T_{flm} \leq h_{fm} \quad \forall f, m \tag{20.12}$$

$$\mu_{fm} \leq h_{fm} \quad \forall f, m \tag{20.13}$$

$$\sum_{c=1}^{C} V_{lcm} \leq e_{lm} \quad \forall l, m \tag{20.14}$$

$$\sum_{k=1}^{K} Q_{kfm} \geq DM_{fm} \quad \forall f, m \tag{20.15}$$

$$\mu_{fm} \geq DM_{fm} \quad \forall f, m \tag{20.16}$$

$$\sum_{f=1}^{F} T_{flm} \geq DD_{lm} \quad \forall l, m \tag{20.17}$$

$$\sum_{l=1}^{L} V_{lcm} \geq DC_{cm} \quad \forall c, m \tag{20.18}$$

$$\sum_{r=1}^{R} \rho_{rwm} \geq DMA_{wm} \quad \forall w, m \tag{20.19}$$

$$\sum_{j=1}^{J} O_{jrm} + \sum_{k=1}^{K} S_{krm} \geq DRCM_{rm} \quad \forall r, m \tag{20.20}$$

$$\tau_{rm} \geq DRCM_{rm} \quad \forall r, m \tag{20.21}$$

$$\sum_{i=1}^{I} \sum_{k=1}^{K} \delta_{ikp} \geq \sum_{r=1}^{R} DRCP_{rp} \quad \forall p \tag{20.22}$$

$$\sum_{j=1}^{J} \sum_{k=1}^{K} G_{jkm} \geq \sum_{f=1}^{F} DM_{fm} \quad \forall m \tag{20.23}$$

$$\sum_{j=1}^{J} \sum_{k=1}^{K} G_{jkm} \leq n_{mp} \left(\sum_{i=1}^{I} \sum_{j=1}^{J} \Phi_{ijp} \right) \quad \forall m, p \tag{20.24}$$

$$\sum_{j=1}^{J} \sum_{r=1}^{R} O_{jrm} \leq n_{mp} \left(\sum_{i=1}^{I} \sum_{j=1}^{J} \Phi_{ijp} \right) \quad \forall m, p \tag{20.25}$$

$$NRS_{\min} \leq \sum_{i=1}^{I} \gamma_{ip} \leq NRS_{\max} \quad \forall p \tag{20.26}$$

$$NDS_{\min} \leq \sum_{j=1}^{J} \alpha_{jm} \leq NDS_{\max} \quad \forall m \tag{20.27}$$

$$NPS_{\min} \leq \sum_{k=1}^{K} \beta_{km} \leq NPS_{\max} \quad \forall m \tag{20.28}$$

$$NRCS_{\min} \leq \sum_{r=1}^{R} \lambda_{rm} \leq NRCS_{\max} \quad \forall m \tag{20.29}$$

$$\sum_{f=1}^{F} T_{flm} = \sum_{c=1}^{C} V_{lcm} \quad \forall l, m \tag{20.30}$$

$$\sum_{k=1}^{K} G_{jkm} + \sum_{r=1}^{R} O_{jrm} \leq Y_{jm} \quad \forall j, m \tag{20.31}$$

$$\sum_{j=1}^{J} \Phi_{ijp} + \sum_{k=1}^{K} \delta_{ikp} \leq X_{ip} \quad \forall i, p \tag{20.32}$$

$$\sum_{f=1}^{F} Q_{kfm} + \sum_{r=1}^{R} S_{krm} \leq \theta_{km} \quad \forall k, m \tag{20.33}$$

$$\sum_{w=1}^{W} \rho_{rwm} \leq \tau_{rm} \quad \forall r, m \tag{20.34}$$

$$\sum_{l=1}^{L} T_{flm} \leq \mu_{fm} \quad \forall f, m \tag{20.35}$$

$$\Phi_{ijp}, \delta_{ikp}, G_{jkm}, O_{jrm}, Q_{kfm}, S_{krm}, \rho_{rwm}, T_{flm}, V_{lcm}, \mu_{fm}, X_{ip}, Y_{jm}, \theta_{km}, \tau_{rm} \geq 0 \quad \forall i, j, k, f, r, w, p, m, l, c \tag{20.36}$$

$$\alpha_{jm}, \beta_{km}, \gamma_{ip}, \lambda_{rm} = \{0, 1\} \quad \forall i, j, k, p, m \tag{20.37}$$

20.3.2 Objective Function

We want to demonstrate a model in reverse supply chain is a way to minimize the chain costs. We should introduce a model which minimizes the transportation cost of products and parts between centers and at the same time minimizes the fixed opening cost of sites and operation's cost on parts and supply maintenance costs and remanufacturing costs. By attention to the definition of indices, parameters and decision variables, the objective function will be defined, which consists of: minimizing the costs of transportation of products and parts, the fixed opening cost of centers and operations costs on parts and the supply maintenance costs, remanufacturing costs in reverse supply chain (Constraint 20.1).

20.3.3 Constraints

(20.2) and (20.3): These constraints are stating that the amount of shipping products from any returning center (if it is opened) into the disassembly, processing centers for each product should be equal or smaller than the capacity of that returning center.

(20.4): This constraint is stating that the amount of products that will be collected in the returning center should be equal or smaller than the capacity of that returning center.

(20.5) and (20.6): These constraints are stating that the amount of sent parts from any disassembly centers and recycling centers should be equal or smaller than the capacity of the same disassembly center for each part.

(20.7): This constraint is stating that the amount of a part which is in the disassembly center should be equal or smaller than the capacity of the same disassembly center.

(20.8) and (20.9): These constraints are stating that the amount of shipping parts from any processing centers (if it is opened) into the manufacturing centers and recycling centers should be equal or smaller than the capacity of the same processing centers for each parts.

(20.10): This constraint is stating that the amount of a part which is in the processing center should be equal or smaller than the capacity of the same processing center.

(20.11): This constraint is stating that the amount of the parts which shipping from any recycling center (if it is opened) into the material centers should be equal or smaller than the capacity of the same recycling for each part.

(20.12): This constraint states that the amount of sent parts from any manufacturing center into the distribution centers should be equal or smaller than the capacity of the same manufacturing center for each part.

(20.13): This constraint states that the amount of part in each manufacturing center should be equal or smaller than the capacity of the same manufacturing center.

(20.14): This constraint states that the amount of sent parts from any distribution center to the client should be equal or smaller than the capacity of the same distribution center for clients.

(20.15) and (20.16): These constraints state the demand amounts of manufacturing centers for parts.

(20.17): This constraint states the part demand amount of distribution centers.

(20.18): This constraint indicates the client's part demand amount.

(20.19): This constraint states the part demand amount of material center.

(20.20) and (20.21): These constraints state the part demand amount of recycling centers.

(20.22) and (20.23): These constraints state that the manufacturing and recycling center's demand is for products and parts which are transported from the returning and disassembly centers into the processing center.

(20.24) and (20.25): These constraints are related to the balance of parts flow from the disassembly of products.

(20.26), (20.27), (20.28), (20.29): These constraints are stating that the min and max index amount of returning, disassembling, processing and recycling centers.

(20.30): This constraint states that the amount of sent parts from manufacturing centers to the distribution center is equal to the sent parts from distribution centers in to the client.

(20.31): This constraint states that the amount of sent parts from each disassembly center into the processing and recycling centers should be equal or smaller than the parts amount in that disassembly center.

(20.32): This constraint states that the amount of sent products from each returning center into the disassembly, processing centers, should be equal or smaller than the product's amount in that returning center.

(20.33): This constraint states that the amount of sent parts from each of the processing centers into the manufacturing and recycling centers should be equal or smaller than the flow amount of parts in that processing center.

(20.34): This constraint states that the sent parts amount from any recycling center into the material centers should be equal or smaller than the parts amount in that recycling center.

(20.35): This constraint states that the sent parts amount from any manufacturing center into the distribution centers should be equal or smaller than the parts flow amount in that manufacturing center.

(20.36) and (20.37): These constraints enforce the binary and non-negativity restrictions on the corresponding decision variables.

20.4 Analytical Example

We solved the presented mathematic model by using LINGO, an operation research software. In this multi-layer and multi-product model, we are attempting to minimize the costs of fixed opening facilities, transportation and shipping of products and parts between centers and also the operations, supply maintenance and remanufacturing costs, and also the product amount and sending parts into the centers and the amount of it would be calculated. To analyzing the suggested model we create numerical example in small size and then solve the created example by LINGO software.

In small size we consider the index quantities as variables between 3–5 to solve the problem, so we replace the inputs of problem in the model. With respect to the inputs of the model and solving it, the outputs of model and objective function amount and the implementation time has been obtained which are as follow; the obtained objective function is 29653.20 which obtained in zero time. All the variables that were not zero 0 quantities are shown in Table 20.1; after solving the model we will find out that the decision variable $\alpha(1,2)$ gained 1 quantity. This means that the disassembly center 1 should be opened for part 2. The decision variable $\lambda(3,2)$ obtained 1, means that the recycling center 3 would be opened for part 2. Generally when the decision variables α_{jm}, β_{km}, γ_{ip}, λ_{rm} gained 1, it indicates that the considered center to that decision variable will be opened for that part or product. The decision variable $Q(1,4,2)$ is considered 5. This means that the amount of part 2 from processing center 1 into the manufacturing center 4 is 5. The decision variable $\tau(2,3)$ got 15, it means that the amount of part 3 in recycling center 2 is 15. $\rho(3,2,2) = 8$ means that the amount of part 2 from recycling center 3 into the material center 2 is 8.

According to the results that obtained from the mathematical model in different dimension we want to evaluate the importance of decision variables. This importance of decision variables is determined by an elimination method. In this method after elimination of each decision variable and rerunning the model, the importance of the variable would be clear. It is not possible to remove all decision variables in the model because some of the variables in the model are substantial so that the elimination of the variables will cause the closure of facilities in the model. Changes of decision variables after removal variables δ_{ikp}, G_{jkm}, O_{jrm}, S_{krm}, ρ_{rwm} is shown in Table 20.2.

TABLE 20.1

Numerical Results Using LINGO Software

$\varphi(2,1,4)$	1.2	$\rho(3,1,2)$	3	$\beta(4,3)$	1
$\varphi(2,3,2)$	0.9	$\rho(3,2,2)$	8	$\theta(1,1)$	71
$\varphi(3,3,1)$	0.9	$\rho(3,3,1)$	18	$\theta(1,2)$	57
$\varphi(3,3,3)$	1.1	$\rho(3,3,2)$	14	$\theta(2,2)$	16
$\delta(2,1,2)$	34	$T(1,1,3)$	6	$\theta(2,3)$	63
$\delta(2,4,4)$	28	$T(1,2,1)$	8	$\theta(3,1)$	8
$\delta(3,1,1)$	41	$T(1,2,3)$	13	$\theta(4,2)$	20
$\delta(3,4,3)$	8	$T(1,3,2)$	3	$\theta(4,3)$	7
$G(1,3,2)$	62	$T(2,1,2)$	7	$\gamma(2,1)$	1
$G(1,4,3)$	50	$T(2,2,3)$	17	$\gamma(2,2)$	1
$G(2,1,3)$	16	$T(2,3,1)$	24	$\gamma(2,3)$	1
$G(3,2,1)$	59	$T(2,3,3)$	16	$\gamma(2,4)$	1
$O(1,1,3)$	8	$T(3,2,1)$	8	$\gamma(3,1)$	1
$O(1,3,2)$	18	$T(3,2,2)$	8	$\gamma(3,2)$	1
$O(1,3,3)$	4	$T(4,1,1)$	13	$\gamma(3,3)$	1
$O(2,2,3)$	4	$T(4,2,1)$	3	$\gamma(3,4)$	1
$O(2,3,1)$	18	$T(4,2,2)$	5	$X(2,2)$	34.87
$Q(1,1,1)$	9	$T(5,1,1)$	8	$X(2,4)$	29.19
$Q(1,1,2)$	17	$T(5,1,2)$	16	$X(3,1)$	41.94
$Q(1,2,1)$	18	$T(5,3,3)$	17	$X(3,3)$	9.09
$Q(1,2,2)$	16	$V(1,1,2)$	6	$\lambda(1,1)$	1
$Q(1,3,2)$	8	$V(1,1,3)$	6	$\lambda(1,2)$	1
$Q(1,4,1)$	16	$V(1,2,2)$	17	$\lambda(1,3)$	1
$Q(1,4,2)$	5	$V(1,3,1)$	1	$\lambda(2,1)$	1
$Q(1,5,1)$	8	$V(1,4,1)$	20	$\lambda(2,2)$	1
$Q(2,1,3)$	19	$V(2,1,2)$	4	$\lambda(2,3)$	1
$Q(2,3,3)$	14	$V(2,2,3)$	16	$\lambda(3,1)$	1
$Q(2,4,3)$	9	$V(2,3,1)$	19	$\lambda(3,2)$	1
$Q(2,5,2)$	16	$V(2,4,2)$	9	$\tau(1,1)$	32
$Q(2,5,3)$	17	$V(2,4,3)$	14	$\tau(1,2)$	20
$Q(3,3,1)$	8	$V(3,1,1)$	20	$\tau(1,3)$	12
$Q(4,2,3)$	7	$V(3,1,3)$	13	$\tau(2,1)$	17
$S(1,1,1)$	16	$V(3,2,1)$	4	$\tau(2,2)$	11
$S(1,2,1)$	4	$V(3,3,2)$	3	$\tau(2,3)$	15
$S(1,2,2)$	11	$V(3,3,3)$	20	$\tau(3,1)$	18
$S(2,1,3)$	4	$\alpha(1,2)$	1	$\tau(3,2)$	25
$S(4,1,2)$	20	$\alpha(1,3)$	1	$\tau(3,3)$	4
$\rho(1,1,1)$	1	$\alpha(2,1)$	1	$\mu(1,1)$	9
$\rho(1,1,2)$	12	$\alpha(2,2)$	1	$\mu(1,2)$	17
$\rho(1,2,1)$	17	$\alpha(2,3)$	1	$\mu(1,3)$	19
$\rho(1,2,3)$	6	$\alpha(3,1)$	1	$\mu(2,1)$	24
$\rho(1,4,1)$	14	$Y(1,2)$	80	$\mu(2,2)$	16
$\rho(1,4,3)$	2	$Y(1,3)$	62	$\mu(2,3)$	33
$\rho(1,5,2)$	8	$Y(2,1)$	18	$\mu(3,1)$	8

(Continued)

TABLE 20.1 (*Continued*)

Numerical Results Using LINGO Software

$\rho(1,5,3)$	4	$Y(2,3)$	20	$\mu(3,2)$	8
$\rho(2,1,3)$	1	$Y(3,1)$	59	$\mu(3,3)$	14
$\rho(2,3,1)$	1	$\beta(1,1)$	1	$\mu(4,1)$	16
$\rho(2,3,3)$	1	$\beta(1,2)$	1	$\mu(4,2)$	5
$\rho(2,4,2)$	4	$\beta(2,2)$	1	$\mu(4,3)$	9
$\rho(2,5,1)$	16	$\beta(2,3)$	1	$\mu(5,1)$	8
$\rho(2,5,2)$	7	$\beta(3,1)$	1	$\mu(5,2)$	16
$\rho(2,5,3)$	13	$\beta(4,2)$	1	$\mu(5,3)$	17

TABLE 20.2

Changes of Decision Variables after Removing Some Variables

Decision Variables	δ_{ikp}	G_{jkm}		O_{jrm}		S_{krm}		ρ_{rwm}
				Changes of Decision Variables				
φ_{ijp}	✓	$\varphi(2,1,4)$	0.35					
		$\varphi(2,3,2)$	0.21					
		$\varphi(3,3,1)$	0.24					
		$\varphi(3,3,3)$	0.32					
G_{jkm}				✓ $G(1,4,3)$	62	✓ $G(1,4,3)$	46	
				$G(2,1,3)$	4	$G(2,1,3)$	20	
O_{jrm}	✓	$O(1,1,3)$	12			✓ $O(1,1,3)$	12	
						$O(2,2,1)$	4	
		$O(2,2,3)$	0			$O(2,2,2)$	11	
						$O(3,1,1)$	16	
						$O(3,1,2)$	20	
Q_{kfm}				✓ $Q(1,2,1)$	17			
				$Q(1,5,3)$	8			
				$Q(2,2,1)$	1			
				$Q(2,5,3)$	9			
S_{krm}	✓	$S(2,1,3)$	0	✓ $S(1,1,1)$	0			
				$S(1,3,1)$	18			
				$S(1,3,3)$	4			
		$S(4,2,3)$	4	$S(2,1,1)$	16			
				$S(2,1,3)$	12			
				$S(2,3,2)$	18			
				$S(4,2,3)$	4			
T_{flm}				✓ $T(1,3,1)$	1	✓ $T(1,3,1)$	1	
				$T(2,3,1)$	23	$T(2,3,1)$	23	
α_{jm}	✓	$\alpha(2,3)$	0			✓ $\alpha(2,3)$	0	
		$\alpha(3,2)$	0			$\alpha(3,2)$	0	
		$\alpha(3,3)$	1					
β_{km}				✓ $\beta(1,3)$	1	✓ $\beta(4,2)$	0	
				$\beta(2,1)$	1	$\beta(4,3)$	1	

(*Continued*)

TABLE 20.2 (*Continued*)

Changes of Decision Variables after Removing Some Variables

Decision Variables	Changes of Decision Variables				
	δ_{ikp}	G_{jkm}	O_{jrm}	S_{krm}	ρ_{rwm}
λ_{rm}					✓ $\lambda(1,1)$ 0 $\lambda(1,2)$ 0 $\lambda(2,2)$ 0 $\lambda(2,3)$ 0 $\lambda(3,1)$ 0
μ_{fm}				✓ $\mu(2,1)$ 23	
X_{ip}	✓ $X(2,2)$ 0.9 $X(2,4)$ 1.2 $X(3,1)$ 0.9 $X(3,3)$ 1.1				
Y_{jm}		✓ $Y(1,2)$ 18 $Y(1,3)$ 16 $Y(2,1)$ 18 $Y(2,3)$ 0 $Y(3,1)$ 0	✓ $Y(1,2)$ 62 $Y(2,1)$ 0 $Y(2,3)$ 4	✓ $Y(2,1)$ 22 $Y(2,2)$ 11 $Y(2,3)$ 24 $Y(3,1)$ 75 $Y(3,2)$ 20	
θ_{km}			✓ $\theta(1,1)$ 72 $\theta(1,3)$ 12 $\theta(2,1)$ 17 $\theta(2,2)$ 34 $\theta(4,3)$ 11	✓ $\theta(1,1)$ 51 $\theta(1,2)$ 46 $\theta(2,3)$ 59 $\theta(4,2)$ 0	
τ_{rm}					✓ $\tau(1,1)$ 16 $\tau(2,1)$ 4 $\tau(2,3)$ 4 $\tau(3,2)$ 18

For instance after removal of decision variable G_{jkm} that represents the amount shipped from disassembly center j into the processing center k for part m and rerunning the model we find the value of the objective function is changed to 26,622.4 and the value of the decision variables $\varphi_{ijp}, T_{flm}, Y_{jm}, X_{ip}, \mu_{fm}$ are changed.

According to the analysis we perceive that the decision variables $G_{jkm}, O_{jrm}, S_{krm}$ are more important than the decision variables δ_{ikp}, ρ_{rwm}. Just decision variable X_{ip} will be changed after removing the decision variable δ_{ik} and decision variables λ_{rm}, τ_{rm} will be changed after removing the decision variable ρ_{rwm} whereas five decision variables $\varphi_{ijp}, O_{jrm}, S_{krm}, \alpha_{jm}, Y_{jm}$ will be changed after removing the decision variable G_{jkm}. Seven decision variables $G_{jkm}, Q_{kfm}, S_{krm}, T_{flm}, \beta_{km}, Y_{jm}, \theta_{km}$ will be changed after removing the decision variable O_{jrm}. Eight decision variables $G_{jkm}, O_{jrm}, T_{flm}, \alpha_{jm}, \beta_{km}, \mu_{fm}, Y_{jm}, \theta_{km}$ will be changed after removing the decision variable S_{krm}. The result of performed analysis has many applications in strategic decision-making.

20.5 Discussions

In this chapter, a reverse supply chain was considered minimizing the total cost of transport, inspection, remanufacture and maintenance. The presented model was an integer linear programming model for multi-layer, multi-product reverse supply chain that minimized the products and parts transportation costs among centers and also sites launch, operation parts, maintenance and remanufacturing costs at the same time. We can solve the proposed model using any optimization software. Also, comprehensive sensitivity analysis can be studied to validate the proposed mathematical model.

References

Amin SH, Zhang G. A proposed mathematical model for closed-loop network configuration based on product life cycle. *The International Journal of Advanced Manufacturing Technology* 2012;58(5):791–801.

Aras N, Aksen D, Tanugur AG. Locating collection centers for incentive- dependent returns under a pick-up policy with capacitated vehicles. *European Journal of Operational Research* 2008;191:1223–1240.

Du F, Evans GW. A bi-objective reverse logistics network analysis for post-sale service. *Computers & Operations Research* 2008;35(8):2617–2634.

Fleischmann M, Beullens P, Bloemhof-Ruwaard JM, Van Wassenhove LN. The impact of product recovery on logistics network design. *Production and Operations Management* 2001;10(2):156–173.

Fuente MVD, Ros L, Cardos M. Integrating forward and reverse supply chains: Application to a metal-mechanic company. *International Journal of Production Economics* 2008;111(2):782–792.

Gupta A, Evans GW. A goal programming model for the operation of closed-loop supply chains. *Engineering Optimization* 2009;41(8):713–735.

He W, Li G, Ma X, Wang H, Huang J, Xu M et al., WEEE recovery strategies and the WEEE treatment status in China. *Journal of Hazardous Materials* 2006;136(3):502–512.

Jayaraman V, Patterson RA, Rolland E. The design of reverse distribution networks: Models and solution procedures. *European Journal of Operational Research* 2003;150:128–149.

Kannan G. Fuzzy approach for the selection of third party reverse logistics provider. *Asia Pacific Journal of Marketing and Logistics* 2009;21(3):397–416.

Kannan G, Noorul Haq A, Devika M. Analysis of closed loop supply chain using genetic algorithm and particle swarm optimization. *The International Journal of Production Research* 2009;47(5):1175–1200.

Kannan G, Sasikumar P, Devika K. 2010. A genetic algorithm approach for solving a closed loop supply chain model: A case of battery recycling. *Applied Mathematical Modelling* 2010;34(3):655–670.

Kim KB, Song IS, Jeong BJ. Supply planning model for remanufacturing system in reverse logistics environment. *Computers & Industrial Engineering* 2006;51(2):279–287.

Ko HJ, Evans GW. A genetic-based heuristic for the dynamic integrated forward/reverse logistics network for 3PLs. *Computers and Operations Research* 2007;34:346–366.

Lee D, Dong M. A heuristic approach to logistics network design for end-of- lease computer products recovery. *Transportation Research Part E* 2008;44:455–474.

Lee J, Gen E, Rhee MKG. Network model and optimization of reverse logistics by hybrid genetic algorithm. *Computers and Industrial Engineering* 2009;56(3):951–964.

Listes O. A generic stochastic model for supply-and-return network design. *Computers and Operations Research* 2007;34(2):417–442.

Liu X, Tanaka M, Matsui Y. Electrical and electronic waste management in China: Progress and the barrier to overcome. *Waste Management and Research* 2006;24:92–101.

Min H, Ko CS, Ko HJ. The spatial and temporal consolidation of returned products in a closed-loop supply chain network. *Computers and Industrial Engineering* 2006;51(2):309–320.

Pati RK, Vrat P, Kumar P. A goal programming model for paper recycling system. *The International Journal of Management Science, Omega,* 2008;36(3):405–417.

Pishvaee MS, Farahani RZ, Dullaert W. A memetic algorithm for bi-objective integrated forward/reverse logistics network design. *Computers and Operations Research* 2010;37(6):1100–1112.

Rogers DS, Tibben-Lembke RS. *Going backward: Reverse logistics trends and practices.* Reno, Nevada: Reverse Logistics Executive Council, 1999.

Salema MIG, Barbosa-Povoa AP, Novais AQ. An optimization model for the design of a capacitated multi-product reverse logistics network with uncertainty. *European Journal of Operational Research* 2007;179: 1063–1077.

Schultmann F, Engels B, Rentz O. Closed-loop supply chains for spent batteries. *Interfaces* 2003;33:57–71.

Sheu JB, Chou YH, Hu CC. An integrated logistics operational model for green-supply chain management. *Transportation Research Part E: The Logistics and Transportation Review* 2005;41(4):287–313.

Shi J, Zhang G, Sha J, Amin SH. Coordinating production and recycling decision with stochastic demand and return. *Journal of Systems Science and Systems Engineering* 2010;19(4):385–407.

Stock JK. *Reverse logistics. White Paper, Oak Brook,* IL: Council of Logistics Management, 1992.

Teunter R, Kaparis K, Tang O. Multi-product economic lot scheduling problem with separate production lines for manufacturing and remanufacturing. *European Journal of Operational Research* 2008;191:1241–1253.

Wang H, Hsu H. Resolution of an uncertain closed-loop logistics model: An application to fuzzy linear programs with risk analysis. *Journal of Environmental Management* 2010;91(11):2148–2162.

Zuidwijk R, Krikke H. Strategic response to EEE returns: Product ecodesign or new recovery processes? *European Journal of Operational Research* 2008;191:1206–1222.

Section III

Uncertain and Intelligent Supply Chain Models

21

Multi-Layer Electronic Supply Chain: Intelligent Information System

SUMMARY In this chapter, we propose a multi-agent information system to cluster the elements of a supply network based on the similarity in information flow. Each layer consists of elements that are differentiated by their performance throughout the supply network. We apply intelligent agents as decision aids in different layers of our supply network. The proposed agents measure and record the performance flow of elements considering their web interactions.

21.1 Introduction

The current business environment is becoming increasingly complex, uncertain, unpredictable, and as a result, more and more competitive. As competition and complexity have increased, flexibility-based supply chain management (SCM) has emerged as an increasingly important issue for companies. The challenge of flexibility in SCM is to identify and implement strategies that minimize cost while maximizing flexibility in an increasingly competitive and complex market (Browne et al., 1997; Wadhwa and Saxena, 2005). Flexibility stands out as the most discussed and applied domain in manufacturing and supply chains (SCs) (Stecke and Solberg, 1981; Browne et al., 1995; Chan et al., 2006). Sushil (2000) while deliberating upon the concept of systematic flexibility has essentially stressed the multiplicity of connotations of flexibility in response to diversity of situations. Wadhwa and Rao (2000) defined flexibility as the ability to deal with change by judiciously providing and exploiting controllable options dynamically. The potential of certain types of flexibility to enhance the overall performance of manufacturing and supply chain system has attracted the attention of many researchers (e.g., Browne et al., 1984; Chan et al., 2004; Wadhwa et al., 2005). Flexibility implications on the SCs performance need to be more closely understood as most researchers have interpreted it differently. Enhanced competitiveness requires that companies ceaselessly integrate within a network of organizations. Firms ignoring this challenge are destined to fall behind their rivals. This integration of companies within a network has led to more emphasis on supply chain management (SCM). "SCM is the management of upstream and downstream relationships in order to deliver superior customer value at less cost to the supply chain as a whole" (Christopher, 1998). The integral value of the SCM philosophy is that "total performance of the entire supply chain is enhanced when we simultaneously optimize all the links in the chain as compared to the resulting total performance when each individual link is separately optimized" (Burke and Vakkaria, 2002).

Recent technological developments in information systems and information technologies have the potential to facilitate this coordination, and this, in turn, allows the virtual integration of the entire supply chain. The focus of this integration in the context of internet-enabled activities is generally referred to as e-SCM. Merging these two fields (SCM and the internet) is a key area of concern for contemporary managers and researchers. Managers have realized that the internet can enhance SCM decision-making by providing real-time information and enabling collaboration between trading partners. Many companies have implemented point-of-sales scanners, which read, on real time, what is being sold. These companies do not only collect information on real time to make decisions about what to order or how to replenish the stores; they also send this information, through the internet, to their suppliers in order to make them able to synchronize their production to actual sales.

Following the definition of SCM of Cooper et al. (1997), we define e-SCM as the impact that the internet has on the integration of key business processes from end-user through original suppliers that provides products, services and information that add value for customers and other stakeholders. The main objective of this chapter is to identify the major issues surrounding the impact of the internet on SCM, focusing on supply chain processes. The internet can have three main impacts on the supply chain. One of the most covered topics in the literature is the impact of e-commerce, which refers mainly to how companies can respond to the challenges posed by the internet on the fulfillment of goods sold through the net. Another impact refers to information sharing: how the internet can be used as a medium to access and transmit information among supply chain partners. However, the internet not only enables supply chain partners to access and share information, but also to access data analysis and modeling to jointly make a better planning and decision-making. This joint planning and decision-making is the third type of impact of the internet on SCM and we refer to it as knowledge sharing. A configuration of e-SCM is presented in Figure 21.1.

As Croom (2005) pointed out very recently, there is some debate about the scope of SCM. For example, Oliver and Webber (1992) and Houlihan (1984) used the term SCM for the internal supply chain that integrates business functions involved in the flow of materials and information from inbound to outbound ends of the business. Ellram (1991) viewed SCM as an alternative to vertical integration. Cooper et al. (1997) defined SCM as "the integration of key business processes from end-user through original suppliers that provides products, services, and information that add value for customers and other stakeholders." And, Christopher (1998) defined SCM as the management of upstream

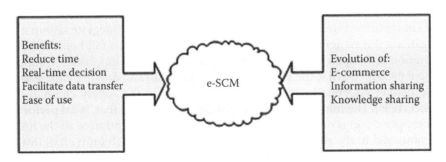

FIGURE 21.1
A configuration of e-SCM.

and downstream relationships. Croom (2005) suggested that one way of dealing with the diversity of SCM definitions is to concentrate on the core processes and functions relating to the management of supply chains (e.g., fulfillment, operations planning and procurement).

In the literature there is a diversity of models suggesting the main supply chain processes. For example, the Supply Chain Operations Reference (SCOR) model developed in 1996 focuses on five key processes: plan, source, make, deliver and return. Cooper et al. (1997) defined SCM taking into account the eight supply chain processes identified by the International Centre for Competitive Excellence (now named Global Supply Chain Forum): customer relationship management (CRM), customer service management, demand management, fulfillment, procurement, manufacturing flow management, product development and commercialization and reverse logistics. Hewitt (1994) found that executives identify up to 14 business processes. As a result, a definition comprising a number of processes closer to 14 might provide more detailed information for practitioners and researchers. Accordingly, from the two previous models we decided to adopt the definition of SCM provided by Cooper et al. (1997). This definition has been widely referred to (Romano and Vinelli, 2001; Cagliano et al., 2003; Mills et al., 2004; Cousins, 2005; Danese et al., 2006).

Here, we consider a five-layer supply chain network including supplier, manufacturer, distributor, retailer, and customer. In each layer we propose the corresponding intelligent information systems based on intelligent agents in virtual environment through internet. The aim to propose such agent-based systems is to cluster the network elements using their information flows.

21.2 Problem Definition

Data mining processes can be divided to six sequential, iterative steps:

1. Problem definition
2. Data acquisition
3. Data preprocessing and survey
4. Data modeling
5. Evaluation
6. Knowledge deployment

Each step is essential: The problem defines what data are used and what a good solution is. Modeling makes it possible to apply the results to new data. On the other hand, data modeling without good understanding and careful preparation of the data leads to problems. Finally, the whole mining process is meaningless if the new knowledge will not be used (Pyle, 1999).

Given the rate of growth of the Web, proliferation of e-commerce, Web services, and Web-based information systems, the volumes of clickstream and user data collected by Web-based organizations in their daily operations has reached huge proportions. Meanwhile, the substantial increase in the number of websites presents a challenging task for webmasters to organize the contents of the websites to cater to the needs of users (Cooley et al., 1997). Modeling and analyzing web navigation behavior is helpful in understanding what information of online users demand. Following that, the analyzed results can be seen as knowledge to be used in intelligent online applications, refining website maps, Web-based personalization system and improving searching accuracy when seeking information. Nevertheless, an online navigation behavior grows each passing day, and thus extracting information intelligently from it is a difficult issue (Wang and NetLibrary, 2000). Web usage mining refers to the automatic discovery and analysis of patterns in clickstream and associated data collected or generated as a result of user interactions with Web resources on one or more websites (Srivastava et al., 2000).

In this research, supply network elements in a multi-layer structure are considered to have online interactions. Since many elements in various layers may have common interests up to a point during their online interactions, information flow patterns should capture the overlapping interests or the information needs of these elements. In this study we advance a model for clustering the supply network elements.

21.3 Intelligent Model

An agent is anything that can be viewed as perceiving its environment through sensors and acting upon that environment through effectors. A human agent has eyes, ears, and other organs for sensors, and hands, legs, mouth, and other body parts for effectors. A robotic agent substitutes cameras and infrared range finders for the sensors and various motors for the effectors. A software agent has encoded bit strings as its percepts and actions.

According to the definitions provided in previous sections, each component of SC is involved in various activities such as: planning and controlling stock, quality control, procurement, marketing, relationship with customers, sale, distribution, etc. Therefore, concerning multi-agent system definitions and concepts, an SC can be considered as multi-agent system in which each element of chain has the nature of an agent.

On the other hand, each component of SC can be viewed as multi-agent system in which every agent communicates with other multi-agent systems of SC in addition to their interaction with other agents of the multi-agent system of which they are members. So, each SC can be seen as a system of n agents in which the magnitude of n depends on the number of activities, the nature of performance and complexity of SC in question.

Clustering can be considered the most important unsupervised learning problem; so, as every other problem of this kind, it deals with finding a structure in a collection of unlabeled data.

A loose definition of clustering could be "the process of organizing objects into groups whose members are similar in some way."

A cluster is therefore a collection of objects that are "similar" between them and are "dissimilar" to the objects belonging to other clusters.

In this case we easily identify the four clusters into which the data can be divided; the similarity criterion is distance: two or more objects belong to the same cluster if they are "close" according to a given distance (in this case geometrical distance). This is called distance-based clustering.

Another kind of clustering is conceptual clustering: two or more objects belong to the same cluster if this one defines a concept common to all that objects. In other words, objects are grouped according to their fit to descriptive concepts, not according to simple similarity measures.

So, the goal of clustering is to determine the intrinsic grouping in a set of unlabeled data. But how to decide what constitutes a good clustering? It can be shown that there is no absolute "best" criterion that would be independent of the final aim of the clustering. Consequently, it is the user who must supply this criterion, in such a way that the result of the clustering will suit their needs.

For instance, we could be interested in finding representatives for homogeneous groups (data reduction), in finding "natural clusters" and describe their unknown properties ("natural" data types), in finding useful and suitable groupings ("useful" data classes) or in finding unusual data objects (outlier detection).

Clustering algorithms can be applied in many fields, for instance:

Marketing: finding groups of customers with similar behavior given a large database of customer data containing their properties and past buying records

Biology: classification of plants and animals given their features

Libraries: book ordering

Insurance: identifying groups of motor insurance policy holders with a high average claim cost; identifying frauds

City planning: identifying groups of houses according to their house type, value and geographical location

Earthquake studies: clustering observed earthquake epicenters to identify dangerous zones

WWW: document classification; clustering weblog data to discover groups of similar access patterns

21.3.1 Clustering Algorithms

Clustering algorithms may be classified as listed below:

- Exclusive clustering
- Overlapping clustering
- Hierarchical clustering
- Probabilistic clustering

In the first case, data are grouped in an exclusive way, so that if a certain datum belongs to a definite cluster then it could not be included in another cluster.

On the contrary the second type, the overlapping clustering, uses fuzzy sets to cluster data, so that each point may belong to two or more clusters with different degrees of membership. In this case, data will be associated to an appropriate membership value.

Instead, a hierarchical clustering algorithm is based on the union between the two nearest clusters. The beginning condition is realized by setting every datum as a cluster. After a few iterations it reaches the final clusters wanted.

Finally, the last kind of clustering uses a completely probabilistic approach.

Here, four of the most used clustering algorithms are considered as follows:

- K-means
- Fuzzy C-means

- Hierarchical clustering
- Mixture of Gaussians

Each of these algorithms belongs to one of the clustering types listed above. So that, K-means is an exclusive clustering algorithm, fuzzy C-means is an overlapping clustering algorithm, hierarchical clustering is obvious and lastly mixture of Gaussian is a probabilistic clustering algorithm.

21.3.2 Distance Measure

An important component of a clustering algorithm is the distance measure between data points. If the components of the data instance vectors are all in the same physical units then it is possible that the simple Euclidean distance metric is sufficient to successfully group similar data instances. However, even in this case the Euclidean distance can sometimes be misleading. Despite both measurements being taken in the same physical units, an informed decision has to be made as to the relative scaling. Notice however that this is not only a graphic issue. The problem arises from the mathematical formula used to combine the distances between the single components of the data feature vectors into a unique distance measure that can be used for clustering purposes: Different formulas lead to different clustering.

Again, domain knowledge must be used to guide the formulation of a suitable distance measure for each particular application. For higher dimensional data, a popular measure is the Minkowski metric,

$$d_p(x_i, x_j) = \left(\sum_{k=1}^{d} |x_{i,k} - x_{j,k}|^p \right)^{1/p},$$

where d is the dimension of the data. The Euclidean distance is a special case, where $p = 2$, while Manhattan metric has $p = 1$. However, there are no general theoretical guidelines for selecting a measure in any given application. A clustering Q means partitioning a data set into a set of clusters Q_i, $i = 1, \ldots, C$. In crisp clustering, each data sample belongs to exactly one cluster (Bezdek and Pal, 1992). Fuzzy clustering is a generalization of crisp clustering where each sample has a varying degree of membership in all clusters. Clustering can also be based on mixture models (McLahlan and Basford, 1987). In this approach, the data are assumed to be generated by several parameterized distributions (typically Gaussians). Distribution parameters are estimated using, for example, the expectation-maximization algorithm. A widely adopted definition of optimal clustering is a partitioning that minimizes distances within and maximizes distances among clusters. However, this leaves much room for variation: within- and between-cluster distances can be defined in several ways; see Table 21.1. The selection of the distance criterion depends on the application. The distance norm $\|.\|$ is yet another parameter to consider. Here, we use the Euclidean norm. We utilize local criteria in clustering data. Thus, S_{nn} and d_s in Table 21.1 are based on distance to nearest neighbor. In Table 21.1, $x_i, x_{i'} \in Q_k$, for $i \neq i'$, $x_j \in Q_l$, $k \neq l$, N_k is the number of samples in cluster Q_k, and $c_k = (1/N_k)\sum_{x_i \in Q_k} x_i$. However, the problem is that they are sensitive to noise and outliers. Addition of a single sample to a cluster can radically change the distances (Bezdek, 1998). To be more robust, the local criterion should depend on collective features of a local data set (Blatt et al., 1996). Solutions include using more than one neighbor (Karypis et al., 1999) or a weighted sum of all distances.

TABLE 21.1

Within-Cluster and Between-Clusters Distances

Within-cluster distance	$S(Q_k)$
Average distance	$S_a = \dfrac{\sum_{i,i'} \lVert x_i - x_{i'} \rVert}{N_k(N_k - 1)}$
Nearest neighbor distance	$S_{nn} = \dfrac{\sum_i \min_{i'}\{\lVert x_i - x_{i'} \rVert\}}{N_k}$
Centroid distance	$S_c = \dfrac{\sum_i \lVert x_i - c_k \rVert}{N_k}$
Between-clusters distance	$D(Q_k, Q_l)$
Single linkage	$d_s = \min_{i,j}\{\lVert x_i - x_j \rVert\}$
Complete linkage	$d_{co} = \max_{i,j}\{\lVert x_i - x_j \rVert\}$
Average linkage	$d_a = \dfrac{\sum_{i,j} \lVert x_i - x_j \rVert}{N_k . N_l}$
Centroid linkage	$d_{ce} = \lVert c_k - c_l \rVert$

Generally speaking, agents are active, persistent (software) components with the abilities of perceiving, reasoning, acting and communicating (Fung and Chen, 2005). The agent may follow a set of rules predefined by the user and then applies them. The intelligent agent will learn and be able to adapt to the environment in terms of user requests consistent with the available resources (Papazoglou, 2001). The key aspects of agents are their autonomy and abilities to reason and act in their environment, as well as to interact and communicate with other agents to solve complex problems (Jain et al., 1999). Autonomy means that the agent can act without the direct intervention of humans or other agents and that it has control over its own actions and internal state. The agent must communicate with the user or other agents to receive instructions and provide results. An essential quality of an agent is the amount of learned behavior and possible reasoning capacity that it has.

As the market needs are extremely various and fashion updates quickly, the supply chain member usually cannot make decisions immediately because of the inaccurate or incomplete information. The decision delay in the supply chain prolongs the process time and causes a company to lose competence. In order to reduce this delay, the supply chain member needs to give quick response. Thus, a supply chain can be characterized as a logistic network of partially autonomous decision-makers. Supply chain management has to do with the coordination of decisions within the network. Different segments of the network are communicating with one another through flows of material and information, being controlled and coordinated by the activities of supply chain management.

Here, we describe our multi-layer agent-based electronic supply network. As stated before, our proposed network consists of five layers. Suppliers should provide raw materials for factories to make a product and present at the markets. In our proposed model, all suppliers present their goods in their websites and any manufacturer visit the websites and collect the information about the suppliers. Then, using the proposed intelligent agent the ranking of suppliers, data analysis and data saving is performed.

The next layer is for manufacturers. Manufacturers exchange their manufacturing data with each other through World Wide Web. These data are collected and analyzed via an intelligent agent. The intelligent agent work out as a decision aid and provide the information such as machine information, production processes, depot analysis and optimization, and manufacturing optimizations. The results are saved in the corresponding database and viewed for public visit in an internet website. The third and fourth layers are distributors and retailers. Here, an interaction between distributor and retailer is considered. Distributors conduct their depot information, due dates, order list and etc., to the internet using an information sharing mechanism. The vehicles of distributors are connected to a server and report real-time information about the delivery of products to retailers. Retailers' interests, need and orders are given in their corresponding websites and deposited in a database.

Here, the intelligent systems collect the information from different servers, analyze them and provide a report containing orders in transit, orders delivered, orders sent and distributor depot inventory control. The last layer of our proposed multi-layer e-SCM contains an interaction between retailer and customer. Customers present their interest and needs in a local server. The data are saved in the market databases and transferred to the World Wide Web. At the same time, retailers show their products and their specifications in another local server, and therefore transfer them to retailer database. The intelligent agent collects the information from both sides and help retailers in decision-making about the customer relationship management through web, electronic quality function deployment. For customers the decisions may be about modifying the interests due to product specifications, and procurement through web.

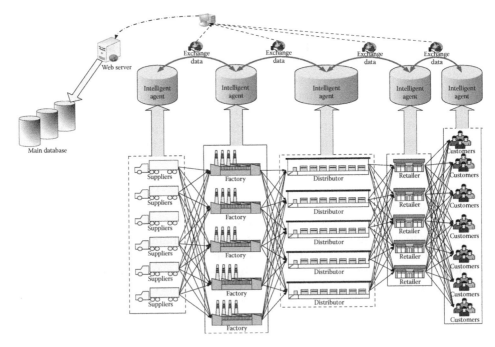

FIGURE 21.2
A configuration of the proposed intelligent agent-based e-SCM.

Considering the interactions among elements of our proposed multi-layer e-SCM, an information flow exists which implies the information and data exchanges between any two layers via intelligent agents. We want to develop a new clustering methodology to segment the elements in different layers. This method is based on the information flow between any two elements in virtual environment. The aim of this segmentation is to improve the serviceability of the network and to increase the flexibility of the network for presenting multi-commodity markets. A configuration of the proposed intelligent agent-based e-SCM is shown in Figure 21.2. The intelligent agents record and trace the data transfer and exchange between and within the layers. All the information are transferred and saved in a main data base. The clustering is performed based on the data of the main database.

21.4 Discussions

We proposed an information system to integrate a multi-layer multi-agent e-SCM. The information systems for each layer are designed and the information flow is traced. The intelligent agents are suitable decision aid for our proposed e-SCM. In each layer, the intelligent agents collect information and present a report that includes different optimization, planning and control decisions. A new approach for clustering the e-SCM elements is developed due to similarities of the information flow during interactions. The advantages of such a system are real-time decision-making and saving time due to online information flow.

References

Bezdek JC. Some new indexes of cluster validity. *IEEE Transactions on Systems, Man, and Cybernetics, Part B (Cybernetics)* 1998;28:301–315.

Bezdek JC, Pal SK. *Fuzzy Models for Pattern Recognition: Methods that Search for Structures in Data.* IEEE, New York, 1992

Blatt M, Wiseman S, Domany E. Super paramagnetic clustering of data. *Physical Review Letters* 1996;76(18): 3251–3254.

Browne J, Dubois D, Rathmill K, Sethi P, Steke KE. Classification of flexible manufacturing systems. *FMS Magazine* 1984;7:114–117.

Browne J, Hunt J, Zhang J. *The Extended Enterprise. The Handbook of Lifecycle Engineering Concepts, Models and Technologies.* UK: Chapman & Hall, 1997.

Browne J, Sackett PJ, Wortmann JC. Future manufacturing system—Towards the extended enterprise. *Computers in Industry* 1995;25:235–254.

Burke GJ, Vakkaria AJ. *Supply Chain Management.* Internet Encyclopedia. New York, NY: Wiley, 2002.

Cagliano R, Caniato F, Spina G. E-business strategy: How companies are shaping their supply chain through the Internet. *International Journal of Operations & Production Management* 2003;23(10):1142–1162.

Chan FTS, Bhagwat R, Wadhwa S. Increase in flexibility: Productive or counterproductive? A study on the physical and operating characteristics of a flexible manufacturing system. *International Journal of Production Research* 2006;44(7):1431–1445.

Chan FTS, Chung SH, Wadhwa S. A heuristic methodology for order distribution in a demand driven collaborative supply chain. *International Journal of Production Research* 2004;42:1–19.

Christopher M. *Logistics & supply chain management: Strategies for reducing cost and improving service.* London: Financial Times Pitman Publishing, 1998.

Cooley R, Mobasher B, Srivastava J. Web mining: Information and pattern discovery on the World Wide Web. *Ninth IEEE International Conference on Tools with Artificial Intelligence*, Newport Beach, CA, 1997:558–567.

Cooper MC, Lambert DM, Pagh JD. Supply chain management: More than a new name for logistics. *The International Journal of Logistics Management* 1997;8(1):1–13.

Cousins PD. The alignment pf appropriate firm and supply strategies for competitive advantages. *International Journal of Operations & Production Management* 2005;25(5):403–428.

Croom SR. The impact of e-business on supply chain management. *International Journal of Operations & Production Management* 2005;25(1):55–73.

Danese P, Romano P, Vinelli A. Sequences of improvements in supply networks: Case studies from pharmaceutical industry. *International Journal of Operations & Production Management* 2006;26(11): 1199–222.

Ellram LM. Supply chain management: The industrial organization perspective. *International Journal of Physical Distribution & Logistics Management* 1991;21(1):13–22.

Fung RYK, Chen T. A multiagent supply chain planning and coordination architecture. *International Journal of Advanced Manufacturing Technology* 2005;25:811–819.

Hewitt F. Supply chain redesign. *The International Journal of Logistics Management* 1994;5(2):1–9.

Houlihan J. Supply chain management. *Proceedings of the 19th International Technical Conference*, BPICS, 1984:101–110.

Jain AK, Aparico M, Singh MP. Agents for process coherence in virtual enterprises. *Communications of the ACM* 1999;42(3):62–69.

Karypis G, Han EH, Kumar V. Chameleon: Hierarchical clustering using dynamic modeling. *IEEE Computer* 1999;32(Aug):68–74.

McLahlan GJ, Basford KE. *Mixture Models: Inference and Applications to Clustering.* Marcel Dekker, New York, 1987;vol 84.

Mills J, Schmitz J, Frizelle G. A strategic review of supply networks. *International Journal of Operations & Production Management* 2004;24(10):1012–1036.

Oliver RK, Webber MD. Supply chain management: Logistics catches up with strategy. In: Christopher, M. (ed.), *Logistics: The Strategic Issues.* London: Chapman & Hall, 1992:63–75.

Papazoglou MP. Agent-oriented technology in support of e-business. *Communications of the ACM* 2001;44(4):71–77.

Pyle D. *Data Preparation for Data Mining.* San Francisco, CA: Morgan Kaufmann, 1999.

Romano P, Vinelli A. Quality management in a supply chain perspective: Strategic and operative choices in a textile-apparel network. *International Journal of Operations & Production Management* 2001;21(4):446–460.

Srivastava J, Cooley R, Deshpande M, Tan PN. Web usage mining: Discovery and applications of usage patterns from Web data. *ACM SIGKDD Explorations Newsletter* 2000;1:12–23.

Stecke KE, Solberg JJ. Loading and control policies for flexible manufacturing systems. *International Journal of Production Research* 1981;19(5):481–490.

Sushil S. Concept of systemic flexibility. *Global Journal of Flexible Systems Management* 2000;1:77–80.

Wadhwa S, Rao KS. Flexibility: An emerging meta-competence form an aging high technology. *International Journal of Technology Management* 2000;19(7):820–845.

Wadhwa S, Rao KS, Chan FTS. Flexibility enabled lead time reduction in flexible systems. *International Journal of Production Research* 2005;43(15):3131–3162.

Wadhwa S, Saxena A. Knowledge management based supply chain: An evolution perspective. *Global Journal of e-Business and Knowledge Management* 2005;2(2):13–29.

Wang J, NetLibrary I. *Encyclopedia of Data Warehousing and Mining.* Idea Group, New York, USA, 2000.

22

Multi-Layer Electronic Supply Chain: Agent-Based Model

SUMMARY   This chapter is concerned with analyzing the interaction of supply chains (SC) and the internet. Combining these two fields is a focal point of concern for contemporary decision-makers and researchers. They have found that the internet can enhance SC by making real-time information available and enabling collaboration among trading partners. Here, we aim to propose a management information system to integrate the segmented factors of a supply network. We apply intelligent agents as decision supports in various layers of our supply network.

22.1 Introduction

Enhanced competitiveness requires that companies ceaselessly integrate within a network of organizations. Firms ignoring this issue are destined to fall behind their rivals. This integration of companies within a network has led to more emphasis on supply chain management (SCM). "SCM is the management of upstream and downstream relationships in order to deliver superior customer value at less cost to the supply chain as a whole" (Christopher, 1998). The integral value of the SCM philosophy is that "total performance of the entire supply chain is enhanced when we simultaneously optimize all the links in the chain as compared to the resulting total performance when each individual link is separately optimized" (Burke and Vakkaria, 2002).

Recent technological expansions in information systems and information technologies have the potential to facilitate this coordination, and this, in turn, lets the virtual integration of the entire supply chain. The focus of this integration in the context of internet-enabled activities is generally referred to as e-SCM. Merging these two fields (SCM and the internet) is a key area of concern for contemporary decision-makers and researchers. Managers have understood that the internet can enhance SCM decision-making by providing real-time information and enabling collaboration between trading partners. Many companies have implemented point-of-sales scanners, which read, on real time, what is being sold. These companies do not only gather information on real time to make decisions about what to order or how to replenish the stores; they also send this information, through the internet, to their suppliers in order to enable them to synchronize their production to actual sales.

The internet can have three main impacts on the supply chain. One of the most covered topics in the literature is the impact of e-commerce, which refers basically to how companies can respond to the challenges posed by the internet on the fulfillment of goods sold through the net. Another impact refers to information sharing, how the internet can be employed as a medium to access and transmit information among supply chain partners. However, the internet not only enables supply chain partners to access and share information, but also to obtain data analysis and modeling to jointly make a better planning and decsion-making. This jointly planning and decision-making is the third type of impact of the internet on SCM and we refer to it as knowledge sharing.

In the literature there are a variety of models suggesting the general supply chain processes. For example, the Supply Chain Operations Reference (SCOR) model developed in 1996 focuses on five key processes: plan, source, make, deliver and return. Cooper et al. (1997) defined SCM taking into account the eight supply chain processes introduced by the International Centre for Competitive Excellence (now named Global Supply Chain Forum): customer relationship management (CRM), customer service management, demand management, fulfillment, procurement, manufacturing flow management, product development and commercialization and reverse logistics.

Hewitt (1994) found that executives identify up to 14 business processes. As a result, a definition comprising a number of processes closer to 14 might provide more detailed information for practitioners and researchers. Also, from the two previous models we decided to adopt the definition of SCM provided by Cooper et al. (1997). This definition has been widely referred to (Romano and Vinelli, 2001; Cagliano et al., 2003; Mills et al., 2004; Cousins, 2005; Danese et al., 2006).

Here, we consider a five-layer supply network including supplier, manufacturer, distributor, retailer, and customer. In each layer we propose the corresponding management information systems based on intelligent agents in virtual environment using internet. As a result, we configure a multi-layer, multi-agent system (MAS) of electronic supply chain.

22.2 Problem Definition

An agent is anything that can be considered as perceiving its atmosphere through sensors and acting upon that environment through effectors. A human agent has eyes, ears, and other organs for sensors, and hands, legs, mouth, and other body parts for effectors. A robotic agent substitutes cameras and infrared range detectors for the sensors and various motors for the effectors. A software agent has encoded bit strings as its percepts and actions.

According to the definitions provided in last sections, each component of SC is involved in various activities such as: planning and controlling stock, quality control, procurement, marketing, relationship with customers, sale, distribution, etc. Therefore, concerning multi-agent system (MAS) definitions and concepts, an SC can be considered as MAS in which each element of chain has the nature of an agent.

On the other hand, each component of SC can be viewed as MAS in which every agent communicates with other MAS of SC in addition to their interaction with other agents of MAS of which they are a member. So each SC can be seen as a system of n agents in which the magnitude of n depends on the number of activities, the nature of performance and complexity of SC in question.

As mentioned before, this article aims at investigating the possibility of modeling a distribution system of multi-level SC of physical goods in the case of orders received from customers outside the organization. It is also assumed that orders are sent from external agents (companies or distribution institutes) to customers. The structure of this SC is one of the most common that is found in SC of different goods such as oil productions, medicine and food.

An attribute of this type of SC structure is the competition among external agents involved in transportation and physical distribution. Due to the competition, these agents make effort to maximize their profits, which can be in conflict with previous SC elements or the whole SC objectives. For instance, at the time of resource allocation these agents try to assume responsibility for distribution of goods with minimum cost and maximum profit. Therefore it is quite clear that customer's order distribution that is of less attraction to distribution agents is disturbed.

In most supply chains, due to the complication of resource allocation and categorizing the solution to this problem in NP-Hard problems, distribution agents job allocation and distribution are based on negotiations between SC allocation operators and agents or so-called bargaining mechanisms to reach the utility point that each one's profits, not comprehensively but relatively are made. This issue is another reason for consistency of distribution modeling system with MAS.

What is done in bargaining mechanisms in distribution systems consists of announcing a number of priorities presentable to the agents and picking out one of them by the agent on the basis of maximizing policy. Of course, the rules and regulations governing the distribution system, customers' orders, technical and quality conditions of agent are effective in making a list of priorities

22.3 Agent-Based Model

Here, we configure the agent-based management information systems for all layers of our proposed e-SCM.

22.3.1 Agent-Based Supplier Layer

Supplier should provide raw materials for factories to make a product and present at the markets. In our proposed model, all suppliers present their goods in their websites and any manufacturer visit the websites and collect the information about the suppliers. Then, using the proposed intelligent agent the ranking of suppliers, data analysis and data saving is performed. A configuration of agent-based supplier is shown in Figure 22.1.

22.3.2 Agent-Based Manufacturer Layer

Manufacturers exchange their manufacturing data with each other through World Wide Web. These data are collected and analyzed via an intelligent agent. The intelligent agent works out as a decision aid and provides the information such as machine information, production processes, depot analysis and optimization, and manufacturing optimizations. The results are saved in the corresponding database and viewed for public visit in an internet website. A configuration of agent-based manufacturer is shown in Figure 22.2.

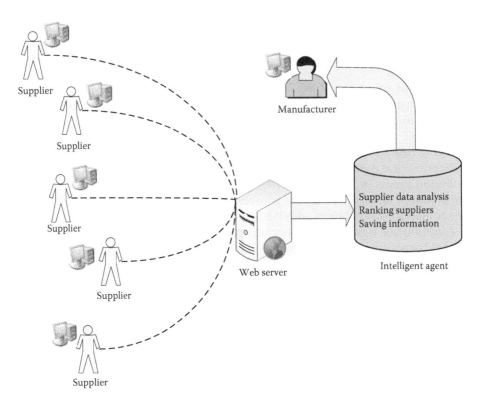

FIGURE 22.1
A configuration of agent-based supplier.

22.3.3 Agent-Based Distributor-Retailer Layer

In this layer an interaction between distributor and retailer is considered. Distributors conduct their depot information, due dates, order list, etc., to the internet using an information sharing mechanism. The vehicles of distributors are connected to a server and report real-time information about the delivery of products to retailers. Retailers' interests, need and orders are given in their corresponding websites and deposited in a database. Here, the intelligent systems collect the information from different servers, analyze them and provide a report containing orders in transit, orders delivered, orders sent and distributor depot inventory control. A configuration of agent-based distributor-retailer is shown in Figure 22.3.

22.3.4 Agent-Based Retailer-Customer Layer

The last layer of our proposed multi-layer e-SCM contains an interaction between retailer and customer. Customers present their interest and needs in a local server. The data are saved in the market databases and transferred to the World Wide Web. At the same time, retailers show their products and their specifications in another local server, and transfer them to retailer databases. The intelligent agent collects the information from both sides and help retailers in decision-making about the customer relationship management through web, electronic quality function deployment. For customers the decisions may be about modifying the interests due to product specifications, and procurement through web. A configuration of agent-based retailer-customer is shown in Figure 22.4.

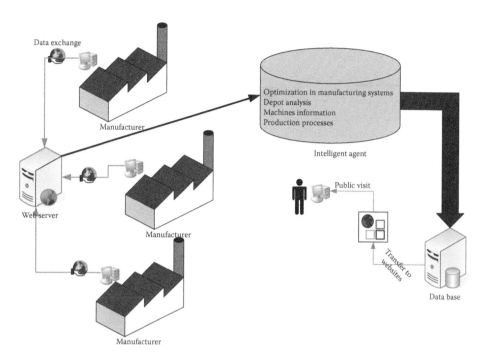

Data exchange

Manufacturer

Web server

Manufacturer

Manufacturer

Optimization in manufacturing systems
Depot analysis
Machines information
Production processes

Intelligent agent

Public visit

Transfer to websites

Data base

FIGURE 22.2
A configuration of agent-based manufacturer.

22.4 Discussions

We proposed an information system to make an integration of a multi-layer, multi-agent e-SCM. The information systems for each layer were designed and the information flow was traced. The intelligent agents were suitable decision aid for our proposed e-SCM. In each layer the intelligent agents collected information and presented a report including different optimization, planning and control decisions. The advantages of such a system were real-time decision-making and saving time due to online information flow. Also in comparison with ordinary supply chains the proposed e-SCM includes the following advantages:

e-SCM covers all aspects of a business, from the stage of raw materials right on to the end user. Each and every aspect of the cycle is covered by the e-SCM be it sourcing, product design, production planning, order processing, inventory management, transportation, warehousing and customer service. The e-SCM manages the flow between the different cycles and spans across the different departments and companies involved and the applications used by these departments and companies should be able to talk to each other and understand each other for the e-SCM to work properly.

In a traditional company that does not employ e-commerce, 17%–50% of the price of its products is from moving the products from their manufacturing plant to shop shelves. This includes the margin of the retailer and of the distributors. Most of the cost is attributed to logistics and holding inventory. An efficient e-SCM can bring down the prices of products by as high as 40% and it does so by eliminating overstocking by reducing the average inventory levels to what is needed and by so doing lowering warehousing costs and transport costs since there won't be any unnecessary trips when every stage of the supply chain is in synch with each other. This will not only give the company a cost benefit

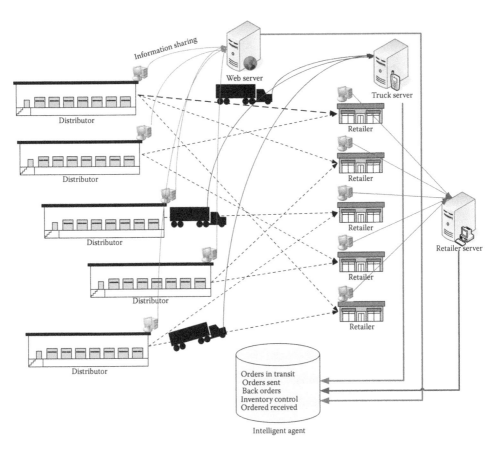

FIGURE 22.3
A configuration of agent-based distributor-retailer.

but will also result in improved customer service levels, improved competitiveness and an overall gain in profitability for the organization.

In an e-SCM application system communication between the different departments or different companies is in real time and data can be integrated with back office systems thus reducing paperwork. Using the Web to eliminate paper transactions can generate substantial savings of cost and time. It facilitates the removal of purchase orders, delivery confirmations, bills of material and invoices. The switch away from paper can also speed up response and improve communications with those in different time zones or who work outside normal office hours. Another significant potential benefit is a reduction in the errors associated with activities such as re-keying data and receiving orders by telephone calls and handwritten faxes.

To leverage the full benefits of e-logistics in an e-SCM and achieve full customer satisfaction visibility throughout the entire supply chain must be completely transparent. This is achieved through the movement of information in tandem with goods and services. Customers thus have complete real-time consignment status information over the Web, while at the same time suppliers and delivery companies can save on the salary previously devoted to employees answering queries on order status.

e-SCM's main strategic advantage lies in its ability to allow real-time exchange of information to take place between the company's employees and their trading partners, namely

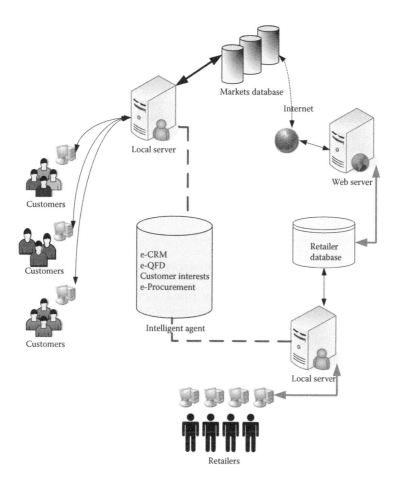

FIGURE 22.4
A configuration of agent-based retailer-customer.

customers, distributors and manufacturers, regarding product configuration, order status, pricing and inventory availability. Such functions improve order accuracy and provide 100% order fulfillment through accurate inventory information. These real-time data enable users to make informed ordering, purchasing and inventory decisions, thereby enhancing the quality and scope of customer service.

In addition to increasing productivity and reducing overall operating expenses, e-SCM maximizes selling opportunities by capturing valuable customer information-buying patterns, frequency of visits, preferences, order history and then uses this information for up-selling, cross-selling and promotional opportunities. e-SCM provides the tool sets to get new business by reaching out to customers that you never could before.

References

Burke GJ, Vakkaria AJ. Supply chain management. In: *Internet Encyclopedia*, Bidgoli H (ed.), New York, NY: Wiley, 2002.

Cagliano R, Caniato F, Spina G. E-business strategy: How companies are shaping their supply chain through the internet. *International Journal of Operations & Production Management* 2003;23(10):1142–1162.

Christopher M. *Logistics & Supply Chain Management: Strategies for Reducing Cost and Improving Service.* London: Financial Times Pitman Publishing, 1998.

Cooper MC, Lambert DM, Pagh JD. Supply chain management: More than a new name for logistics. *The International Journal of Logistics Management* 1997;8(1):1–13.

Cousins PD. The alignment pf appropriate firm and supply strategies for competitive advantages. *International Journal of Operations & Production Management* 2005;25(5):403–428.

Danese P, Romano P, Vinelli A. Sequences of improvements in supply networks: Case studies from pharmaceutical industry. *International Journal of Operations & Production Management* 2006;26(11):1199–1222.

Hewitt F. Supply chain redesign. *The International Journal of Logistics Management* 1994;5(2):1–9.

Mills J, Schmitz J, Frizelle G. A strategic review of supply networks. *International Journal of Operations & Production Management* 2004;24(10):1012–1036.

Romano P, Vinelli A. Quality management in a supply chain perspective: Strategic and operative choices in a textile-apparel network. *International Journal of Operations & Production Management* 2001;21(4):446–460.

Wadhwa S, Saxena A. Knowledge management based supply chain: An evolution perspective. *Global Journal of e-Business and Knowledge Management* 2005;2(2):13–29.

23

Multi-Layer Electronic Supply Chain: Dynamic Route Selection in an Agent Model

SUMMARY In this chapter, we develop an intelligent information system in a multi-layer electronic supply chain network. It has been realized that the internet can facilitate SCM by making real-time information available and enabling collaboration between trading partners. Here, we propose a multi-agent system to analyze the performance of the elements of a supply network based on the attributes of the information flow. Each layer consists of elements that are differentiated by their performance throughout the supply network. A dynamic programming approach is applied to determine the optimal route for a customer in the end-user layer.

23.1 Introduction

The internet can have three main impacts on the supply chain. Most interest in the literature is on the impact of e-commerce, referring mainly to how companies can respond to the challenges posed by the internet in fulfilling the goods sold through the net. Another impact refers to information sharing, that is, how the internet can be used as a medium to access and transmit information among supply chain partners. However, the internet enables supply chain partners not only to access and share information, but also to access data analysis and modeling in order to make a better planning and decision-making. The joint planning and decision-making is the third impact of the internet on SCM and we refer to it as knowledge sharing.

As pointed out in Croom (2005), there is some debate about the scope of SCM. For example, in Houlihan (1984) and Oliver and Webber (1992), the authors used the term SCM for the internal supply chain integrating business functions involved in the flow of materials and information from inbound to outbound ends of the business. Ellram (1991) viewed SCM as an alternative to vertical integration. Cooper et al. (1997) defined SCM as "the integration of key business processes from end-user through original suppliers that provides products, services, and information that add value for customers and other stakeholders." And Christopher (1998) defined SCM as the management of upstream and downstream relationships. Croom (2005) suggested that one way of dealing with the diversity of SCM definitions was to concentrate on the core processes and functions relating to the management of supply chains (e.g., fulfillment, operations planning and procurement).

The Supply Chain Operations Reference (SCOR) model developed focusing on five key processes: plan, source, make, deliver and return. Cooper et al. (1997) defined SCM taking account of the eight supply chain processes identified by the International Centre

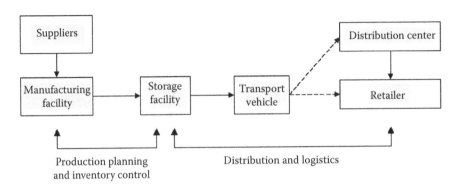

FIGURE 23.1
A supply chain process.

for Competitive Excellence (renamed Global Supply Chain Forum): customer relationship management (CRM), customer service management, demand management, fulfillment, procurement, manufacturing flow management, product development and commercialization, and reverse logistics. A supply chain process is depicted in Figure 23.1.

Fung and Chen (2005) found out that executives identified up to 14 business processes. As a result, a definition encompassing a number of processes closer to the 14 might provide more detailed information for practitioners and researchers. Accordingly, we consider using the definition of SCM given in (Cooper et al., 1997), as it being widely adopted (Romano and Vinelli, 2001; Cagliano et al., 2003; Mills et al., 2004; Cousins, 2005; Danese et al., 2006).

23.2 Problem Definition

Here, we describe a multi-layer, agent-based electronic supply network. As stated before, the proposed network consists of five layers. At one layer, suppliers provide raw materials for factories to make a product and present it at the markets. All suppliers present their goods in their websites and manufacturers collect the information about the suppliers by visiting the websites. Then, using the proposed intelligent agent, the ranking of suppliers, data analysis and data saving are performed.

The next layer is for manufacturers. Manufacturers exchange their manufacturing data with one other through the World Wide Web. The data are collected and analyzed via an intelligent agent. The intelligent agent works out to be a decision aid and provides information about vehicles, production processes, depot analysis and optimization, and manufacturing optimization. The results are saved in the corresponding database and viewed by the public on a website. The third and fourth layers are distributors and retailers. Here, interaction between distributor and retailer is allowed. Distributors share their depot information, due dates, order list, etc., in the internet using an information sharing mechanism. The vehicles of distributors are connected to a server and report real-time information about the delivery of products to retailers. Retailers' interests, needs and orders are specified in their corresponding websites, deposited in a database.

The intelligent system collects the information from different servers, analyzes them and provides a report containing orders in transit, orders delivered, orders sent and the distributor depot inventory control. The last layer of the proposed multi-layer e-SCM is composed of customers having interactions with retailers. Customers present their interests and needs in a local server. The data are saved in the market databases and transferred to the World Wide Web. At the same time, retailers show their products and their specifications in another local server, and then transfer them to the retailer database. The intelligent agent collects the information from both sides and helps retailers in decision-making about the customer relationship management through the Web and electronic quality function deployment. For customers, the decisions may concern modifying the interests due to product specifications, and procurement through the Web.

23.3 Route Selection Model

Considering the interactions among elements of the proposed multi-layer e-SCM, an intelligent agent exists which flows the information and exchanges data between any two layer. Here, we develop a new methodology for the integration of elements in different layers. The aim of the integration is to improve the serviceability of the network and increase the flexibility in presenting multi-commodity markets' network. Intelligent agents record and trace the data transfer and exchange between and within the layers. All the information is deposited in a main database. The route selection is performed using the data of the main database.

23.3.1 Information Flow Interaction

The interaction between any two elements of the multi-layer e-SCM depends on time periods. The information flow is different in different time periods, and thus we need to take time into account. We consider the information flow between any two elements as some attributes affecting the interactions. A configuration of the interaction among elements is presented in Figure 23.2. Considering that the information sharing takes place among

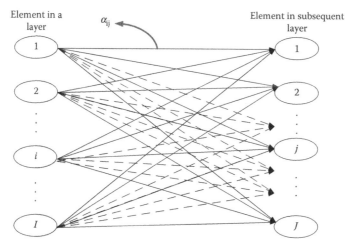

FIGURE 23.2
A configuration of the interaction between elements.

all the elements in different layers, any element in a layer is considered to get a score in relation to the elements in the preceding layer with respect to the knowledge gained from information sharing. The mathematical notations are as follows:

Mathematical Notations

i index for element in a layer; $i = 1, \dots, I$
j index for element in a subsequent layer; $j = 1, \dots, J$
t index for time period; $t = 1, \dots, T$
m index for attribute; $m = 1, \dots, M$
A_m the mth attribute
α_{ij} the score of interaction between ith element in a layer with jth element in the subsequent layer
S^{it} matrix with S^{it}_{jm} values for interactions α_{ij} with attributes m (extracted from Table 23.1), for $j = 1, \dots, J, m = 1, \dots, M$ $(i = 1, \dots, I, t = 1, \dots, T)$

Considering the above notations, the matrix in Table 23.2 is configured for scores of interactions with respect to attributes. In this matrix, we consider a numerical preference value chosen from Table 23.1, for each interaction with respect to an attribute.

Note that the scores presented in Table 23.2 are only related to the time period 1 and the interaction is between two layers; for other time periods and other pairs of layers, the same scoring process is performed (the same applies to Tables 23.3 and 23.4 as well). A process of scoring in different time periods and the continuous improvement are depicted in Figure 23.3. As depicted in Figure 23.3, the interactions among elements of two sequential layers are computed using the proposed matrices and gathered via an intelligent agent and transferred to the database in one period. Due to weaknesses of the supply network resulting from the computations, modifications are performed in the next period. The dynamic structure is continued until the end of the periods.

TABLE 23.1

Arc Preferences with Their Numerical Values

Preference	Numerical Value
Extremely preferred	9
Very strongly preferred	7
Strongly preferred	5
Moderately preferred	3
Equally preferred	1
Preferences in between the above preferences	2,4,6,8

TABLE 23.2

Interaction Scoring

$i = 1, t = 1$	1	2	...	M
1	S^{11}_{11}	S^{11}_{12}	...	S^{11}_{1M}
2	S^{11}_{21}	S^{11}_{22}	...	S^{11}_{2M}
$\vdots$	$\vdots$	$\vdots$	$\vdots$	$\vdots$
J	S^{11}_{J1}	S^{11}_{J2}	...	S^{11}_{JM}

TABLE 23.3

The Normalized Values

$i = 1, t = 1$	1	2	...	M
1	$\bar{S}_{11}^{11}$	$\bar{S}_{12}^{11}$	...	$\bar{S}_{1M}^{11}$
2	$\bar{S}_{21}^{11}$	$\bar{S}_{22}^{11}$	...	$\bar{S}_{2M}^{11}$
⋮	⋮	⋮	⋮	⋮
J	$\bar{S}_{J1}^{11}$	$\bar{S}_{J2}^{11}$	...	$\bar{S}_{JM}^{11}$

TABLE 23.4

The Weighted Scores

$i = 1, t = 1$	1	2	...	M
1	$\beta_1 . \bar{S}_{11}^{11}$	$\beta_2 . \bar{S}_{12}^{11}$	...	$\beta_M . \bar{S}_{1M}^{11}$
2	$\beta_1 . \bar{S}_{21}^{11}$	$\beta_2 . \bar{S}_{22}^{11}$	...	$\beta_M . \bar{S}_{2M}^{11}$
⋮	⋮	⋮	⋮	⋮
J	$\beta_1 . \bar{S}_{J1}^{11}$	$\beta_2 . \bar{S}_{J2}^{11}$	...	$\beta_M . \bar{S}_{JM}^{11}$

To obtain the α values considering the attributes, a normalization process is performed as follows:

$$\bar{S}_{jm}^{it} = \frac{S_{jm}^{it}}{\sqrt{\sum_{k=1} \left(S_{km}^{it} \right)^2}},$$

where $\bar{S}_{jm}^{it}$ is the normalized vector for S_{jm}^{it}. Using the normalized values, we can configure the normalized table (Table 23.3).

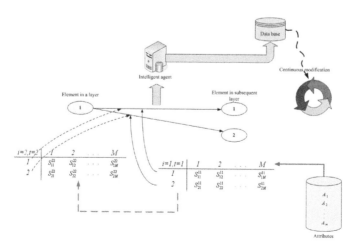

FIGURE 23.3
A process of scoring in different time periods and the continuous improvement.

TABLE 23.5

The Element-to-Element Scores

$t=1$	Element	Element	...	Element
Element	α_{11}	α_{12}	...	α_{1J}
Element	0	α_{22}	...	α_{2J}
.	0	0	.	.
.	0	0	.	.
.	0	0	.	.
Element	0	0	...	α_{IJ}

Since the weight of an attribute should be based on the decision-maker's interest and its own significance, we multiply the values in Table 23.3 by a weight coefficient β_m for the mth attribute, with $\sum_{m=1}^{M} \beta_m = 1$, which is given by decision-maker. Thus the weighted table is configured as Table 23.4.

Computing the arithmetic mean corresponding to the rows in Table 23.4, we obtain the corresponding α values. Hence, we configure an *element-to-element* matrix, which is being filled with the normalized weighted values of α (see Table 23.5). Each element in the table corresponds to a preceding layer and the one immediately after.

We use a threshold value to evaluate the quality of the obtained α scores. The proposed threshold is Z_p; that is, the pth percentile of standard normal distribution. Next, we describe the proposed threshold.

23.3.2 Element Scoring Threshold

Here, we discuss a threshold value applied to assess the scores obtained for each interaction. The aim of using the threshold is to improve the performance of the proposed e-SCM. The threshold helps identifying the weaknesses between interactions of any two layers in the proposed network. Assume that α_{ij} is the score of each interaction for each ith element in any layer. Then μ_{io} and σ_{io} are mean and standard deviation of scores between any pairs of layers for one element, respectively. Thus, we have,

$$\mu_{io} = \frac{\sum_{j=1}^{J} \alpha_{ij}}{J}, \quad i=1,\ldots,I,$$

$$\sigma_{io} = \sqrt{\frac{\sum_{j=1}^{J} (\alpha_{ij} - \mu_{io})^2}{J-1}}, \quad i=1,\ldots,I,$$

where J is the number of interactions for one element (or elements of subsequent layer).

Applying the mean and standard deviation, we define the proposed threshold value as follows:

$$Z_{ij} = \frac{\alpha_{ij} - \mu_{io}}{\sigma_{io}/\sqrt{J}}.$$

For any confidence level, we can decide upon the appropriateness of α using the proposed threshold value Z. For instance, if the confidence level is 95%, then using standard normal distribution tables we obtain $Z_{95\%} = 1.96$, and thus any value of α lower than 1.96 is considered to be inappropriate and hence omitted from the process.

As a result, the intelligent agent omits the α values lower than Z. Then, using the remaining α values, we determine the optimal route in the multi-layer network for the customer using a backward dynamic program. The details of the proposed dynamic program are given in the next section.

23.3.3 A Dynamic Program for the Optimal Route

Dynamic programming is a technique widely used for multi-stage decision processes. A given problem is subdivided into smaller subproblems, which are sequentially solved until the original problem is solved by the aggregation of the subproblem solutions. In each stage, a set of states is defined. The states would describe all possible conditions of the process in the current decision stage, which corresponds to every feasible partial solution. The set of all possible states is known as the state space. The states of a stage u can be transformed to states of a stage $u + 1$ by using a transition. A transition indicates the decisions adopted in a stage, and a sequence of transitions taken to reach a state starting from another state is known as a policy. Dynamic programming approaches can be seen as transformations of the original problem to one associated with the exploration of a multi-stage graph $G(S, T)$, where the vertices in S correspond to the state space and the arcs in T correspond to the set of transitions, leading to an optimal policy.

The optimality principle states that an optimal policy should be constituted by optimal policies from every state of the decision chain to the final state. Here, we make use of a dynamic programming approach in our proposed network to identify the optimal route for the customers or any other element of the multi-layer e-SCM. The dynamic model would be defined as follows:

Indices

n Number of layer; $n = 1, 2, 3, 4, 5$.
i Start node number corresponding to a layer; $i = 1, 2, \ldots, I$.
j End node number corresponding to subsequent layer; $j = 1, 2, \ldots, J$.

Notations

$\varphi_n(i)$ The maximum value of moving from an element i in a layer n to an element j in layer $n + 1$.
α_{ij} Numerical value of an arc between two elements.

Optimal Policy

$$\varphi_n(i) = \underset{j \text{ in layer } n+1}{Max} \{\varphi_{n+1}(j) + \alpha_{ij}\}, \quad n = 1, 2, 3, 4, \quad \forall i \text{ in layer } n.$$

Boundary condition: $\varphi_5(i) = 0, \quad \forall i$.

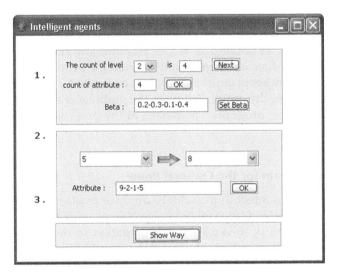

FIGURE 23.4
A user interface for the proposed intelligent agent.

Answer: $\varphi^* = \varphi_1(i), \quad \forall i.$

Using the answer φ^*, we can identify the optimal route.

All computations and configurations of the proposed e-SCM can be carried out using an intelligent agent encoded in JAVA programming language. The user interface for the designed intelligent agent is shown in Figure 23.4.

23.4 Numerical Example

Here, we illustrate the proposed model by an example. We assume that a five-layer supply chain and four attributes are considered to evaluate the interactions between elements of each pair of layers. The weights for the stated attributes are assumed to be $\beta_1 = 0.2$, $\beta_2 = 0.3$, $\beta_3 = 0.1$, and $\beta_4 = 0.4$, respectively. The interactions and corresponding scores, normalized values, and weighted normalized values are shown in Table 23.6.

Using the information in Table 23.6, the element-to-element matrix is configured. Then, using the element-to-element matrix, we compute the threshold values corresponding to the interactions.

Considering a confidence level of 95% and using standard normal distribution tables, we obtain $Z_{95\%} = 1.96$. Thus, any α value lower than 1.96 is considered to be inappropriate and hence omitted from the process. After eliminating the inappropriate scores and applying the dynamic program, we identify the optimal route in the proposed multi-layer supply chain. In Figure 23.5, the blue lines are the remaining interactions after the threshold analysis and the bold red lines show the optimal route obtained using the dynamic program. The optimal route is 2→5→10→11→16 and the corresponding optimal value is: $2.867 + 1.969 + 2.755 + 2.751 = 10.342$.

TABLE 23.6

Data Entries for the Example

	A_1	A_2	A_3	A_4	N_1	N_2	N_3	N_4	β_1*N_1	β_2*N_2	β_3*N_3	β_4*N_4
$a_{1,4}$	5	3	2	7	0.1326	0.0828	0.0527	0.1627	0.0265	0.0248	0.0052	0.0651
$a_{1,5}$	1	2	7	9	0.0265	0.0552	0.1846	0.2093	0.0053	0.0165	0.0184	0.0837
$a_{1,6}$	4	3	5	8	0.1061	0.0828	0.1318	0.1860	0.0212	0.0248	0.0131	0.0744
$a_{1,7}$	1	2	5	7	0.0265	0.0552	0.1318	0.1627	0.0053	0.0165	0.0131	0.0651
$a_{2,4}$	6	4	3	7	0.1592	0.1105	0.0791	0.1627	0.0318	0.0331	0.0079	0.0651
$a_{2,5}$	2	5	4	1	0.0530	0.1381	0.1055	0.0232	0.0106	0.0414	0.0105	0.0093
$a_{2,6}$	3	4	5	6	0.0796	0.1105	0.1318	0.1395	0.0159	0.0331	0.0131	0.0558
$a_{2,7}$	7	8	9	1	0.1857	0.2210	0.2374	0.0232	0.0371	0.0663	0.0237	0.0093
$a_{3,4}$	3	5	2	7	0.0796	0.1381	0.0527	0.1627	0.0159	0.0414	0.0052	0.0651
$a_{3,5}$	6	8	2	8	0.1592	0.2210	0.0527	0.1860	0.0318	0.0663	0.0052	0.0744
$a_{3,6}$	5	4	1	3	0.1326	0.1105	0.0263	0.0697	0.0265	0.0331	0.0026	0.0279
$a_{3,7}$	6	2	4	5	0.1592	0.0552	0.1055	0.1162	0.0318	0.0165	0.0105	0.0465
$a_{4,8}$	7	9	5	1	0.1857	0.2486	0.1318	0.0232	0.0371	0.0745	0.0131	0.0093
$a_{4,9}$	4	1	8	3	0.1061	0.0276	0.2110	0.0697	0.0212	0.0082	0.0211	0.0279
$a_{4,10}$	9	1	2	4	0.2388	0.0276	0.0527	0.0930	0.0477	0.0082	0.0052	0.0372
$a_{5,8}$	9	2	1	5	0.2388	0.0552	0.0263	0.1162	0.0477	0.0165	0.0026	0.0465
$a_{5,9}$	9	1	7	5	0.2388	0.0276	0.1846	0.1162	0.0477	0.0082	0.0184	0.0465
$a_{5,10}$	4	9	1	6	0.1061	0.2486	0.0263	0.1395	0.0212	0.0745	0.0026	0.0558
$a_{6,8}$	2	1	3	9	0.0530	0.0276	0.0791	0.2093	0.0106	0.0082	0.0079	0.0837
$a_{6,9}$	6	7	1	3	0.1592	0.1934	0.0263	0.0697	0.0318	0.0580	0.0026	0.0279
$a_{6,10}$	8	9	2	4	0.2122	0.2486	0.0527	0.0930	0.0424	0.0745	0.0052	0.0372
$a_{7,8}$	1	6	2	8	0.0265	0.1657	0.0527	0.1860	0.0053	0.0497	0.0052	0.0744
$a_{7,9}$	9	1	7	6	0.2388	0.0276	0.1846	0.1395	0.0477	0.0082	0.0184	0.0558
$a_{7,10}$	4	1	9	6	0.1061	0.0276	0.2374	0.1395	0.0212	0.0082	0.0237	0.0558
$a_{8,11}$	2	8	6	4	0.0530	0.2210	0.1582	0.0930	0.0106	0.0663	0.0158	0.0372
$a_{8,12}$	1	3	5	7	0.0265	0.0828	0.1318	0.1627	0.0053	0.0248	0.0131	0.0651
$a_{8,13}$	9	5	1	3	0.2388	0.1381	0.0263	0.0697	0.0477	0.0414	0.0026	0.0279
$a_{8,14}$	2	4	6	1	0.0530	0.1105	0.1582	0.0232	0.0106	0.0331	0.0158	0.0093
$a_{9,11}$	6	5	9	8	0.1592	0.1381	0.2374	0.1860	0.0318	0.0414	0.0237	0.0744
$a_{9,12}$	1	2	4	6	0.0265	0.0552	0.1055	0.1395	0.0053	0.0165	0.0105	0.0558
$a_{9,13}$	3	5	4	1	0.0796	0.1381	0.1055	0.0232	0.0159	0.0414	0.0105	0.0093
$a_{9,14}$	2	1	8	9	0.0530	0.0276	0.2110	0.2093	0.0106	0.0082	0.0211	0.0837
$a_{10,11}$	8	7	9	9	0.2122	0.1934	0.2374	0.2093	0.0424	0.0580	0.0237	0.0837
$a_{10,12}$	1	7	6	2	0.0265	0.1934	0.1582	0.0465	0.0053	0.0580	0.0158	0.0186
$a_{10,13}$	3	5	4	1	0.0796	0.1381	0.1055	0.0232	0.0159	0.0414	0.0105	0.0093
$a_{10,14}$	1	1	1	1	0.0265	0.0276	0.0263	0.0232	0.0053	0.0082	0.0026	0.0093
$a_{11,15}$	5	2	9	8	0.1326	0.0552	0.2374	0.1860	0.0265	0.0165	0.0237	0.0744
$a_{11,16}$	1	6	1	1	0.0265	0.1657	0.0263	0.0232	0.0053	0.0497	0.0026	0.0093
$a_{11,17}$	1	2	8	8	0.0265	0.0552	0.2110	0.1860	0.0053	0.0165	0.0211	0.0744
$a_{11,18}$	3	2	1	5	0.0796	0.0552	0.0263	0.1162	0.0159	0.0165	0.0026	0.0465
$a_{11,19}$	4	3	7	9	0.1061	0.0828	0.1846	0.2093	0.0212	0.0248	0.0184	0.0837
$a_{12,15}$	2	4	1	7	0.0530	0.1105	0.0263	0.1627	0.0106	0.0331	0.0026	0.0651
$a_{12,16}$	1	6	5	3	0.0265	0.1657	0.1318	0.0697	0.0053	0.0497	0.0131	0.0279
$a_{12,17}$	2	3	2	8	0.0530	0.0828	0.0527	0.1860	0.0106	0.0248	0.0052	0.0744

(Continued)

TABLE 23.6 (*Continued*)

Data Entries for the Example

	A_1	A_2	A_3	A_4	N_1	N_2	N_3	N_4	β_1*N_1	β_2*N_2	β_3*N_3	β_4*N_4
$a_{12,18}$	9	4	5	6	0.2388	0.1105	0.1318	0.1395	0.0477	0.0331	0.0131	0.0558
$a_{12,19}$	1	3	2	7	0.0265	0.0828	0.0527	0.1627	0.0053	0.0248	0.0052	0.0651
$a_{13,15}$	2	5	6	1	0.0530	0.1381	0.1582	0.0232	0.0106	0.0414	0.0158	0.0093
$a_{13,16}$	3	4	2	8	0.0796	0.1105	0.0527	0.1860	0.0159	0.0331	0.0052	0.0744
$a_{13,17}$	9	5	5	5	0.2388	0.1381	0.1318	0.1162	0.0477	0.0414	0.0131	0.0465
$a_{13,18}$	3	3	3	4	0.0796	0.0828	0.0791	0.0930	0.0159	0.0248	0.0079	0.0372
$a_{13,19}$	2	5	6	1	0.0530	0.1381	0.1582	0.0232	0.0106	0.0414	0.0158	0.0093
$a_{14,15}$	8	9	2	3	0.2122	0.2486	0.0527	0.0697	0.0424	0.0745	0.0052	0.0279
$a_{14,16}$	7	4	5	1	0.1857	0.1105	0.1318	0.0232	0.0371	0.0331	0.0131	0.0093
$a_{14,17}$	1	3	5	5	0.0265	0.0828	0.1318	0.1162	0.0053	0.0248	0.0131	0.0465
$a_{14,18}$	2	5	3	3	0.0530	0.1381	0.0791	0.0697	0.0106	0.0414	0.0079	0.0279
$a_{14,19}$	8	7	7	9	0.2122	0.1934	0.1846	0.2093	0.0424	0.0580	0.0184	0.0837

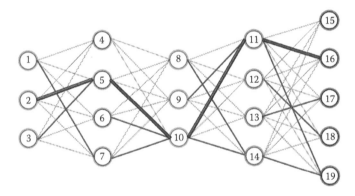

FIGURE 23.5
The obtained optimal route.

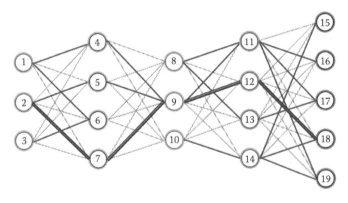

FIGURE 23.6
The configuration of optimal route in the second period.

We performed the same process for a different period. After the corresponding computations, another route was obtained as an optimal route. The difference between the optimal routes corresponding to the two periods is due to variety in requirements and interests of the elements of the layers in different time periods. The configuration of optimal route in the new period is shown in Figure 23.6.

The optimal route is 2→7→9→12→18 with the corresponding optimal value of 2.679 + 1.998 + 2.662 + 3.695 = 11.034.

23.5 Discussions

We proposed an information system to integrate the elements of a multi-layer, agent-based e-SCM. We showed how to design the information systems for the layers and trace the information flows. In each layer, the intelligent agents collect information and produce a report that includes the flow of multi-attribute decision-making. A new method was presented for the information flow of interactions using a dynamic programming approach. The advantages of the system are real-time decision-making and optimal route selection.

References

Cagliano R, Caniato F, Spina G. E-business strategy: How companies are shaping their supply chain through the internet. *International Journal of Operations & Production Management* 2003;23(10):1142–1162.

Christopher M. *Logistics & Supply Chain Management: Strategies for Reducing Cost and Improving Service.* London: Financial Times Pitman Publishing, 1998.

Cooper MC, Lambert DM, Pagh JD. Supply chain management: More than a new name for logistics. *The International Journal of Logistics Management* 1997;8(1):1–13.

Cousins PD. The alignment of appropriate firm and supply strategies for competitive advantages. *International Journal of Operations & Production Management* 2005;25(5):403–428.

Croom SR. The impact of e-business on supply chain management. *International Journal of Operations & Production Management* 2005;25(1):55–73.

Danese P, Romano P, Vinelli A. Sequences of improvements in supply networks: Case studies from pharmaceutical industry. *International Journal of Operations & Production Management* 2006;26(11): 1199–1222.

Ellram LM. Supply chain management: The industrial organization perspective. *International Journal of Physical Distribution & Logistics Management* 1991;21(1):13–22.

Fung RYK, Chen T. A multiagent supply chain planning and coordination architecture. *International Journal of Advanced Manufacturing Technology* 2005;25:811–819.

Houlihan J. Supply chain management. In: *Proceedings of the 19th International Technical Conference,* BPICS, 1984:101–110.

Mills J, Schmitz J, Frizelle G. A strategic review of supply networks. *International Journal of Operations & Production Management* 2004;24(10):1012–1036.

Oliver RK, Webber MD. Supply chain management: Logistics catches up with strategy. In: Christopher M. (ed.), *Logistics: The Strategic Issues.* London: Chapman & Hall, 1992:63–75.

Romano P, Vinelli A. Quality management in a supply chain perspective: Strategic and operative choices in a textile-apparel network. *International Journal of Operations & Production Management* 2001;21(4):446–460.

24

Electronic Supply Chain Management System: Electronic Market, Customer Satisfaction, and Logistic Model

SUMMARY This chapter considers an electronic supply chain system to be capable in an electronic market. A supply chain composed of supplier, plant, and customer is proposed. The aim is to optimize a real-time, web-based order-delivery system in which customer satisfaction is noted. As such, a comprehensive web-based order-delivery system in an electronic market is proposed and optimized applying mathematical programming.

24.1 Introduction

Being a complex network of suppliers, factories, warehouses, distribution centers, and retailers, the success of any supply chain management system (SCMS) depends on how well these system components are handled and integrated (Zhao et al., 2008). Recently, information has become a significant item in determining the productivity of a complex enterprise. The enterprise's ability to process information and make rapid but accurate decisions leads to growth (Halldorsson et al., 2003). In such a scenario it is necessary to forecast and estimate the demand, supply raw materials to the point of sale locations, and reorganize the business structure if necessary (Simchi-Levi et al., 2007). To realize these goals a system must seamlessly integrate both information and material flow. Such a system can provide access to information, aid decision-making and execution (Halldorsson et al., 2007).

Supply chain management (SCM) is an integration of materials, information, and finances in a link among supplier, manufacturer, wholesaler, retailer and consumer (Cooper et al., 1997). Supply chain management consists of coordinating and integrating these flows both within and among companies. It is considered that the final goal of any effective SCMS is to reduce inventory (with the assumption that products are available when needed). As a solution for successful supply chain management, complicated software systems with web interfaces are competing with web-based application service providers (ASP) who facilitate to provide part or all of the SCM service for companies who rent their service (Ketchen and Hult, 2006).

Supply chain management flows can be divided into three main flows:

- The product flow
- The information flow
- The finances flow

The product flow includes the movement of goods from a supplier to a customer, at the same time, any customer returns or service requirements. The information flow consists of transmitting orders and updating the status of delivery. The financial flow involves credit terms, payment schedules, and consignment and title ownership arrangements. There are two main types of SCM software: planning applications and execution applications. Planning applications use advanced algorithms to identify the best way to fill an order. Execution applications track the physical status of goods, the handling of materials, and financial information involving all parties. Some SCM applications are based on open data models that support the sharing of data both inside and outside the enterprise (this is called the extended enterprise, and includes key suppliers, manufacturers, and end customers of a specific firm). This shared data may reside in diverse database systems, or data warehouses, at the websites of the enterprises. By sharing this data "upstream" (with a company's suppliers) and "downstream" (with a company's customer), SCM applications have the potential to improve the time-to-market of products, reduce costs, and allow all parties in the supply chain to better manage current resources and plan for future needs (Larson and Halldorsson, 2004).

Increasing numbers of companies are turning to websites and web-based applications as part of the SCM solution. A number of major websites offer e-procurement marketplaces where manufacturers can trade and even make auction bids with suppliers (Haag et al., 2006).

E-supply chain management (e-SCM) refers to the flow of physical goods and associated information from the source to the consumer (Tanriverdi, 2006). Key e-supply chain activities include purchasing, materials management, distribution, customer service, and inventory forecasting. Effectively managing these processes is critical to the success of any online operation (Chen et al., 2007).

In commerce, a *retailer* buys goods or products in large quantities from manufacturers or importers, either directly or through a wholesaler, and then sells individual items or small quantities to the general public or end user customers, usually in a *shop*, also called *store*. Retailers are at the end of the supply chain (Lavassani et al., 2008b).

Many shops are part of a *chain*: a number of similar shops with the same name selling the same products in different locations. The shops may be owned by one company, or there may be a franchising company that has franchising agreements with the shop owners (see also: restaurant chains).

Traditionally, marketing, distribution, planning, manufacturing, and the purchasing organizations along the supply chain operated independently. These organizations have their own aims and these are often conflicting. Clearly, there is a need for a mechanism through which different functions can be integrated together. Supply chain management is a strategy where such integration can be obtained (Lavassani et al., 2008a).

Based on our study three important components are deemed essential for a web-based generic SCMS—the supplier, the plants, and customers. As retail and wholesale organizations have stores scattered across many places the Web is utilized as the media for information exchange (Li and Lin, 2006).

24.2 Problem Definition

The World Wide Web has changed the traditional landscape of the business environment from that of being a market *place* to one that is more of a market *space*. This market space is an information- and communication-based electronic exchange environment occupied by sophisticated computer and telecommunication technologies and digitized offerings. The impact of this digitization is quite evident in the following:

1. The content of transaction is different—information about a product often replaces the product itself.
2. The context of transaction is different—an electronic screen replaces the face-to face transaction.
3. The enabling infrastructure of transactions is different—computers ad communications infrastructure may replace typical physical resources especially if the offering lends itself to a digitized format.

The Internet, or World Wide Web precisely, allows the supplying enterprise from the smallest enterprise to largest corporation to establish its global presence. Hence, a supplying enterprise now has the opportunity to reach geographically dispersed markets that would otherwise be cost prohibitive to access. Purchasing enterprises also now have the opportunity to select the best suppliers, by utilizing suppliers' bids on the Web, and thus avert the time consuming and costly outside sources (middlemen like professional import brokers). The possibilities of automating supplier selection procedures as an e-commerce application offer several advantages such as automatically searching supplier bids (product data) on web servers, and filtering data to find the best supplier. By automation, an enterprise can seek potential suppliers from all over the world, faster and cheaper. Moreover, an enterprise can seek suppliers often, for an ongoing project or for future projects. Thus, automation of supplier selection procedures enables the enterprise to maintain its key attributes for survival, namely agility and dynamic collaboration (virtual enterprising). For illustration, an electronic market is represented in Figure 24.1.

As seen in Figure 24.1, customers exist that reflect their preferences to suppliers. Those preferences would be stored in the suppliers' database. Then, the preferences are transferred to the Web. At the same time, the suppliers would communicate with the plant database and try to provide the products based on the aforementioned preferences. After that, the plants introduce the products to the Web.

In this real-time system, a customer orders online, via a web-based mechanism, one type of a product to a supplier. This order is based on the list of products viewed in the supplier website. In this website the specifications of each product such as price, plants produce the same products, brands, logistical considerations, etc. After that the supplier receives the order and analyzes it through a logistical intelligent agent. The logistical intelligent agent performs the following investigations:

1. Distance analysis
2. Routes selection
3. Amount of order
4. Vehicle capacity analysis
5. Vehicle delivery time
6. Vehicle transferring cost
7. Vehicle allocation
8. Final price of each product

The first factor that the logistical intelligent agent considers is the distance between the customer (who gives an order) and different plants that provide the ordered products.

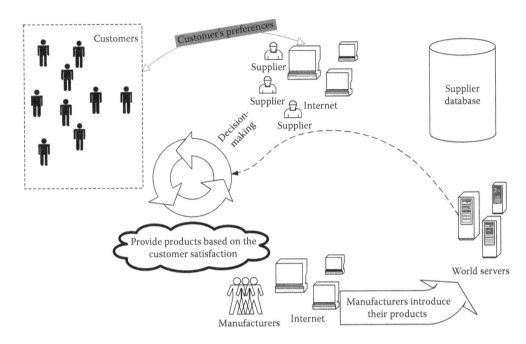

FIGURE 24.1
A proposed electronic market.

Then the agent presents some candidate routes to supply the order. After that the amount of the order is investigated considering the production capacity of the plants. Also the vehicles are checked due to their capacity. Here, the delivery time is significant for both supplier and customer. The supplier should apply a procedure which satisfies the customer due date. This way, the procedure that shortens the delivery time is more optimal, based on customer satisfaction. Because each vehicle is associated with its corresponding cost, therefore allocating a vehicle is depending on the transferring cost, too. The final price of each product is another factor that affects the customer satisfaction. Customers are looking for lower prices accompanying with the shorter delivery time.

Final price is composed of lean production cost and transferring cost. The lean production cost is the material cost and the cost of set of all operating costs that should be spent to produce a product. The final price is set considering four elements: competitive market, maximum profit of plants, minimum price for satisfying customers, and the lower level for finished product price. The supplier should offer a tradeoff between these elements and provide a unique price. At the end, the report is given to the customer and customer would decide whether to process the order or not. The proposed logistical intelligent agent is shown in Figure 24.2.

24.3 Web-Based Model

As stated in previous sections, the logistical agent is the core of decision-making process in our e-SCM. We proposed a mathematical model to formulate the procedures of decision-making in the logistical intelligent agent. The notations, parameters and decision variable are as follows:

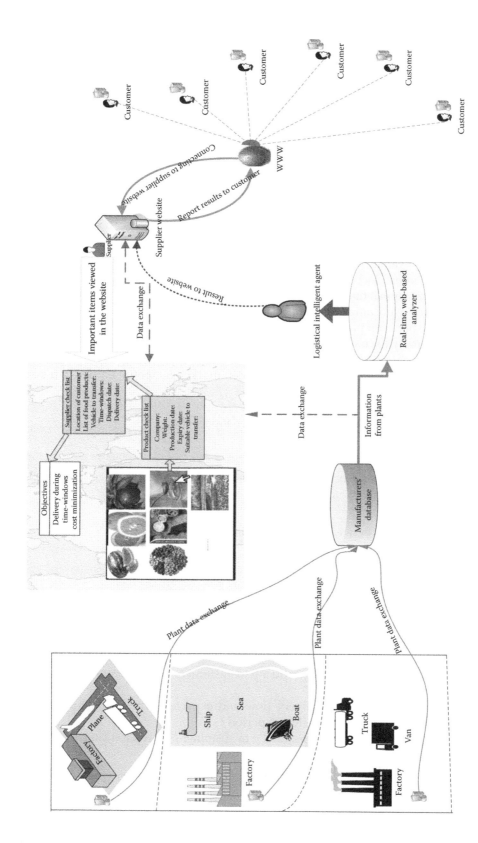

FIGURE 24.2
The proposed logistical intelligent agent.

Notations

p set of product
i set of plants
j set of customer
v set of vehicles
t unit of period

Parameters

D_{tjp} demand of customer j for product p of period t
ET_{tjp} the expected time of customer j for product p of period t
PC_{tip} production cost of product p in plant i of period t
C_{ip} the capacity of product p in plant i
V_{tiv} the number of vehicles v in plant i at the end of period t
VC_{vp} the capacity of product p in vehicle v
TC_v the cost of transportation per unit of distance by vehicle v
TT_v the time of transportation per unit of distance by vehicle v
DI_{ij} the distance between plant i and customer j
DP the fixed penalty for tardiness of products delivery to customer
SC the fixed cost of stored product in each period
CO the fixed coefficient for difference of between market competitive price and evaluated price
M the large number
MP_{tp} the price of product p in competitive market at period t

Decision Variables

$$Z_{tip} : \begin{cases} 1 & \text{if plant } i \text{ produces product } p \text{ at time } t \\ 0 & \text{O.W} \end{cases}$$

$$X_{tijpv} : \begin{cases} 1 & \begin{array}{l} \text{there is a path between plant } i \text{ and customer } j \\ \text{for product } p \text{ by vehicle } v \text{ at time } t \end{array} \\ 0 & \text{O.W} \end{cases}$$

g_{tip} The amount of produced product p in plant i at period t

S_{tip} The amount of stored product p in plant i at the end of period t
N_{tijpv} The Number of required vehicles v for transferring product p between plant i and customer j at period t
FP_{tijpv} The final price of delivered product p to customers j from plant i by vehicle v at period t
DDE_{tijpv} The difference between delivery time and expected time for customer j from plant i for product p by vehicle v at period t
QP_{tijp} The quantity of transferred product p to customer j from plant i at period t

The mathematical model is as follows:

$$Max \quad F = f_1 + f_2 + f_3 + f_4 + f_5$$

$$f_1 = \sum_{t \in T} \sum_{i \in I} \sum_{j} \sum_{p} \sum_{v} QP_{tijp} \cdot FP_{tijpv}, \tag{24.1}$$

$$f_2 = \sum_{t} \sum_{i} \sum_{j} \sum_{p} \sum_{v} (MP_{tp} - FP_{tijpv}) \cdot CP, \tag{24.2}$$

$$f_3 = -\sum_{t} \sum_{i} \sum_{j} \sum_{p} \sum_{v} N_{tijpv} \cdot d_{ij} \cdot TC_v, \tag{24.3}$$

$$f_4 = -\sum_{t} \sum_{i} \sum_{p} S_{tip} \cdot SC, \tag{24.4}$$

$$f_5 = -\sum_{t} \sum_{i} \sum_{j} \sum_{p} \sum_{v} DDE_{tijpv} \cdot DP, \tag{24.5}$$

s.t.

$$g_{tip} \le M \cdot Z_{tip}, \quad \forall t \in T, \ \forall i \in I, \ \forall p \in P, \tag{24.6}$$

$$g_{tip} \ge Z_{tip}, \quad \forall t \in T, \ \forall i \in I, \ \forall p \in P, \tag{24.7}$$

$$\sum_{i} QP_{tijp} = D_{tjp}, \quad \forall t, \ \forall j, \ \forall p, \tag{24.8}$$

$$\sum_{j} QP_{tijp} \le g_{tip} + s_{t-1ip}, \quad \forall t, \ \forall i, \ \forall p, \tag{24.9}$$

$$s_{t-1ip} + g_{tip} - \sum_{j} QP_{tijp} = s_{tip}, \quad \forall t, \ \forall i, \ \forall p, \tag{24.10}$$

$$g_{tip} \le C_{ip}, \quad \forall t, \ \forall i, \ \forall p, \tag{24.11}$$

$$QP_{tijp} \le M \cdot \sum_{v} X_{tijpv}, \quad \forall t, \ \forall i, \ \forall p, \ \forall j, \tag{24.12}$$

$$\sum_{v} x_{tijpv} \le 1, \quad \forall t, \ \forall i, \ \forall p, \ \forall j, \tag{24.13}$$

$$\left(\frac{QP_{tijp}}{VC_{vp}} \right) - M \cdot (1 - x_{tijpv}) \le N_{tijpv}, \quad \forall t, \ \forall i, \ \forall p, \ \forall j, \ \forall v, \tag{24.14}$$

$$N_{tijpv} \leq M \cdot X_{tijpv}, \quad \forall t, \forall i, \forall p, \forall j, \forall v \tag{24.15}$$

$$\sum_{j}\sum_{p} N_{tijpv} \leq V_{t-1iv}, \quad \forall t, \forall i, \forall v, \tag{24.16}$$

$$V_{t-1iv} - \sum_{j}\sum_{p} N_{tijpv} = V_{tiv}, \quad \forall t, \forall i, \forall v, \tag{24.17}$$

$$X_{tijpv}((d_{ij} \cdot TT_v) - ET_{tjp}) = DDE_{tijpv}, \quad \forall t, \forall i, \forall p, \forall j, \forall v, \tag{24.18}$$

$$FP_{tijpv} \leq X_{tijpv}((1.2 \cdot PC_{tip}) + (d_{ij} \cdot TC_v)) \cdot M, \quad \forall t, \forall i, \forall p, \forall j, \forall v, \tag{24.19}$$

$$FP_{tijpv} \geq X_{ijpv}((1.2 \cdot PC_{tip}) + (d_{ij} \cdot TC_v)), \quad \forall t, \forall i, \forall p, \forall j, \forall v, \tag{24.20}$$

$$FP_{tijpv} \leq MP_{tp}, \quad \forall t, \forall i, \forall p, \forall j, \forall v, \tag{24.21}$$

$$g_{tip}, S_{tip}, FP_{tijpv}, QP_{tip} \geq 0, \quad \forall t, \forall i, \forall p, \forall j, \forall v, \tag{24.22}$$

$$N_{tijpv} \in \text{Integer}, \quad \forall t, \forall i, \forall p, \forall j, \forall v, \tag{24.23}$$

$$DDE_{tijpv} \text{ is free}, \quad \forall t, \forall i, \forall p, \forall j, \forall v, \tag{24.24}$$

$$Z_{tip}, X_{tijpv} \in \{0,1\}, \quad \forall t, \forall i, \forall p, \forall j, \forall v. \tag{24.25}$$

Equations 24.1 and 24.2 are the objective functions that consider tradeoff between maximization of price of each product on manufacturers' viewpoint and minimization of price on customers' viewpoint. Equation 24.3 is the objective functions that minimize total cost of forward distance. Equation 24.4 is the objective function that minimizes total cost of storage. Equation 24.5 is the objective function that penalizes total tardiness of each product from expected delivery time. The constraints 24.6 and 24.7 show that each plant can produce an amount of product just after it is selected. The constraint 24.8 guarantees that all customer demands are met for all products in all periods. The constraint 24.9 is the flow conservation at depots. The constraint 24.10 shows the amount of stored products at the end of period. The constraint 24.11 represents capacity restriction. The constraints 24.12 and 24.13 ensure that delivery is accomplished by only one vehicle. The number of required vehicles for transferring products between both of plant and customer has been shown in constraints 24.14 and 24.15. The constraint 24.16 requires that the number of traveled vehicles from depot is lower than or equal to its stationed vehicles. The constraint 24.17 represents the number of remained vehicles at the end of period. Tardiness of delivery time from expected time is shown in constraint 24.18. The constraints 24.19, 24.20 and 24.21 represent lower and upper bounds of price of each product. Upper bound is a parameter that is determined by competitive markets. Lower bound is acquired by both functions of production cost and cost of distance, which are considered as coefficients of production cost. This coefficient is set to be 1.2 as lean production cost. The constraints 24.22 through 24.25 show the signs and kinds of the decision variables.

24.4 Numerical Illustration

The accuracy of the proposed model is validated by a comprehensive numerical example. Our model is tested in small scale of data. The number of plants, customers, products and periods are set to be three. There are two kinds of different vehicles in the logistic network. The amount of customers' demands for each product in various periods is shown in Table 24.1. Table 24.2 represents distance between plants and customers.

24.4.1 Input Data

The restriction of capacity for each product on both vehicles and plants viewpoint are shown in Tables 24.3 and 24.4, respectively. The costs, which include transferring costs and production costs, are recognized by Tables 24.5 and 24.6.

Relevant data for time consideration whether expected time for delivering product on customers' opinions and transferring time per unit of distance are given in Tables 24.7 and 24.8, respectively.

TABLE 24.1

Customers' Demands for Each Product

Order	Product 1	Product 2	Product 3
First Period			
Customer 1	10	5	10
Customer 2	12	10	15
Customer 3	10	12	11
Second Period			
Customer 1	15	10	10
Customer 2	14	12	15
Customer 3	10	10	14
Third Period			
Customer 1	8	14	15
Customer 2	15	10	14
Customer 3	14	10	15

TABLE 24.2

Distance between Plants and Customers

Distance	Customer 1	Customer 2	Customer 3
Plant 1	10	16	11
Plant 2	8	12	13
Plant 3	13	8	12

The price of each product in market environment with respect to competitive levels of market is shown in Table 24.9. Table 24.10 shows the number of stationed vehicles at the beginning of period.

So far, we present the required data for processing the results. To facilitate the computations the mathematical model is encoded in LINGO. The suitable path to deliver product to customer from plant using fitting vehicles is reported in Table 24.11. As well as, the number of responsible vehicles for carrying product and the corresponding amount of product are shown in it. Meanwhile, the final price of product and the tardiness/earliness of delivered product are represented in Table 24.11. The negative sign in *DDE* column implies the delivery with earliness while positive sign shows the tardiness for delivered product. Also the (−) sign certifies the on-time delivery. The unit of time is considered to be minute. Table

TABLE 24.3

The Vehicle Capacity

Capacity of Vehicle	Product 1	Product 2	Product 3
Vehicle 1	20	25	15
Vehicle 2	10	8	5

TABLE 24.4

The Plant Capacity

Plant Capacity	Product 1	Product 2	Product 3
Plant 1	100	85	90
Plant 2	90	80	70
Plant 3	80	75	70

TABLE 24.5

The Transferring Cost

Transferring Cost Per Unit of Distance	Vehicle 1	Vehicle 2
	10	5

TABLE 24.6

The Production Cost

Production Cost	Product 1	Product 2	Product 3
First Period			
Plant 1	20	30	20
Plant 2	40	30	50
Plant 3	40	20	30
Second Period			
Plant 1	40	30	20
Plant 2	40	20	30
Plant 3	30	40	40
Third Period			
Plant 1	20	40	40
Plant 2	50	60	30
Plant 3	60	20	40

TABLE 24.7

The Expected Time for Delivery

Expected Time	Product 1	Product 2	Product 3
First Period			
Customer 1	200	300	200
Customer 2	250	150	200
Customer 3	200	150	200
Second Period			
Customer 1	200	100	150
Customer 2	200	250	300
Customer 3	200	200	250
Third Period			
Customer 1	100	150	200
Customer 2	200	250	200
Customer 3	150	200	200

TABLE 24.8

The Transferring Time

Transferring Time Per Unit of Distance	Vehicle 1	Vehicle 2
	10	20

TABLE 24.9

The Market Price

Market Price	Product 1	Product 2	Product 3
Period 1	140	120	135
Period 2	130	135	140
Period 3	160	140	130

TABLE 24.10

The Number of Stationed Vehicles

Number of Vehicles	Vehicle 1	Vehicle 2
Plant 1	50	50
Plant 2	50	50
Plant 3	50	50

24.12 shows the amount of produced product in each plant. The remaining vehicles at the end of each period are represented in Table 24.13.

Also, the storage is set to be zero for all plants in all periods. The value of objective function is 2173690 unit of price.

24.5 Discussions

Considering the fact of huge expansion of supply network, this chapter proposed an electronic supply chain to overcome the confusion of order-delivery process. Here, we

TABLE 24.11

The Path Output

X	Plant	Customer	Product	Vehicle	Period	N	QP	$FP	DDE
1	1	3	2	2	1	2	12	120	70
1	1	3	3	1	1	1	11	135	−90
1	2	1	1	1	1	1	10	128	−120
1	2	1	2	1	1	1	5	116	−220
1	2	1	3	2	1	2	10	100	−40
1	3	2	1	2	1	2	12	100	−90
1	3	2	2	1	1	1	10	111	−70
1	3	2	3	1	1	1	15	135	−120
1	3	3	1	2	1	1	10	140	40
1	1	1	1	2	2	2	15	130	–
1	1	3	1	2	2	1	10	103	20
1	1	3	2	2	2	2	10	91	20
1	1	3	3	1	2	1	14	140	−140
1	2	1	2	1	2	1	10	111	−20
1	2	1	3	1	2	1	10	126	−70
1	3	2	1	1	2	1	14	130	−120
1	3	2	2	1	2	1	12	135	−170
1	3	2	3	1	2	1	15	140	−220
1	1	1	2	2	3	2	14	140	50
1	1	3	1	1	3	1	14	160	−40
1	1	3	2	2	3	2	10	103	20
1	1	3	3	2	3	3	15	130	20
1	2	1	1	1	3	1	8	140	−20
1	2	1	3	1	3	1	15	130	−120
1	3	2	1	1	3	1	15	160	−120
1	3	2	2	1	3	1	10	104	−170
1	3	2	3	1	3	1	14	130	−120

considered an electronic market in which three components of plant, supplier, and customer exist. A supplier would present its products specifications in the corresponding website. The customer views the website and based on his requirements order a list of products. The supplier analyzes the orders via a logistical intelligent agent. We proposed a mathematical model to perform the real-time, web-based investigations. The core contributions of the proposed approach are highlighted below:

- Collecting customers' preferences in virtual environment
- Designing a real-time supply chain process using information technology
- Proposing a comprehensive logistical intelligent agent for suppliers in an electronic market
- Considering prices and delivery time as customer satisfaction measures
- Presenting an integrated customer relationship management system

TABLE 24.12

The Amount of Produced Products

G	Product 1	Product 2	Product 3
First Period			
Plant 1	0	12	11
Plant 2	10	5	10
Plant 3	22	10	15
Second Period			
Plant 1	25	10	14
Plant 2	0	10	10
Plant 3	14	12	15
Third Period			
Plant 1	14	24	15
Plant 2	8	0	15
Plant 3	15	10	14

TABLE 24.13

The Number of Remained Vehicles at the End of Periods

Number of Left Vehicles	Vehicle 1	Vehicle 2
At the End of Period 1		
Plant 1	49	48
Plant 2	48	48
Plant 3	48	47
At the End of Period 2		
Plant 1	48	43
Plant 2	46	48
Plant 3	45	47
At the End of Period 3		
Plant 1	47	36
Plant 2	44	48
Plant 3	42	47

The results of the analysis are reported to the customer and the customer would decide to accept or refuse the order to that supplier.

References

Chen M, Zhang D, Zhou L. Empowering collaborative commerce with Web services enabled business process management systems. *Decision Support Systems* 2007;43(2):530–546.

Cooper MC, Lambert DM, Pagh J. Supply chain management: More than a new name for logistics. *The International Journal of Logistics Management* 1997;8(1):1–14.

Haag S, Cummings M, McCubbrey D, Pinsonneault A, Donovan R. *Management Information Systems for the Information Age*, 3rd Canadian edition. McGraw Hill Ryerson, Canada, 2006.

Halldorsson A, Kotzab H, Mikkola JH, Skjoett-Larsen T. Complementary theories to supply chain management. *Supply Chain Management: An International Journal* 2007;12(4):284–296.

Halldorsson A, Herbert K, Skjøtt-Larsen T. Interorganizational theories behind supply chain management – discussion and applications. In: *Strategy and Organization in Supply Chains*, Seuring S, Müller M, Goldbach M, Schneidewind U (eds), Physica Verlag, Heidelberg, Germany, 2003:31–63.

Ketchen DJ Jr., Hult TM. Bridging organization theory and supply chain management: The case of best value supply chains. *Journal of Operations Management* 2006;25(2):573–580.

Larson PD, Halldorsson A. Logistics versus supply chain management: An international survey. *International Journal of Logistics: Research & Application* 2004;7(1):17–31.

Lavassani MK, Movahedi B, Kumar V. Transition to B2B e-marketplace enabled supply chain: Readiness assessment and success factors. *Information Resources Management (Conf-IRM)*, Niagara, Canada, 2008a.

Lavassani MK, Movahedi B, Kumar V. Historical developments in theories of supply chain management: The case of B2B e-marketplaces. *Administrative Science Association of Canada (ASAC)*, Halifax, Canada, 2008b.

Li S, Lin B. Accessing information sharing and information quality in supply chain management. *Decision Support Systems* 2006;42(3):1641–1656.

Simchi-Levi D, Kaminsky P, Simchi-Levi E. *Designing and Managing the Supply Chain*, 3rd edition. McGraw Hill, New York, USA, 2007.

Tanriverdi H. Performance effects of information technology synergies in multibusiness firms. *MIS Quarterly* 2006;30(1):57–77.

Zhao X, Huo B, Flynn BB, Yeung JHY. The impact of power and relationship commitment on the integration between manufacturers and customers in a supply chain. *Journal of Operations Management* 2008;26:368–388.

25

Electronic Supply Chain System:
Fuzzy Logistic Model

SUMMARY   In this chapter, we consider a supply chain composed of supplier, plant, and customer. The aim is to optimize a real-time, web-based fuzzy order-delivery system for which customer satisfaction is emphasized. As such, a comprehensive, web-based order-delivery system in an electronic market is proposed and optimized applying fuzzy mathematical programming.

25.1 Introduction

Here, we present a primal framework of a fuzzy variable linear programming (FVLP) problem with fuzzy cost coefficients, fuzzy coefficient matrix and fuzzy right hand side of the general constraints. Consider the FVLP problem,

$$
\begin{aligned}
\max \quad & z = \tilde{c}x \\
\text{s.t.} \quad & \tilde{A}x \underset{R}{\leq} \tilde{b} \\
& x \geq 0,
\end{aligned}
$$

where $\tilde{c}$, $\tilde{A}$ and $\tilde{b}$ have fuzzy trapezoidal components. A fuzzy trapezoidal number is shown to be $\tilde{a} = (a^L, a^U, \alpha, \beta)$, with its configuration as given in Figure 25.1.

25.1.1 Ranking Function

Some early methods for solving fuzzy linear programming problems have been reported in studies by Fang and Hu (1999), Lai and Hwang (1992), Maleki et al. (2000), Shoacheng (1994), and Tanaka and Ichihashi (1984). For comprehensive treatments using ranking functions, in the context of fuzzy linear programming, see Mahdavi-Amiri and Nasseri (2006, 2007). Here, we also use a ranking function, R, for defuzzifications as proposed by Yager (1981):

$$
R(\tilde{a}) = \frac{a^L + a^U}{2} + \frac{(\beta - \alpha)}{4}.
$$

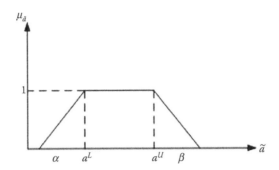

FIGURE 25.1
A trapezoidal fuzzy number.

25.2 Problem Definition

Here, the delivery time is significant for both supplier and customer. The supplier should apply a procedure which satisfies the customer's due date. This way, the procedure that shortens the delivery time should be preferred based on customer satisfaction. Because each vehicle has its own cost, allocating a vehicle is thus dependent on the transferring cost, as well. The final price of each product is another factor affecting customer satisfaction. Customers are looking for lower prices accompanied by shorter delivery time. The demand is considered to be fuzzy. Also, the cost of transportation per unit of distance by a vehicle and the time of transportation per unit of distance by a vehicle are considered to be fuzzy.

25.3 Fuzzy Logistic Model

The fuzzy linear mathematical program is given below. In our proposed model, demand is a fuzzy parameter, and costs are assumed to be fuzzy as well. As stated in the previous section, the logistic agent constitutes the core of the decision-making process in our e-SCM. We propose a mathematical model formulating the decision-making process in the logistic intelligent agent. The notations, parameters and decision variables are

Notations

P set of products p
I set of plants i
J set of customers j
V set of vehicles v
T set of units of period t

Parameters

$\tilde{D}_{tjp}$ fuzzy demand of customer j for product p at period t
ET_{tjp} Expected time of customer j for product p at period t

PC_{tip} Production cost of product p in plant i at period t
C_{ip} Capacity of product p in plant i
V_{tiv} Number of vehicles v in plant i at the end of period t
VC_{vp} Capacity of product p for vehicle v
$\tilde{TC}_v$ Fuzzy transportation cost per unit of distance for vehicle v
$\tilde{TT}_v$ Fuzzy transportation time per unit of distance for vehicle v
DI_{ij} Distance between plant i and customer j
DP Fixed penalty for tardiness of product delivery to customer
SC Fixed cost for product storage in each period
CO Fixed coefficient for the difference between market competitive price and evaluated price
M A large number
MP_{tp} Price of product p in competitive market at period t
CP Fixed cost

Decision Variables

$$Z_{tip} = \begin{cases} 1 & \text{if plant } i \text{ produces product } p \text{ at time } t \\ 0 & \text{otherwise} \end{cases}$$

$$X_{tijpv} = \begin{cases} 1 & \text{if there is a path between plant } i \text{ and customer } j \text{ for} \\ & \text{product } p \text{ for vehicle } v \text{ at time } t \\ 0 & \text{Otherwise} \end{cases}$$

g_{tip} Amount of produced product p in plant i at period t
S_{tip} Amount of stored product p in plant i at the end of period t
N_{tijpv} Number of required vehicles v for transferring product p between plant i and customer j at period t
FP_{tijpv} Final price of delivered product p to customers j from plant i by vehicle v at period t
DDE_{tijpv} Difference between delivery time and expected time for customer j from plant i for product p by vehicle v at period t
QP_{tijp} Quantity of transferred product p to customer j from plant i at period t
QV_{tjiv} Number of vehicles v delivering to plant i by customer j at time t
D_{tiv} Demand of vehicle v for plant i at time t
RV_{tjv} Number of existing vehicles v for customer j at the end of time t

25.3.1 Mathematical Model

$$Max \quad F = f_1 + f_2 + f_3 + f_4 + f_5,$$

where,

$$f_1 = \sum_t \sum_i \sum_j \sum_p \sum_v QP_{tijp} \cdot FP_{tijpv}, \tag{25.1}$$

$$f_2 = \sum_t \sum_i \sum_j \sum_p \sum_v (MP_{tp} - FP_{tijpv}) \cdot CP, \tag{25.2}$$

$$f_3 = -\sum_t \sum_i \sum_j \sum_p \sum_v N_{tijpv} \cdot d_{ij} \cdot \tilde{T}C_v, \tag{25.3}$$

$$f_4 = -\sum_t \sum_i \sum_p S_{tip} \cdot SC, \tag{25.4}$$

$$f_5 = -\sum_t \sum_i \sum_j \sum_p \sum_v DDE_{tijpv} \cdot DP, \tag{25.5}$$

$$f_6 = -\sum_j \sum_i \sum_v \sum_t QV_{tjiv} \cdot d_{ij} \cdot \tilde{T}C_v, \tag{25.6}$$

s.t.

$$g_{tip} \le M \cdot Z_{tip}, \quad \forall t \in T, \ \forall i \in I, \ \forall p \in P, \tag{25.7}$$

$$g_{tip} \ge Z_{tip}, \quad \forall t \in T, \ \forall i \in I, \ \forall p \in P, \tag{25.8}$$

$$\sum_i QP_{tijp} = \tilde{D}_{tjp}, \quad \forall t, \ \forall j, \ \forall p, \tag{25.9}$$

$$\sum_j QP_{tijp} \le g_{tip} + s_{(t-1)ip}, \quad \forall t, \ \forall i, \ \forall p, \tag{25.10}$$

$$s_{(t-1)ip} + g_{tip} - \sum_j QP_{tijp} = s_{tip}, \quad \forall t, \ \forall i, \ \forall p, \tag{25.11}$$

$$g_{tip} \le C_{ip}, \quad \forall t, \ \forall i, \ \forall p, \tag{25.12}$$

$$QP_{tijp} \le M \cdot \sum_v X_{tijpv}, \quad \forall t, \ \forall i, \ \forall p, \ \forall j, \tag{25.13}$$

$$\sum_v x_{tijpv} \le 1, \quad \forall t, \ \forall i, \ \forall p, \ \forall j, \tag{25.14}$$

$$\left(\frac{QP_{tijp}}{VC_{vp}} \right) - M \cdot (1 - x_{tijpv}) \le N_{tijpv}, \quad \forall t, \ \forall i, \ \forall p, \ \forall j, \ \forall v, \tag{25.15}$$

$$N_{tijpv} \le M \cdot X_{tijpv}, \quad \forall t, \ \forall i, \ \forall p, \ \forall j, \ \forall v, \tag{25.16}$$

$$D_{tiv} = \sum_j \sum_p N_{(t-1)ijpv}, \quad \forall t, \forall i, \forall v, \tag{25.17}$$

$$\sum_j QV_{tjiv} = D_{tiv}, \quad \forall t, \forall i, \forall v, \tag{25.18}$$

$$RV_{(t-1)jv} = \sum_i QV_{tjiv}, \quad \forall t, \forall j, \forall v, \tag{25.19}$$

$$V_{(t-1)iv} - \sum_j \sum_p N_{tijpv} + \sum_j QV_{tjiv} = V_{tiv}, \quad \forall t, \forall i, \forall v, \tag{25.20}$$

$$\sum_j \sum_p N_{tijpv} \le V_{(t-1)iv}, \quad \forall t, \forall i, \forall v, \tag{25.21}$$

$$RV_{(t-1)jv} + \sum_i \sum_p N_{tijpv} - \sum_i QV_{tjiv} = RV_{tjv}, \quad \forall t, \forall j, \forall v, \tag{25.22}$$

$$X_{tijpv}((d_{ij} \cdot \tilde{T}T_v) - ET_{tjp}) = DDE_{tijpv}, \quad \forall t, \forall i, \forall p, \forall j, \forall v, \tag{25.23}$$

$$FP_{tijpv} \le X_{tijpv}((1.2 \cdot PC_{tip}) + (d_{ij} \cdot \tilde{T}C_v)) \cdot M, \quad \forall t, \forall i, \forall p, \forall j, \forall v, \tag{25.24}$$

$$FP_{tijpv} \ge X_{ijpv}((1.2 \cdot PC_{tip}) + (d_{ij} \cdot \tilde{T}C_v)), \quad \forall t, \forall i, \forall p, \forall j, \forall v, \tag{25.25}$$

$$FP_{tijpv} \le MP_{tp}, \quad \forall t, \forall i, \forall p, \forall j, \forall v, \tag{25.26}$$

and Yager's constraints for fuzzy considerations:

$$\left| \frac{(\tilde{D}_{tjpa^l} + \tilde{D}_{tjpa^u})}{2} + \frac{(\tilde{D}_{tjp\beta} - \tilde{D}_{tjp\alpha})}{4} \right| = D_{tjp}, \quad \forall j, \forall p, \forall t, \tag{25.27}$$

$$\left| \frac{(\tilde{T}C_{va^l} + \tilde{T}C_{va^u})}{2} + \frac{(\tilde{T}C_{v\beta} - \tilde{T}C_{v\alpha})}{4} \right| = TC_v, \quad \forall v, \tag{25.28}$$

$$\left| \frac{(\tilde{T}T_{va^l} + \tilde{T}T_{va^u})}{2} + \frac{(\tilde{T}T_{v\beta} - \tilde{T}T_{v\alpha})}{4} \right| = TT_v, \quad \forall v, \tag{25.29}$$

and

$$g_{tip}, S_{tip}, FP_{tijpv}, QP_{tijp}, QV_{tjiv}, RV_{tjv}, D_{tivt} \ge 0, \quad \forall t, \forall i, \forall p, \forall j, \forall v, \tag{25.30}$$

$$Z_{tip}, X_{tijpv} \in \{0,1\}, \quad \forall t, \forall i, \forall p, \forall j, \forall v, \tag{25.31}$$

$$N_{tijpv} \text{ integer}, \quad \forall t, \forall i, \forall p, \forall j, \forall v. \tag{25.32}$$

Equations 25.1 and 25.2 are the objective functions considering the tradeoff between maximization of price for the products from manufacturer's viewpoint and minimization of price from customer's viewpoint. Equation 25.3 is the objective function minimizing total cost of forward distance; Equation 25.4 is the objective function minimizing total cost of storage; Equation 25.5 is the objective function penalizing total tardiness of the products from expected delivery time; and Equation 25.6 is the objective function minimizing total cost of backward distance. The constraints 25.7 and 25.8 show that each plant can produce an amount of product just after it is selected. The constraint 25.9 guarantees that all customer demands are met for all products in all periods. The constraint 25.10 imposes the flow conservation at plants. The constraint 25.11 shows the amount of stored products at the end of any period. The constraint 25.12 represents the capacity restriction. The constraints 25.13 and 25.14 ensure that delivery is accomplished by only one vehicle. The number of required vehicles for transferring products between both plant and customer are shown by constraints 25.15 and 25.16. The constraint 25.17 certifies the demand of vehicles for each plant. The constraint 25.18 guarantees that the sum of delivered vehicles from customers to each plant is equal to the number of dispatched vehicles. The flow conservation of vehicles from customer is presented by constraint 25.19. Constraint 25.20 updates the number of plant's vehicles. Constraint 25.21 imposes that the sum of the dispatched vehicles for product delivery from each plant must be lower than the available vehicles. Constraint 25.22 updates the number of customer's vehicles. Tardiness of delivery time from expected time is shown by constraint 25.23. The constraints 25.24, 25.25 and 25.26 represent lower and upper bounds of price for each product. Upper bound is a parameter that is determined by competitive markets. Lower bound is acquired by both functions of production cost and cost of distance, which are considered as coefficients of production cost. This coefficient is set to be 1.2 as lean production cost. Equations 25.27 through 25.29 represent the defuzzified demand of product for customer at each period, the defuzzified traveling cost per unit of distance using vehicle, the defuzzified traveling time per unit of distance using vehicle, respectively. The constraints 25.30 through 25.32 show the signs and types of the decision variables.

25.4 Discussions

We proposed an electronic supply chain for order-delivery process. We considered an electronic market where three elements of plant, supplier, and customer exist. A supplier presents its product specifications in the appropriate website. The customer views the website. and based on his requirements orders a list of products. The supplier analyzes the orders via a logistic intelligent agent. We proposed a mathematical model to perform the real-time, web-based investigations. Two elements of price and delivery time are considered for customer satisfaction. The results of the analysis made by the system are reported to the customer to decide whether to give or refuse to give the order to the supplier.

References

Fang SC, Hu CF. Linear programming with fuzzy coefficients in constraints. *Computers & Mathematics with Applications* 1999;37:63–76.

Lai YJ, Hwang CL. *Fuzzy Mathematical Programming Methods and Applications*. Berlin: Springer, 1992.

Mahdavi-Amiri N, Nasseri SH. Duality in fuzzy number linear programming by use of a certain linear ranking function. *Applied Mathematics and Computation* 2006;180(1):206–216.

Mahdavi-Amiri N, Nasseri SH. Duality results and a dual simplex method for linear programming problems with trapezoidal fuzzy variables. *Fuzzy Sets and Systems* 2007;158(17):1961–1978.

Maleki HR, Tata M, Mashinchi M. Linear programming with fuzzy variables. *Fuzzy Sets and Systems* 2000;109:21–33.

Shoacheng T. Interval number and fuzzy number linear programming. *Fuzzy Sets and Systems* 1994;66:301–306.

Tanaka H, Ichihashi H. A formulation of fuzzy linear programming problem based on comparison of fuzzy numbers. *Control and Cybernet* 1984;13:185–194.

Yager RR. A procedure for ordering fuzzy subsets of the unit interval. *Information Sciences* 1981;24:143–161.

26

Multiple Supply Network: Fuzzy Mathematical Programming Model

SUMMARY   In this chapter, we propose a fuzzy mathematical programming model for a supply chain that considers multiple depots, multiple vehicles, multiple products, multiple customers, and different time periods. In this work, not only demand and cost but also decision variables are considered to be fuzzy. We apply two ranking functions for solving the model. The aim of the fuzzy mathematical program is to select the appropriate depots among candidate depots, the allocation of orders to depots and vehicles, also the allocation of the returning vehicles to depots, to minimize the total costs.

26.1 Introduction

Over the last decade or so, supply chain management has emerged as a key area of research among the practitioners of operations research. In today's increasingly global and competitive market, it is imperative that enterprises work together to achieve common goals such as minimizing the delay of deliveries, the holding and the transportation costs (Roy et al., 2004). A supply chain can be defined as a network consisting of suppliers, manufacturers, wholesales, distributors, retailers, and customers through which material and products are acquired, transformed, and delivered to consumers in markets (Hyung and Sung, 2003). Thus, more and more companies adopt and explore better supply chain management (SCM) to improve the overall efficiency. A successful SCM requires a change from managing individual functions to integrating activities into key supply chain processes.

Owing to the high complexity and uncertainty of the supply chain in industry, a traditional centralized decisional system seems unable to manage easily all the information flows and actions. The decision delay in the supply chain prolongs the process time and causes a company to lose competence. In order to reduce this delay, the supply chain member needs to give quick response. Thus, a supply chain can be characterized as a logistic network of partially autonomous decision-makers. Supply chain management has to do with the coordination of decisions within the network.

In the supply chain, ordering decision and inventory decision are two critical decisions supply chain managers have to face. The orders are usually made based on the forecasted customer demand without considering the uncertain factors in industry.

Mathematical programming models have proven their usefulness as analytical tools to optimize complex decision-making problems such as those encountered in supply chain planning. After that, a diversity of deterministic mathematical programming models dealing with the design of supply chain networks can be found in the literature (see e.g., Geoffrion and Powers, 1995; Goetschalckx et al., 2002; Yan et al., 2003; Amiri, 2006).

Under most circumstances, the critical design parameters for the supply chain, such as customer demands, prices and resource capacities are generally uncertain. Uncertain supply chain design has been one of the promising subjects. A big amount of stochastic programming models have been proposed for strategic and tactical planning (see e.g., Cheung and Powell, 1996; Owen and Daskin, 1998; Landeghen and Vanmaele, 2002; Min and Zhou, 2002; Guillén et al., 2005). However, in certain situations, the assumption of precise parameters of probability distributions is seriously questioned. The parameters are fixed, statistically estimated using past demand information, while demand does not stay "static" in fact. When the conditioning variables, such as the technological innovations and preferences of consumers, considerably change, the mean and variance of the demand distribution are possible to change. Besides, it is almost impossible to specify exactly the true values to the parameters, especially in the absence of abundant information as in the case of demand of new products. Thus, based on expert experience, fuzzy variables are considered to describe them. In this case, random variables with imprecise parameters are random fuzzy variables.

By random fuzzy programming we mean the optimization theory in random fuzzy environment. Random fuzzy programming makes the supply chain design plan more flexible when the parameters of the coefficients' distribution are uncertain. In practice, the values of the fuzzy parameters can be obtained according to the expert experience. Different numbers of the fuzzy variables reflect different conditions that affect the parameters of the probability distribution. The concept of random fuzzy variables was provided by Liu (2002), which is different from the definition used by Nattier (2001). To the best of our knowledge, there is a little research for the programming and solving supply chain design problems in random fuzzy environment.

Here, we propose a supply chain that considers multiple depots, multiple vehicles, multiple products, multiple customers, and different time periods. The supplier receives the order and forwards it to depots of multiple products. Since, demands are associated with fluctuations, thus an uncertainty should be corporate. As a result, we consider demands as fuzzy numbers. Also, causality of consideration costs as fuzzy numbers is uncertainty of different environmental and geographical conditions. The mentioned reasons endorse uncertainty of output decision variables that are considered as fuzzy numbers. A set of depots should be selected among candidate depots. The depots investigate the capacity level and accept/refuse supplying the order. Considering the location of the customers, the depots decide about sending the suitable vehicles. Each vehicle has its corresponding traveling cost. Also when the vehicles deliver the order to the customers, another allocation for the returning vehicles to depots is set. The aim is to identify the allocation of orders to depots, vehicles, and returning vehicles to depots to minimize the total cost.

26.2 Problem Definition

The proposed problem of this chapter considers different customers that should be serviced with one supplier. The supplier provides various products and keeps them in different depots. The initial problem is choosing the appropriate depots among a set of candidate depots. Each depot uses different types of vehicles to satisfy the orders. All of the depots are already stationed at the related locations. Here, we consider a multi-echelon supply chain network (one supplier, multi-depot, and customers) and multi-commodity with fuzzy demand. Sets of vehicles are stationed at each depot. Each depot can store sets of products. The received order list from customers can be responded to by one or multiple depots at each time. Each selected vehicle to deliver the products can transfer only one product. The returning vehicles are allocated to the depots when depots may not have specific vehicles in a period and should respond to an order. A configuration of the proposed supply chain network is shown in Figure 26.1.

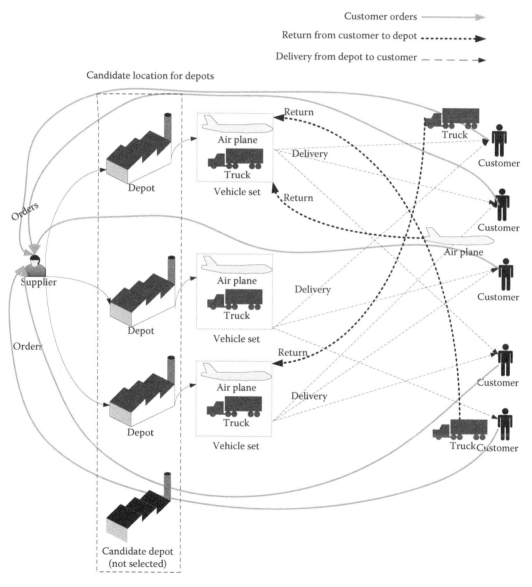

FIGURE 26.1
A configuration of the proposed supply chain network.

26.3 Fuzzy Mathematical Model

The following ranking functions are considered.

1. Bisector of area method (BOA): The proposed method in this category is reported by Yager (1981) which reduces to:

$$R(\tilde{a}) = \frac{a^L + a^U}{2} + \frac{(\beta - \alpha)}{4}$$

2. Mean of maximum method (MOM) which reduces to:

$$R(\tilde{a}) = \frac{a^L + a^U}{2}$$

26.3.1 Fuzzy Mathematical Programming

The fuzzy linear mathematical program is given below. In our proposed problem demand is a fuzzy parameter, also costs are assumed to be fuzzy value. The model is as follows:

Notations

P	set of products
I	set of depots stationed
J	set of customers
T	set of time units
V	set of all vehicles
$\tilde{D}_{jpt}$	the fuzzy demand of product p for customer j at time t
ND_{ivt}	the number of existing vehicles v in depot i at time t
NC_{jvt}	the number of existing vehicles v in customer j at the end of time t
VL_{vp}	capacity of vehicle v for product p
$\tilde{CT}_{ijv}$	the fuzzy traveling cost per mile from depot i to customer j using vehicle v
d_{ij}	distance between depot i and customer j
M	a large number
$\tilde{Ch}_{ip}$	the fuzzy holding cost for product p in depot i
$\tilde{Cs}_{ipt}$	the fuzzy supplying cost for product p in depot i at time t
CO_i	the opening cost of depot i
cap_p_{ip}	maximum capacity of product p in depot i

Decision Variables

x_{ijpvt}	{1, If depot i delivers product p to customer j using vehicle v at time t; 0; o.w}
y_{jipvt}	{1, if customer j delivers vehicle v to depot i at time t; 0; o.w}

z_{ipt} {1, if depot i receives product p at time t;0;o.w}

w_{it} {1, if depot i is active at time t;0;o.w}

TH_{ipt} Amount of received product p in depot i at time t

S_{ipt} Amount of stored product p in depot i at the end of time t

f_{ijpvt} Number of vehicles traveling from depot i to customer j for product p by vehicle v at time t

$\tilde{QP}_{ijpt}$ The fuzzy quantity of product p can be satisfied by depot i to customer j at time t

NR_{jivt} Number of vehicles v delivered to depot i by customer j at time t

DV_{ivt} Demand of vehicle v for depot i at time t

26.3.2 Objective Functions

$$Minimize(F) = Min(f_1 + f_2 + f_3 + f_4 + f_5)$$

$$f_1 = \sum_{i \in I} \sum_{j \in J} \sum_{p \in P} \sum_{v \in V} \sum_{t \in T} x_{ijpvt} \cdot f_{ijpvt} \cdot d_{ij} \cdot \tilde{CT}_{ijv} \tag{26.1}$$

$$f_2 = \sum_{j \in J} \sum_{i \in I} \sum_{v \in V} \sum_{t \in T} y_{jivt} \cdot NR_{jivt} \cdot d_{ij} \cdot \tilde{CT}_{ijv} \tag{26.2}$$

$$f_3 = \sum_{i \in I} \sum_{p \in P} \sum_{t \in T} S_{ipt} \cdot \tilde{Ch}_{ip} \tag{26.3}$$

$$f_4 = \sum_{i \in I} \sum_{p \in P} \sum_{t \in T} TH_{ipt} \cdot \tilde{Cs}_{ipt} \tag{26.4}$$

$$f_5 = \sum_{i \in I} \sum_{t \in T} (w_{it+1} - w_{it}) \cdot CO_i \tag{26.5}$$

26.3.3 Constraints

$$\sum_{p \in P} z_{ipt} \le M \cdot w_{it}, \quad \forall i \in I, \forall t \in T, \tag{26.6}$$

$$\sum_{p \in P} z_{ipt} \ge w_{it}, \quad \forall i \in I, \forall t \in T, \tag{26.7}$$

$$TH_{ipt} \le M \cdot z_{ipt}, \quad \forall i \in I, \forall p \in P, \forall t \in T, \tag{26.8}$$

$$TH_{ipt} \geq z_{ipt}, \quad \forall i \in I, \forall p \in P, \forall t \in T, \tag{26.9}$$

$$w_{it+1} \geq w_{it}, \quad \forall i \in I, \forall t \in T, \tag{26.10}$$

$$\sum_{i \in I} QP_{ijpt} = D_{jpt}, \quad \forall j \in J, \forall p \in P, \forall t \in T, \tag{26.11}$$

$$\sum_{j \in J} QP_{ijpt} \leq TH_{ipt} + S_{ipt-1}, \quad \forall i \in I, \forall p \in P, \forall t \in T, \tag{26.12}$$

$$S_{ipt-1} + TH_{ipt} - \sum_{j \in J} QP_{ijpt} = S_{ipt}, \quad \forall i \in I, \forall p \in P, \forall t \in T, \tag{26.13}$$

$$TH_{ipt} + S_{ipt-1} \leq cap\_p_{ip}, \quad \forall i \in I, \forall p \in P, \forall t \in T, \tag{26.14}$$

$$QP_{ijpt} \cdot \left(1 - \sum_{v \in V} x_{ijpvt}\right) = 0, \quad \forall i \in I, \forall j \in J, \forall p \in P, \forall t \in T, \tag{26.15}$$

$$QP_{ijpt} \geq \sum_{v \in V} x_{ijpvt}, \forall i \in I, \quad \forall j \in J, \forall p \in P, \forall t \in T, \tag{26.16}$$

$$\left\lfloor \left((QP_{ijpt} \div VL_{vp}) \cdot x_{ijpvt}\right) + 0.999 \right\rfloor = f_{ijpvt}, \quad \forall i \in I, \forall j \in J, \forall p \in P, \forall v \in V, \forall t \in T, \tag{26.17}$$

$$ND_{ivt-1} - \sum_{j \in J} \sum_{p \in P} f_{ijpvt} + \sum_{j \in J} NR_{jivt} = ND_{i,v,t}, \quad \forall i \in I, \forall v \in V, \forall t \in T, \tag{26.18}$$

$$\sum_{j \in J} \sum_{p \in P} f_{ijpvt} \leq ND_{ivt-1}, \quad \forall i \in I, \forall v \in V, \forall t \in T, \tag{26.19}$$

$$DV_{ivt} \leq M \cdot w_{it}, \quad \forall i \in I, \forall v \in V, \forall t \in T, \tag{26.20}$$

$$\sum_{j \in J} NR_{jivt} = DV_{ivt}, \quad \forall i \in I, \forall v \in V, \forall t \in T, \tag{26.21}$$

$$NC_{jvt-1} = \sum_{i \in I} NR_{jivt}, \quad \forall j \in J, \forall v \in V, \forall t \in T, \tag{26.22}$$

$$NR_{jivt} \leq M \cdot y_{jivt}, \quad \forall j \in J, \forall i \in I, \forall v \in V, \forall t \in T, \tag{26.23}$$

$$NR_{jivt} \geq y_{jivt}, \quad \forall j \in J, \forall i \in I, \forall v \in V, \forall t \in T, \tag{26.24}$$

$$NC_{jvt-1} + \sum_{i \in I} \sum_{p \in P} f_{ijpvt} - \sum_{i \in I} NR_{jivt} = NC_{jvt}, \quad \forall j \in J, \forall v \in V, \forall t \in T, \tag{26.25}$$

MOM constraints for fuzzy consideration:

$$\left| \left[\frac{\left(\tilde{D}_{jpta^l} + \tilde{D}_{jpta^u} \right)}{2} \right] \right| = D_{jpt}, \quad \forall j \in J, \forall p \in P, \forall t \in T, \tag{26.26}$$

$$\left| \left[\frac{\left(\tilde{C}T_{ijva^l} + \tilde{C}T_{ijva^u} \right)}{2} \right] \right| = CT_{ijv}, \quad \forall i \in I, \forall j \in J, \forall v \in V, \tag{26.27}$$

$$\left| \left[\frac{\left(\tilde{C}s_{ipta^l} + \tilde{C}s_{ipta^u} \right)}{2} \right] \right| = Cs_{ipt}, \quad \forall i \in I, \forall p \in P, \forall t \in T, \tag{26.28}$$

$$\left| \left[\frac{\left(\tilde{C}h_{ipa^l} + \tilde{C}h_{ipa^u} \right)}{2} \right] \right| = Ch_{ip}, \quad \forall i \in I, \forall p \in P, \tag{26.29}$$

$$\left| \left[\frac{\left(\tilde{Q}P_{ijpta^l} + \tilde{Q}P_{ijpta^u} \right)}{2} \right] \right| = QP_{ijpt}, \quad \forall i \in I, \forall j \in J, \forall p \in P, \forall t \in T, \tag{26.30}$$

$$\tilde{Q}P_{ijpta^l} \leq \tilde{Q}P_{ijpta^u}, \quad \forall i \in I, \forall j \in J, \forall p \in P, \forall t \in T, \tag{26.31}$$

Yager constraints for fuzzy consideration:

$$\left| \left[\frac{\left(\tilde{D}_{jpta^l} + \tilde{D}_{jpta^u} \right)}{2} + \frac{\left(\tilde{D}_{jpt\beta} - \tilde{D}_{jpt\alpha} \right)}{4} \right] \right| = D_{jpt}, \quad \forall j \in J, \forall p \in P, \forall t \in T, \tag{26.32}$$

$$\left| \left[\frac{\left(\tilde{C}T_{ijva^l} + \tilde{C}T_{ijva^u} \right)}{2} + \frac{\left(\tilde{C}T_{ijv\beta} - \tilde{C}T_{ijv\alpha} \right)}{4} \right] \right| = CT_{ijv}, \quad \forall i \in I, \forall j \in J, \forall v \in V, \tag{26.33}$$

$$\left| \left[\frac{\left(\tilde{C}s_{ipta^l} + \tilde{C}s_{ipta^u} \right)}{2} + \frac{\left(\tilde{C}s_{ipt\beta} - \tilde{C}s_{ipt\alpha} \right)}{4} \right] \right| = Cs_{ipt}, \quad \forall i \in I, \forall p \in P, \forall t \in T, \tag{26.34}$$

$$\left| \frac{\left(\tilde{C}h_{ipa^l} + \tilde{C}h_{ipa^u} \right)}{2} + \frac{\left(\tilde{C}h_{ip\beta} - \tilde{C}h_{ip\alpha} \right)}{4} \right| = Ch_{ip}, \quad \forall i \in I, \forall p \in P, \tag{26.35}$$

$$\left| \frac{\left(\tilde{Q}P_{ijpta^l} + \tilde{Q}P_{ijpta^u} \right)}{2} + \frac{\left(\tilde{Q}P_{ijpt\beta} - \tilde{Q}P_{ijpt\alpha} \right)}{4} \right| = QP_{ijpt}, \quad \forall i \in I, \forall j \in J, \forall p \in P, \forall t \in T,$$
$$\tag{26.36}$$

$$\tilde{Q}P_{ijpta^l} \leq \tilde{Q}P_{ijpta^u}, \quad \forall i \in I, \quad \forall j \in J, \quad \forall p \in P, \quad \forall t \in T, \tag{26.37}$$

Integrity and non-negativity constraints:

$$x_{ijpvt} \in \{0,1\}, \quad \forall i \in I, \forall j \in J, \forall p \in P, \forall v \in V, \forall t \in T, \tag{26.38}$$

$$y_{jivt} \in \{0,1\}, \quad \forall j \in J, \forall i \in I, \forall v \in V, \forall t \in T, \tag{26.39}$$

$$z_{ipt} \in \{0,1\}, \quad \forall i \in I, \forall p \in P, \forall t \in T, \tag{26.40}$$

$$w_{it} \in \{0,1\}, \quad \forall i \in I, \forall t \in T, \tag{26.41}$$

$$f_{ijpvt} \geq 0, \quad \forall i \in I, \forall j \in J, \forall p \in P, \forall v \in V, \forall t \in T, \tag{26.42}$$

$$\tilde{Q}P_{ijpt}, \quad Integer, \quad \forall i \in I, \forall j \in J, \forall p \in P, \forall t \in T, \tag{26.43}$$

$$NR_{jivt}, \quad Integer, \quad \forall j \in J, \forall i \in I, \forall v \in V, \forall t \in T, \tag{26.44}$$

$$TH_{ipt} \geq 0, \quad i \in I, \forall p \in P, \forall t \in T, \tag{26.45}$$

$$DV_{ivt} \geq 0, \quad \forall i \in I, \forall v \in V, \forall t \in T. \tag{26.46}$$

Equations 26.1 and 26.2 are the objective functions that minimize total cost of both forward and backward distance, respectively. Equation 26.3 is the objective function that minimizes total cost of storage. Equation 26.4 is the objective function that minimizes total cost of supply. Equation 26.5 is the objective function that minimizes cost of opening depot. The constraints 26.6 and 26.7 show that each depot can be supplied when it is activated. The constraints 26.8 and 26.9 ensure that the amount of product each selected depot receives is nonnegative. The constraint 26.10 prevents the depots from changing their status more than once. The constraint 26.11 guarantees that all customer demands are met for all products required at all periods. The constraint 26.12 is the flow conservation at depots. The constraint 26.13 shows amount of stored product at the end of period. The constraint 26.14 represents capacity restriction. The constraints 26.15 and 26.16 ensure that delivery is accomplished by only one vehicle. The number of vehicles traveling between depots and customers has been shown in constraint 26.17. The constraint 26.18 represents the number of remaining vehicles at the end of period. The constraint 26.19 requires that the number of traveled vehicles from depot is lower than or equal to its stationed vehicles. The constraint 26.20 requires that each activated depot can order vehicles. The constraint 26.21 guarantees that all depots' demands of vehicles are met, for all vehicles required and for any period. The constraint 26.22 is the flow balance of stationed vehicles at the end of period. The constraints 26.23 and 26.24 guarantee that delivery of vehicles from customer to depot is accomplished while the corresponded path was selected. The constraint 26.25 represents the number of remaining vehicles stationed at the corresponded customer at the end of period. Equation 26.26 through 26.30 represent the defuzzified demand of product for customer at each period, the defuzzified traveling cost per mile from depot to customer using vehicle, the defuzzified supplying cost for product in depot at each period, the defuzzified holding cost for product in depot, the defuzzified quantity of product can be satisfied by depot to customer at each period, respectively. Note that the all of the used fuzzy numbers in mentioned equations are defuzzified by mean of maximum (MOM) ranking function whereas Equations 26.32 thorugh 26.36 are defuzzified by bisector of area (BOA) ranking function which proposed by Yager. The constraints 26.31 and 26.37 imply that the first fuzzy number of decision variable QP (quantity of product can be satisfied by depot to customer at each time) in trapezoidal form always must be lower or equal than its second fuzzy number. The constraints 26.38 through 26.41 require that this variable is binary. The constraints 26.42 through 26.46 restrict all other variables from taking non-negative values.

26.4 Comprehensive Example

Here, we propose a numerical example to indicate the effectiveness of the proposed fuzzy mathematical model. The number of customers is three, number of products is three, number of candidate depots is seven, and number of vehicles is two. Because of the return of selected vehicles at each period, we must consider additional period in which no demand exists. Then, period four is supposed as additional period. Both of the fuzzy orders and principal orders in different time periods for different customers ($\tilde{D}_{jpt}$) are given in Table 26.1. The mentioned ranking functions in Section 3 are useful tools for defuzzifying fuzzy numbers. While the obtained results of using ranking functions are decimal, considering the real life environment they are estimated to their posterior integer number.

The distance between customers and depots (d_{ij}), capacity of vehicles (VL_{vp}) and capacity of depots (cap_p_{ip}) with respect to products are reported in Tables 26.2 through 26.4.

The transferring fuzzy cost per unit of distance for vehicles in depots ($\tilde{C}T_{ijv}$) is presented in Table 26.5.

The supplying fuzzy cost for products in depots at each period ($\tilde{C}s_{ipt}$) is presented in Table 26.6.

The holding fuzzy cost for products in depots ($\tilde{C}h_{ip}$) is presented in Table 26.7.

The opening costs (CO_i) and number of vehicles in depots at the beginning of first period (ND_{ivt}) are given in Tables 26.8 and 26.9.

To facilitate the computations, LINGO package is applied. The output of forward flow for the decision variables of Yager method are presented in Table 26.10. The amount of

TABLE 26.1

The Fuzzy Orders in Different Time Periods for Different Customers

	Principal	Order	Yager	MOM
First Period				
Customer 1				
Product 1	40	(38,42,2,5)	41	40
Product 2	45	(42,45,3,3)	44	44
Product 3	60	(50,58,4,5)	55	54
Customer 2				
Product 1	70	(65,68,1,3)	67	67
Product 2	30	(28,29,2,2)	29	29
Product 3	50	(45,56,5,3)	50	51
Customer 3				
Product 1	0	(0,0,0,0)	0	0
Product 2	20	(18,22,4,2)	20	20
Product 3	30	(27,31,2,4)	30	29
Second Period				
Customer 1				
Product 1	19	(19,22,3,3)	21	21
Product 2	0	(0,0,0,0)	0	0
Product 3	18	(18,19,2,2)	19	19
Customer 2				
Product 1	0	(0,0,0,0)	0	0
Product 2	0	(0,0,0,0)	0	0
Product 3	13	(12,15,4,4)	14	14
Customer 3				
Product 1	13	(13,14,2,2)	14	14
Product 2	15	(15,18,6,7)	17	17
Product 3	17	(12,15,3,6)	15	14
Third Period				
Customer 1				
Product 1	30	(28,32,5,6)	31	30
Product 2	25	(25,27,2,2)	26	26
Product 3	17	(17,19,6,7)	19	18
Customer 2				
Product 1	16	(10,18,6,8)	15	14
Product 2	20	(17,25,4,6)	22	21
Product 3	18	(15,18,3,6)	18	17
Customer 3				
Product 1	26	(22,25,2,1)	24	24
Product 2	25	(20,30,5,4)	25	25
Product 3	20	(19,20,3,2)	20	20

(*Continued*)

TABLE 26.1 (*Continued*)

The Fuzzy Orders in Different Time Periods for Different Customers

	Principal	Order	Yager	MOM
Fourth Period				
Customer 1				
Product 1	0	(0,0,0,0)	0	0
Product 2	0	(0,0,0,0)	0	0
Product 3	0	(0,0,0,0)	0	0
Customer 2				
Product 1	0	(0,0,0,0)	0	0
Product 2	0	(0,0,0,0)	0	0
Product 3	0	(0,0,0,0)	0	0
Customer 3				
Product 1	0	(0,0,0,0)	0	0
Product 2	0	(0,0,0,0)	0	0
Product 3	0	(0,0,0,0)	0	0

TABLE 26.2

The Distance between Customers and Depots

Distance	Customer 1	Customer 2	Customer 3
Depot 1	20	25	10
Depot 2	10	15	17
Depot 3	14	12	13
Depot 4	10	15	12
Depot 5	16	22	24
Depot 6	13	16	20
Depot 7	14	15	16

TABLE 26.3

The Capacity of Vehicles

Capacity of Vehicle	Product 1	Product 2	Product 3
Vehicle 1	30	50	20
Vehicle 2	10	15	8

TABLE 26.4

The Capacity of Depots

Depot Capacity	Product 1	Product 2	Product 3
Depot 1	100	85	90
Depot 2	90	80	70
Depot 3	80	75	70
Depot 4	90	100	70
Depot 5	85	65	75
Depot 6	80	70	60
Depot 7	100	70	80

TABLE 26.5

The Transferring Fuzzy Cost Per Unit of Distance for Vehicles in Depots

Transferring Cost Per Unit of Distance	Vehicle 1	Yager	MOM	Principal	Vehicle 2	Yager	MOM	Principal
Depot 1								
Customer 1	(45,55,4,6)	51	50	50	(25,32,3,6)	30	29	30
Customer 2	(48,52,3,6)	51	50	50	(28,31,3,4)	30	30	30
Customer 3	(49,57,4,5)	54	53	50	(27,36,4,6)	32	32	30
Depot 2								
Customer 1	(50,55,4,3)	53	53	50	(25,35,3,6)	31	30	30
Customer 2	(44,56,2,6)	51	50	50	(24,30,3,5)	28	27	30
Customer 3	(49,52,2,6)	52	51	50	(26,31,2,4)	29	29	30
Depot 3								
Customer 1	(47,56,3,2)	52	52	50	(25,34,3,3)	30	30	30
Customer 2	(47,55,2,2)	51	51	50	(24,33,5,6)	29	29	30
Customer 3	(40,50,2,1)	45	45	50	(22,32,1,6)	29	27	30
Depot 4								
Customer 1	(46,57,5,4)	52	52	50	(24,37,2,6)	32	31	30
Customer 2	(42,50,3,6)	47	46	50	(25,30,2,5)	29	28	30
Customer 3	(49,56,3,4)	53	53	50	(29,34,2,3)	32	32	30
Depot 5								
Customer 1	(46,52,3,6)	50	49	50	(29,35,2,6)	33	32	30
Customer 2	(49,56,5,5)	53	53	50	(29,36,4,5)	33	33	30
Customer 3	(47,54,5,4)	51	51	50	(28,35,4,6)	32	32	30
Depot 6								
Customer 1	(45,52,3,6)	50	49	50	(25,35,2,4)	31	30	30
Customer 2	(47,57,5,4)	52	52	50	(26,34,5,6)	31	30	30
Customer 3	(48,53,6,5)	51	51	50	(27,36,4,5)	32	32	30
Depot 7								
Customer 1	(48,55,3,2)	52	52	50	(28,34,3,4)	32	31	30
Customer 2	(46,50,2,3)	49	48	50	(26,36,4,6)	32	31	30
Customer 3	(48,50,3,2)	49	49	50	(29,31,4,6)	31	30	30

received (TH) product from supplier in each depot at each period is shown in Table 26.11. Table 26.12 represents amount of storage (S) of products in each depot at the end of each period as ⎯⎯⎯▶ pursues stored status of product in each depot at all time periods and the number poses upon it represents the related amount of storage, whereas, ⎯⎯⎯▶ halts the described status. The backward flow for the decision variables is presented in Table 26.13. The number of remaining vehicles at the end of period (ND) is presented in Table 26.14, and the activation periods of depots are presented in Table 26.15.

TABLE 26.6

The Supplying Fuzzy Cost for Products in Depots at Each Period

Supplying Cost	Product 1				Product 2				Product 3			
	Fuzzy Number	Yager	MOM	Principal	Fuzzy Number	Yager	MOM	Principal	Fuzzy Number	Yager	MOM	Principal
First Period												
Depot 1	(10,17,2,2)	14	14	15	(8,9,2,1)	9	9	8	(9,16,3,5)	13	13	10
Depot 2	(8,14,3,6)	12	11	8	(7,8,2,3)	8	8	7	(9,16,5,2)	12	13	15
Depot 3	(7,14,5,3)	10	11	12	(7,11,5,2)	9	9	10	(15,19,2,2)	17	17	15
Depot 4	(10,16,3,4)	14	13	14	(13,16,3,5)	15	15	15	(8,10,3,3)	9	9	9
Depot 5	(12,16,1,4)	15	14	13	(6,10,2,3)	9	8	7	(9,16,4,5)	13	13	9
Depot 6	(9,16,3,6)	14	13	15	(10,15,3,5)	13	13	12	(13,17,2,6)	16	15	13
Depot 7	(5,8,2,2)	7	7	5	(10,14,3,4)	13	12	10	(7,15,4,3)	11	11	7
Second Period												
Depot 1	(10,19,3,3)	15	15	11	(5,10,2,2)	8	8	6	(14,16,3,3)	15	15	15
Depot 2	(10,14,4,6)	13	12	12	(6,9,2,3)	8	8	7	(10,15,5,3)	12	13	11
Depot 3	(12,14,3,4)	14	13	12	(5,10,5,2)	7	8	6	(13,19,2,2)	16	16	14
Depot 4	(11,16,3,4)	14	14	13	(9,11,3,2)	10	10	10	(8,10,3,3)	9	9	9
Depot 5	(9,16,1,2)	13	13	12	(8,12,2,3)	11	10	10	(12,16,4,5)	15	14	13
Depot 6	(9,14,3,4)	12	12	13	(12,18,3,5)	16	15	14	(8,17,5,6)	13	13	10
Depot 7	(9,10,2,2)	10	10	9	(10,14,3,4)	13	12	11	(8,10,4,3)	9	9	9
Third Period												
Depot 1	(12,16,2,6)	15	14	13	(8,10,2,2)	9	9	9	(8,16,3,3)	12	12	9
Depot 2	(7,14,4,6)	11	11	12	(5,8,2,3)	7	7	6	(9,14,5,3)	11	12	13
Depot 3	(7,14,3,4)	11	11	12	(7,15,5,2)	11	11	8	(10,19,2,2)	15	15	13
Depot 4	(12,16,3,4)	15	14	12	(10,11,3,2)	11	11	11	(7,15,3,3)	11	11	14
Depot 5	(12,15,1,2)	14	14	14	(8,13,2,3)	11	11	8	(14,16,4,5)	16	15	15
Depot 6	(9,18,3,4)	14	14	11	(12,18,3,5)	16	15	13	(15,17,5,6)	17	16	15
Depot 7	(6,9,2,2)	8	8	7	(10,14,3,4)	13	12	11	(9,10,4,3)	10	10	9

(*Continued*)

TABLE 26.6 (Continued)

The Supplying Fuzzy Cost for Products in Depots at Each Period

Supplying Cost	Product 1				Product 2				Product 3			
	Fuzzy Number	Yager	MOM	Principal	Fuzzy Number	Yager	MOM	Principal	Fuzzy Number	Yager	MOM	Principal
Fourth Period												
Depot 1	(10,15,2,3)	13	13	12	(8,12,2,2)	10	10	9	(15,16,3,3)	16	16	16
Depot 2	(10,18,4,6)	15	14	11	(5,9,2,3)	8	7	8	(9,17,5,3)	13	13	12
Depot 3	(9,16,3,4)	13	13	10	(5,9,5,2)	7	7	7	(13,22,2,2)	18	18	15
Depot 4	(10,18,3,4)	15	14	14	(7,11,3,2)	9	9	10	(8,10,3,3)	9	9	9
Depot 5	(12,17,1,2)	15	15	15	(6,15,2,3)	11	11	13	(11,16,4,5)	14	14	11
Depot 6	(9,14,3,4)	12	12	10	(12,18,3,5)	16	15	13	(11,15,5,6)	14	13	14
Depot 7	(8,10,2,2)	9	9	10	(10,14,3,4)	13	12	12	(9,10,4,3)	10	10	10

TABLE 26.7

The Holding Fuzzy Cost for Products in Depots

Holding Cost	Product 1				Product 2				Product 3			
	Fuzzy number	Yager	MOM	Principal	Fuzzy number	Yager	MOM	Principal	Fuzzy number	Yager	MOM	Principal
Depot 1	(10,15,2,3)	13	13	12	(6,9,2,2)	8	8	6	(11,16,3,3)	14	14	11
Depot 2	(10,14,4,6)	13	12	13	(3,8,2,3)	6	6	4	(9,11,5,3)	10	10	10
Depot 3	(9,14,3,4)	12	12	13	(7,9,5,2)	8	8	7	(13,19,2,2)	16	16	14
Depot 4	(10,16,3,4)	14	13	12	(10,11,3,2)	11	11	10	(8,10,3,3)	9	9	9
Depot 5	(12,15,1,2)	14	14	13	(6,9,2,3)	8	8	7	(11,16,4,5)	14	14	15
Depot 6	(9,14,3,4)	12	12	12	(12,15,3,5)	14	14	14	(11,17,5,6)	15	14	16
Depot 7	(4,8,2,2)	6	6	7	(10,14,3,4)	13	12	13	(7,10,4,3)	9	9	7

TABLE 26.8

The Opening Costs

	Opening Cost
Depot 1	2000
Depot 2	2000
Depot 3	2000
Depot 4	2000
Depot 5	2000
Depot 6	2000
Depot 7	2000

TABLE 26.9

The Number of Vehicles in Depots at the Beginning of First Period

Number of Vehicles	Vehicle 1	Vehicle 2
Depot 1	14	12
Depot 2	14	12
Depot 3	14	12
Depot 4	14	12
Depot 5	14	12
Depot 6	14	12
Depot 7	14	12

TABLE 26.10

The Forward Path Output

X	Depot	Customer	Product	Vehicle	Period	f	$\tilde{Q}P$	QP
1	2	2	1	1	1	3	(53.8,53.8,1,53.8)	67
1	2	2	2	1	1	1	(1,57,1,1)	29
1	2	2	3	1	1	3	(40.2,40.2,1,40.2)	50
1	2	3	3	1	1	1	(1,9,1,1)	5
1	7	1	1	1	1	2	(41,41,1,1)	41
1	7	1	2	2	1	3	(35.4,35.4,1,35.4)	44
1	7	1	3	1	1	3	(1,109,1,1)	55
1	7	3	2	2	1	2	(16.2,16.2,1,16.2)	20
1	7	3	3	1	1	2	(1,49,1,1)	25
1	2	1	1	1	2	1	(21,21,1,1)	21
1	2	1	3	1	2	1	(15.4,15.4,1,15.4)	19
1	2	2	3	1	2	1	(11.4,11.4,1,11.4)	14
1	2	3	2	2	2	2	(13.8,13.8,1,13.8)	17
1	2	3	3	2	2	2	(1,29,1,1)	15
1	7	3	1	1	2	1	(11.4,11.4,1,11.4)	14
1	2	1	1	2	3	4	(1,61,1,1)	31
1	2	1	2	2	3	2	(1,51,1,1)	26
1	2	2	1	2	3	1	(1,9,1,1)	5
1	2	2	2	1	3	1	(17.8,17.8,1,17.8)	22
1	2	3	1	1	3	1	(1,47,1,1)	24
1	2	3	2	1	3	1	(1,49,1,1)	25
1	7	1	3	2	3	3	(15.4,15.4,1,15.4)	19
1	7	2	1	2	3	1	(13,13,13,1)	10
1	7	2	3	2	3	3	(14.6,14.6,1,14.6)	18
1	7	3	3	2	3	3	(16.2,16.2,1,16.2)	20

TABLE 26.11

The Amount of Received Product in Each Depot

TH	Product 1	Product 2	Product 3
First Period			
Depot 2	67	29	55
Depot 7	41	64	80
Second Period			
Depot 2	21	17	48
Depot 7	14	1	0
Third Period			
Depot 2	60	73	0
Depot 7	10	0	57

TABLE 26.12

The Amount of Stored Product in Each Depot

```
·············▶ Continue
             │
·············▶ Stop
```

Depot	Product	Period 1	Period 2	Period 3	Period 4
Depot 1					
Depot 2					
Depot 3					
Depot 4					
Depot 5					
Depot 6					
Depot 7	2		$\xrightarrow{1}$	$\xrightarrow{1}$	$\xrightarrow{1}$

TABLE 26.13

The Backward Path Output

Y	Customer	Depot	Vehicle	Period	NR
1	1	2	2	2	2
1	1	7	1	2	5
1	1	7	2	2	1
1	2	7	1	2	7
1	3	7	1	2	3
1	3	7	2	2	2
1	1	2	1	3	2
1	2	7	1	3	1
1	3	7	1	3	1
1	3	7	2	3	4
1	1	2	2	4	9
1	2	2	2	4	5
1	2	3	1	4	1
1	3	1	1	4	2
1	3	1	2	4	3

TABLE 26.14

The Number of Remaining Vehicles at the End of Periods

Number of Left Vehicles	Vehicle 1	Vehicle 2
At the End of Period 1		
Depot 1	14	12
Depot 2	6	12
Depot 3	14	12
Depot 4	14	12
Depot 5	14	12
Depot 6	14	12
Depot 7	7	7
At the End of Period 2		
Depot 1	14	12
Depot 2	3	10
Depot 3	14	12
Depot 4	14	12
Depot 5	14	12
Depot 6	14	12
Depot 7	21	10
At the End of Period 3		
Depot 1	14	12
Depot 2	2	3
Depot 3	14	12
Depot 4	14	12
Depot 5	14	12
Depot 6	14	12
Depot 7	23	4

(Continued)

TABLE 26.14 (*Continued*)

The Number of Remaining Vehicles at the End of Periods

Number of Left Vehicles	Vehicle 1	Vehicle 2
At the End of Period 4		
Depot 1	16	15
Depot 2	2	17
Depot 3	15	12
Depot 4	14	12
Depot 5	14	12
Depot 6	14	12
Depot 7	23	4

TABLE 26.15

The Activation Period of Depots

	Period 1	Period 2	Period 3	Period 4
Depot 1				✓
Depot 2	✓	✓	✓	✓
Depot 3				✓
Depot 4				
Depot 5				
Depot 6				
Depot 7	✓	✓	✓	✓

Whereas activation of depot necessitates returning of vehicles, with a tradeoff between corresponding costs, the model has made decision to activate new depots (1 and 3) at the first of period four.

The output of forward flow for the decision variables of principal method are presented in Table 26.16. The amount of received (TH) product from supplier in each depot at each period is shown in Table 26.17. Table 26.18 represents amount of storage (S) of product in each depot at the end of each period. The backward flow for the decision variables is presented in Table 26.19. The number of remained vehicles at the end of period (ND) is presented in Table 26.20, and the activation periods of depots are presented in Table 26.21.

The output of forward flow for the decision variables of MOM method are presented in Table 26.22. The amount of received (TH) product from supplier in each depot at each period is shown in Table 26.23. Table 26.24 represents amount of storage (S) product in each depot at the end of each period. The backward flow for the decision variables is presented in Table 26.25. The number of remained vehicles at the end of period (ND) is presented in Table 26.26, and activation periods of depots are presented in Table 26.27.

The best objectives of three methods are given in Table 26.28.

The objective values of the two fuzzy ranking methods are very close to each other.

TABLE 26.16

The Forward Path Output

X	Depot	Customer	Product	Vehicle	Period	f	QP
1	2	1	2	1	1	1	45
1	2	1	3	1	1	3	60
1	2	3	2	2	1	1	15
1	2	3	3	2	1	1	6
1	3	2	1	1	1	3	70
1	3	2	2	1	1	1	30
1	3	2	3	1	1	3	50
1	4	1	1	1	1	2	40
1	4	3	2	2	1	1	5
1	4	3	3	2	1	3	24
1	2	1	3	1	2	1	18
1	2	3	3	2	2	1	1
1	4	1	1	2	2	2	19
1	4	2	3	1	2	1	13
1	4	3	1	2	2	1	3
1	4	3	2	2	2	1	15
1	4	3	3	2	2	2	16
1	7	3	1	2	2	1	10
1	2	2	1	2	3	1	6
1	2	3	1	2	3	3	26
1	2	3	2	2	3	2	25
1	2	3	3	2	3	3	20
1	4	1	1	1	3	1	30
1	4	1	2	2	3	1	10
1	4	1	3	1	3	1	17
1	4	2	1	2	3	1	10
1	5	1	2	2	3	1	15
1	5	2	2	2	3	1	15
1	7	2	2	2	3	1	5
1	7	2	3	1	3	1	18

TABLE 26.17

The Amount of Received Product in Each Depot

TH	Product 1	Product 2	Product 3
First Period			
Depot 2	0	60	66
Depot 3	70	30	50
Depot 4	40	5	24
Depot 7	1	0	1

(Continued)

TABLE 26.17 (Continued)

The Amount of Received Product in Each Depot

TH	Product 1	Product 2	Product 3
Second Period			
Depot 2	1	0	19
Depot 4	22	15	29
Depot 7	9	0	0
Third Period			
Depot 1	1	0	0
Depot 2	31	25	20
Depot 4	40	10	17
Depot 5	0	30	0
Depot 7	0	5	17

TABLE 26.18

The Amount of Stored Product in Each Depot

Depot / Product	Period 1	Period 2	Period 3	Period 4
Depot 1 / 1			$\xrightarrow{1}$	$\xrightarrow{1}$
Depot 2 / 1		$\xrightarrow{1}$		
Depot 3				
Depot 4				
Depot 5				
Depot 6				
Depot 7 / 1	$\xrightarrow{1}$			
Depot 7 / 3	$\xrightarrow{1}$	$\xrightarrow{1}$		

TABLE 26.19

The Backward Path Output

Y	Customer	Depot	Vehicle	Period	NR
1	1	4	1	2	6
1	2	3	1	2	7
1	3	1	2	2	6
1	1	2	1	3	1
1	1	2	2	3	2
1	2	7	1	3	1
1	3	3	2	3	6
1	1	4	1	4	2
1	1	4	2	4	2
1	2	3	1	4	1
1	2	3	2	4	4
1	3	1	2	4	8

TABLE 26.20

The Number of Remaining Vehicles at the End of Periods

Number of Left Vehicles	Vehicle 1	Vehicle 2
At the End of Period 1		
Depot 1	14	12
Depot 2	10	10
Depot 3	7	12
Depot 4	12	8
Depot 5	14	12
Depot 6	14	12
Depot 7	14	12
At the End of Period 2		
Depot 1	14	18
Depot 2	9	9
Depot 3	14	12
Depot 4	17	2
Depot 5	14	12
Depot 6	14	12
Depot 7	14	11
At the End of Period 3		
Depot 1	14	18
Depot 2	10	2
Depot 3	14	18
Depot 4	15	0
Depot 5	14	10
Depot 6	14	12
Depot 7	14	10

(Continued)

TABLE 26.20 (*Continued*)

The Number of Remaining Vehicles at the End of Periods

Number of Left Vehicles	Vehicle 1	Vehicle 2
At the End of Period 4		
Depot 1	14	26
Depot 2	10	2
Depot 3	15	22
Depot 4	17	2
Depot 5	14	10
Depot 6	14	12
Depot 7	14	10

TABLE 26.21

The Activation Period of Depots

	Period 1	Period 2	Period 3	Period 4
Depot 1		✓	✓	✓
Depot 2	✓	✓	✓	✓
Depot 3	✓	✓	✓	✓
Depot 4	✓	✓	✓	✓
Depot 5	✓	✓	✓	✓
Depot 6				
Depot 7	✓	✓	✓	✓

TABLE 26.22

The Forward Path Output

X	Depot	Customer	Product	Vehicle	Period	f	$\tilde{Q}P$	QP
1	1	1	2	2	1	1	(14,14,1,1)	14
1	1	2	3	1	1	3	(51,51,1,1)	51
1	1	3	3	2	1	1	(8,8,1,1)	8
1	2	1	2	2	1	2	(1,59,1,1)	30
1	2	1	3	2	1	5	(1,79,1,1)	40
1	2	2	1	1	1	3	(1,133,1,1)	67
1	2	2	2	1	1	1	(1,57,1,1)	29
1	2	3	2	2	1	2	(20,20,1,1)	20
1	2	3	3	2	1	1	(8,8,1,1)	8
1	3	1	1	2	1	1	(10,10,1,1)	10
1	3	1	3	2	1	2	(14,14,1,1)	14
1	3	3	3	2	1	1	(1,1,1,1)	1
1	6	1	1	2	1	3	(1,59,1,1)	30
1	6	3	3	1	1	1	(1,23,1,1)	12
1	1	1	3	1	2	1	(11,11,1,1)	11
1	1	3	1	2	2	1	(1,7,1,1)	4
1	2	1	1	1	2	1	(1,39,1,1)	20

(*Continued*)

TABLE 26.22 (*Continued*)

The Forward Path Output

X	Depot	Customer	Product	Vehicle	Period	f	$\tilde{Q}P$	QP
1	2	1	3	2	2	1	(1,15,1,1)	8
1	2	3	1	2	2	1	(1,19,1,1)	10
1	3	2	3	2	2	1	(1,11,1,1)	6
1	3	3	2	2	2	1	(1,29,1,1)	15
1	6	1	1	2	2	1	(1,1,1,1)	1
1	6	2	3	2	2	1	(8,8,1,1)	8
1	6	3	2	1	2	1	(1,3,1,1)	2
1	6	3	3	1	2	1	(1,27,1,1)	14
1	2	1	1	1	3	1	(1,59,1,1)	30
1	2	1	2	1	3	1	(1,51,1,1)	26
1	2	1	3	1	3	1	(1,35,1,1)	18
1	2	2	2	2	3	1	(1,29,1,1)	15
1	2	2	3	2	3	1	(1,15,1,1)	8
1	2	3	1	2	3	1	(10,10,1,1)	10
1	2	3	2	1	3	1	(1,47,1,1)	24
1	3	2	1	1	3	1	(1,25,1,1)	13
1	3	2	2	2	3	1	(1,11,1,1)	6
1	3	3	2	2	3	1	(1,1,1,1)	1
1	3	3	3	1	3	1	(1,39,1,1)	20
1	6	2	1	2	3	1	(1,1,1,1)	1
1	6	2	3	2	3	2	(1,17,1,1)	9
1	6	3	1	1	3	1	(1,27,1,1)	14

TABLE 26.23

The Amount of Received Product in Each Depot

TH	Product 1	Product 2	Product 3
First Period			
Depot 1	0	14	59
Depot 2	67	79	48
Depot 3	10	0	15
Depot 6	30	0	12
Second Period			
Depot 1	4	1	11
Depot 2	30	1	8
Depot 3	0	15	6
Depot 6	1	2	22
Third Period			
Depot 2	40	64	26
Depot 3	13	7	20
Depot 6	15	0	9

TABLE 26.24

The Amount of Stored Product in Each Depot

Depot	Product	Period 1	Period 2	Period 3	Period 4
Depot 1	2		$\xrightarrow{\;1\;}$	$\xrightarrow{\;1\;}$	$\xrightarrow{\;1\;}$
Depot 2	2		$\xrightarrow{\;1\;}$		
Depot 3					
Depot 4					
Depot 5					
Depot 6					
Depot 7					

TABLE 26.25

The Backward Path Output

Y	Customer	Depot	Vehicle	Period	NR
1	1	2	2	2	3
1	1	6	2	2	11
1	2	3	1	2	7
1	3	1	2	2	5
1	3	3	1	2	1
1	1	3	1	3	2
1	1	6	2	3	2
1	2	3	2	3	2
1	3	1	1	3	2
1	3	3	2	3	3
1	1	2	1	4	3
1	2	2	2	4	1
1	2	3	1	4	1
1	2	3	2	4	5
1	3	1	1	4	3
1	3	1	2	4	2

TABLE 26.26

The Number of Remaining Vehicles at the End of Periods

Number of Left Vehicles	Vehicle 1	Vehicle 2
At the End of Period 1		
Depot 1	11	10
Depot 2	10	2
Depot 3	14	8
Depot 4	14	12
Depot 5	14	12
Depot 6	13	9
Depot 7	14	12
At the End of Period 2		
Depot 1	10	14
Depot 2	9	3
Depot 3	22	6
Depot 4	14	12
Depot 5	14	12
Depot 6	11	18
Depot 7	14	12
At the End of Period 3		
Depot 1	12	14
Depot 2	5	0
Depot 3	22	9
Depot 4	14	12
Depot 5	14	12
Depot 6	10	17
Depot 7	14	12
At the End of Period 4		
Depot 1	15	16
Depot 2	8	1
Depot 3	23	14
Depot 4	14	12
Depot 5	14	12
Depot 6	10	17
Depot 7	14	12

TABLE 26.27

The Activation Period of Depots

	Period 1	Period 2	Period 3	Period 4
Depot 1	✓	✓	✓	✓
Depot 2	✓	✓	✓	✓
Depot 3	✓	✓	✓	✓
Depot 4				
Depot 5				
Depot 6	✓	✓	✓	✓
Depot 7				

TABLE 26.28

The Best Objectives of Three Methods

	Principal	Yager	MOM
Best objective	61081	67924	67853

26.5 Discussions

Because of the complexity of real-life environment and existing constraints to attain useful data, presentation of a deterministic mathematical program is not sufficient in SCM. Therefore, we proposed a fuzzy mathematical program. In this way, we used ranking function approach to handle such problems.

References

Amiri A. Designing a distribution network in a supply chain system: Formulation and efficient solution procedure. *European Journal of Operational Research* 2006;171(2):567–576.

Cheung RK-M, Powell BW. Models and algorithms for distribution problems with uncertain demands. *Transportation Science* 1996;30:822–844.

Geoffrion AM, Powers RF. Twenty years of strategic distribution system design: An evolution perspective. *Interfaces* 1995;25:105–128.

Goetschalckx M, Vidal CJ, Dogan K. Modeling and design of global logistics systems: A review of integrated strategic and tactical models and design algorithms. *European Journal of Operational Research* 2002;143:1–18.

Guillén G, Mele FD, Bagajewicz MJ, Espuñ a A, Puigjaner L. Multi objective supply chain design under uncertainty. *Chemical Engineering Science* 2005;60:1535–1553.

Hyung JA, Sung JP. Modeling of a multi-agent system for coordination of supply chains with complexity and uncertainty. In: *PRIMA 2003, LNAI 2891*, Lee J, Barley M (eds.), Springer-Verlag Berlin Heidelberg, 2003:13–24.

Landeghen HV, Vanmaele H. Robust planning: A new paradigm for demand chain planning. *Journal of Operations Management* 2002;20:769–783.

Liu B. *Theory and Practice of Uncertain Programming*. Physica-verlag, Springer Berlin Heidelberg, 2002.

Min H, Zhou G. Supply chain modeling: Past, present and future. *Computers and Industrial Engineering* 2002;43:231–249.

Nattier W. Random fuzzy variables of second order and applications to statistical inference. *Information Sciences* 2001;133:69–88.

Owen SH, Daskin MS. Strategic facility location: A review. *European Journal of Operational Research* 1998;111:423–447.

Roy D, Anciaux D, Monteiro T, Ouzizi L. Multi-agent architecture for supply chain management. *Journal of Manufacturing Technology Management* 2004;15(8):745–755.

Yager RR. A procedure for ordering fuzzy subsets of the unit interval, *Information Sciences*, 1981;24:143–161.

Yan H, Yu Z, Cheng TCE. A strategic model for supply chain design with logical constraints: Formulation and solution. *Computers and Operations Research* 2003;30:2135–2155.

27

Facilities Relocation Problem in Supply Chain: Genetic Optimization Model

SUMMARY   In this chapter a relocation problem in supply chain is modeled mathematically and a genetic algorithm is proposed for optimization. During programming horizon some variations happen in some of the primary conditions of the problem. These changes motivate to the possibility of supply chain facilities location replacements, because of cost reduction, rival increase, and services improvement. Genetic algorithm's operators are fully discussed for optimization purpose.

27.1 Introduction

Location in any time is corresponding to that specific time conditions, and a location that is appropriate for one of the facilities, because of the condition's change may not be appropriate for the same facilities after years. Location problems contained large spectrum of OR problems that were always attractive for researchers, and also many varied researches have been done like facility location problem where new facilities or close down already existing facilities at two different distribution levels over a given time horizon (Yapicioglua et al., 2007). Facility layout problems (FLPs) concerning space layout optimization have been investigated in depth by researchers in many engineering fields. Recent advances in computing science and increased understanding of methods for developing mathematical models have helped with layout design investigations. The FLP has applications in both manufacturing and the service industry now. The FLP is a common industrial problem of allocating facilities to minimize the total cost of transporting materials or to maximize adjacency requirement (Koopmans and Beckmann, 1957) or to both minimize the cost of transporting materials and maximize adjacency requirement between the facilities (Meller and Gau, 1996).

The FLP can be classified in two categories according to the arrangement method of facilities, either an equal area layout problem or an unequal area layout problem. The unequal area layout problem can be classified primarily into two categories depending on the plan type that the facility layout is to be drawn, either a grid-based block plan layout problem or a continual block plan layout problem. In the grid-based block plan layout problem the facility layout is constructed on the grid plan, called the grid-based block plan and divided into squares or rectangles having a unit area. In continual block plan layout problem the facility layout is constructed on the continual plan.

Relocating a production site is a difficult industrial project and companies are often reluctant to get into this kind of trouble, especially small or medium-sized companies that have to go on operating with the same machines. A simple solution consists in removing the machines during paid holidays: it is then possible to close the firm for the duration of the removal process. However, a firm working on orders needs not only to be continuously present on the market but also to be highly reactive, and therefore this strategy is not conceivable for this type of firms. Although similar situations are commonplace, this problem has not been dealt with yet. We considered the case of relocating the current facilities of a company to a nearby site. Two companies that were faced with the problem of removing a line of machines from one site to another without interrupting the production twice approached us. The constraints of the removal project were the same for both companies:

- The production could not be entirely stopped during the removal of the machines.
- The removal process was of flow-shop type.
- The budget of the removal operation was limited, which forbade any solution of flash removal type (removal of the whole of the production system overnight or over a weekend).
- The global production process remained identical, that is, the logical chain of operations remained the same (the same machines would continue to be used one after the other).

This unusual problem is not dealt with in reference books about production organization (Heizer and Render, 1999) or facilities planning (Tompkins and White, 1984). Even the fundamental book on the methods of facilities layout (Muther, 1969) does not deal with the organization of a removal faced with such constraints. Recent papers have focused on facilities relocation. In fact, these papers deal with location or layout strategies, by considering mathematical approaches (Batta and Huang, 1989; Huang et al., 1990; Lin and Tseng, 1993; Brimberg and Wesolowsky, 2000) or by proposing oriented management methods (Rheault et al., 1996; Nozick and Turnquist, 1998; Moon and Kim, 2001; Nozick, 2001). However, the removal organization problem is not dealt with in these papers. In this context, and as a consequence, we have elaborated an organizational method that allows production to keep on going during the removal of a flow-shop.

The true problem lies in the choice of the right parameters to balance the relocation organization. A great number of solutions can be planned to segment the totality of the production line. The use of simulations in the field of manufacturing systems to simplify the complexity and reduce the problem dimension is recognized by scientists and industrial managers. In our case, the simulation aims at determining the best combination of input parameter values, given an output criterion (Pierreval and Paris, 2003).

Relocation is relatively a new branch of location. In relocation problems, according to the changes that take place on parameters of a problem, in a time period, new locations are suggested for facilities and the time of those location changes are also determined. In this chapter supply chain warehouse and manufacturing facilities relocation problem and the allocation of them to customers during time periods is being studied. Most traditional location models are analytical and while the parameters of the models are clear, the optimum answer of the problem is determinable by filling the parameters in mathematical equations related to the model.

In supply chain location problems, the relationships among components become more complicated, because at the same time optimization of more components is desirable. Hence, the traditional models, which are based on simplifying assumptions, cannot provide acceptable and correct answers for those problems. Also in such complicated problems, in general, rarely analytical optimum answer could be found. Thus, in these cases usually mathematical programming models are being used (like linear programming models, integer, mixed, etc.). To find optimum answers by those models, different solver software is being used (like GAMS, LINGO). The new ways of solving such complicated facility location problems are Meta heuristic solutions like polynomial algorithm (Hinojosaa et al., 2006) and a single-objective particle swarm optimizer (PSO) and a bi-objective PSO are devised to solve the problem (Colebrook and Siciliaa, 2007). In this chapter, the proposed model is a genetic algorithm (GA).

Recently, artificial intelligence-based methods have been applied to solving facility layout problems. For example, knowledge-based systems, which have been developed to provide users with problem-specific heuristic knowledge so that facilities can be allocated accordingly (Rad and James, 1983; Tommelein et al., 1991). Moreover, Yeh (1995) applied annealed neural networks to solve construction site level facility layout problems.

Although no studies of applying genetic algorithms in solving site-level facility layout problems have been reported, GAs have been applied to a diverse range of engineering and construction management searching problems, which include:

- Structural optimization (Koumousis and Georgiou, 1994; Nagendra, et al., 1996)
- Resource scheduling (Chan et al., 1996; Li and Love, 1997)
- Optimizing labor and equipment assignment (Li et al., 1998)
- Determination of laying sequence for a continuous girder reinforced concrete floor system, and space allocation (Gero and Kazakov, 1997)

Genetic algorithm systems have also been applied to solve space layout planning problems. For example, Gero and Kazakov (1997) incorporated the concept of genetic engineering into the GA system for solving building space layout problems.

27.2 Problem Definition

In the proposed problem of this chapter we encounter a supply chain that includes varied facilities. In the mentioned chain, some factories exist that have warehouses too; that is, each factory is a combination of two facilities—warehouse facilities and manufacturing facilities. Each factory prepares its required materials and parts whether from local provider or imports them from foreign countries. The imported parts and materials by seaports, airports, or other places of delivering imports are being sent to the factories of chain. Each factory can send the products to customers (distributors) directly from its warehouse or first send the products to transshipment points (or transshipment terminals) for sending them to costumers (retailers). Also each factory can send the products to other existed warehouses directly or by transshipment points, instead of keeping the products in its own warehouse, for delivering them to customers. A figuration of supply chain network is being shown in Figure 27.1.

During programming horizon some variations happen in some of the primary conditions of the problem. These changes motivate to the possibility of supply chain facilities

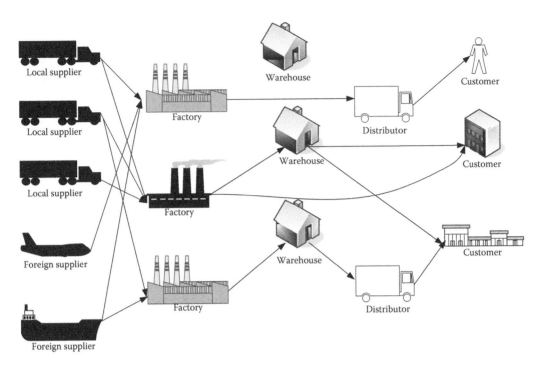

FIGURE 27.1
Supply chain network.

location replacements, because of cost reduction, rival increase, and services improvement. In any of these locations, a factory or warehouse will be built, or a warehouse is rented, or a contractor will be in charge of it. Some candidate points are considered for new factories and warehouses locations. Each of the new locations is one of the implemental decisions. Maybe in those locations new capacities are added or the existing facilities' capacity would transfer to them. But it is assumed if no decision on closing a factory or warehouse is received, no capacity would be transferred from it to other factories or warehouses. Furthermore, new facilities' capacity can be created independently and there is no obligation to transfer that capacity from the existed facilities to them.

Deciding from which transshipment terminal each customer could be serviced is not the desired decision of this problem; but by solving this problem the customers who are covered by each factory or warehouse will be identified. In each time period, each customer is being serviced by one factory or warehouse. The amount of customers' demands in each period depends on which supply chain facilities their required goods are from; but this amount is not stochastic and it is deterministic. In this problem, it is assumed that just one product is manufactured and distributed.

In this problem, the distance between the existed supply chain facilities and candidate locations for building facilities and also their related transportation costs are cleared. Also variable costs of inventory manufacturing and holding in each period, in both existed facilities and new facilities are identified. The fix costs of closing or opening facilities, related costs to capacity transfer from existed facilities to new one and the annual facilities operational costs are in hand. Each existed facilities is revealed and in any time period only a limited segment of each facilities capacity could be added or declined. Operational budget is limited. It is not necessary that the whole facilities locations be transferred at the end of the programming horizon. In this problem of location, the objective is reducing the whole operational costs and investment during programming horizon.

27.3 Genetic Algorithm Model

In this part, it is assumed if a factory or warehouse opens during programming horizon, it won't be closed. The model is described as follows:

Indices

h Existed factories locations ($h = 1 \dots H$)
i New factories locations ($i = 1 \dots I$)
n Existed warehouses locations ($n = 1 \dots N$)
p New warehouses locations ($p = 1 \dots P$)
j Customers ($j = 1 \dots J$)
k Local suppliers and seaports and other import quitting locations ($k = 1 \dots K$)
m Transportation facilities like transshipment terminal ($m = 1 \dots M$)
t Time periods ($t-1 \dots T$)

Parameters

D_{hj}^t Demand of customer j in time period t, if covered by existed factory h
D_{ij}^t Demand of customer j in time period t, if covered by factory in new location i
D_{nj}^t Demand of customer j in time period t, if covered by existed warehouse n
D_{pj}^t Demand of customer j in time period t, if covered by warehouse in new location p
R_h^t Net present value of unit income of product produced in existed factory h in period t
R_i^t Net present value of unit income of product produced in factory in new location i in period t
Q_h Production capacity of existed factory h
H_h Warehouse capacity of existed factory h
H_n Capacity of existed warehouse n
u_h^t Upper bound of part of production capacity that can be transferred from existed factory h in period t
u_i^t Upper bound of production capacity that can be transferred to new factory i in period t
B^t Maximum operational budget in period t (that is spent on operational costs and facilities holding)
S_{hj}^t Net present value of unit transportation cost from existed factory h to customer j in period t
S_{ij}^t Net present value of unit transportation cost from new factory i customer j in period t
S_{nj}^t Net present value of unit transportation cost from existed warehouse n to customer j in period t
S_{pj}^t Net present value of unit transportation cost from new warehouse p to customer j in period t

S_{hn}^t Net present value of unit transportation cost from existed factory h to existed warehouse n in period t

S_{hp}^t Net present value of unit transportation cost from existed factory h to new warehouse p in period t

S_{in}^t Net present value of unit transportation cost from new factory i to existed factory n in period t

S_{ip}^t Net present value of unit transportation cost from new factory i to new warehouse p in period t

c_h^t Net present value of production variable cost of unit product in existed factory h in period t

c_i^t Net present value of production variable cost of unit product in new factory i in period t

w_h^t Net present value of inventory holding variable cost of unit product in existed factory's warehouse h in period t

w_i^t Net present value of inventory holding variable cost of unit product in new factory's warehouse i in period t

w_n^t Net present value of inventory holding variable cost of unit product in existed warehouse n in period t

w_p^t Net present value of inventory holding variable cost of unit product in new warehouse p in period t

r_{hi}^t Net present value of manufacturing capacity unit location changes cost from existed factory h to new factory i in period t (contain man power, equipments, machines, and … transfer costs) $(i = 1 … I–1)$

e_i^t Net present value of a new manufacturing capacity unit creation variable cost in factory location i in period t

ei_i^t Net present value of a new warehouse capacity unit creation variable cost in factory location i in period t

ei_p^t Net present value of a new warehouse capacity unit creation variable cost in warehouse location p in period t

F_i^t Net present value of new factory creation fix cost in location i in period t

F_p^t Net present value of new warehouse creation fix cost in location p in period t

FR_p^t Net present value of new warehouse renting cost in location p in period t

F_h^t Net present value of existed factory h closing fix cost in period t

F_n^t Net present value of existed warehouse n closing fix cost in period t

Decision Variables

x_h^t Number of produced product unit by existed factory h in period t

x_i^t Number of produced product unit by new factory i in period t

y_{hj}^t Number of carried products unit from existed factory h to customer j in period t

y_{ij}^t Number of carried products unit from new factory i to customer j in period t

y_{nj}^t Number of carried products unit from existed warehouse n to customer j in period t

y_{pj}^t Number of carried products unit from new warehouse p to customer j in period

yi_{hn}^t Number of carried products unit from existed factory h to existed warehouse n in period t

yi_{in}^t Number of carried products unit from new factory i to existed warehouse n in period t

yi^t_{hp} Number of carried products unit from existed factory h to new warehouse p in period t

yi^t_{ip} Number of carried products unit from new factory i to new warehouse p in period t

$$v^t_{hj} = \begin{cases} 1 & \text{If customer } j \text{ allocated to existed factory } h \text{ in period } t \\ 0 & \text{o.w} \end{cases}$$

$$v^t_{ij} = \begin{cases} 1 & \text{if customer } j \text{ allocated to new factory } i \text{ in period } t \\ 0 & \text{o.w} \end{cases}$$

$$v^t_{nj} = \begin{cases} 1 & \text{if customer } j \text{ allocated to existed warehouse } n \text{ in period } t \\ 0 & \text{o.w} \end{cases}$$

$$v^t_{pj} = \begin{cases} 1 & \text{if customer } j \text{ allocated to new warehouse } p \text{ in period } t \\ 0 & \text{o.w} \end{cases}$$

$$z^t_i = \begin{cases} 1 & \text{if a factory is opened in location } i \text{ at the beginning of period } t \\ 0 & \text{o.w} \end{cases}$$

$$z^t_p = \begin{cases} 1 & \text{if a warehouse is opened in location } p \text{ at the beginning of period } t \\ 0 & \text{o.w} \end{cases}$$

$$zr^t_p = \begin{cases} 1 & \text{if a warehouse is rented in location } p \text{ in period } t \\ 0 & \text{o.w} \end{cases}$$

$$z^t_h = \begin{cases} 1 & \text{if factory } h \text{ is closed at the end of period } t \\ 0 & \text{o.w} \end{cases}$$

$$z^t_n = \begin{cases} 1 & \text{if warehouse } n \text{ is closed at the end of period } t \\ 0 & \text{o.w} \end{cases}$$

π^t_{hi} The amount of production capacity of existed factory h which transfer to new factory i in period t

ψ^t_i Created production capacity in new factory i in period t

ψi^t_i Created inventory holding capacity in new factory i in period t

ψ^t_h Created inventory holding capacity in new warehouse h in period t

27.3.1 Objective Function

TP(X, Y, TI, Z, ZR, Π, Ψ)   Net present value of total benefit

$TP(X, Y, YI, Z, ZR, \Pi, \Psi)$

$$= \sum_t \left[\sum_h (R_h^t - c_h^t + w_h^t) x_h^t + \sum_i (R_i^t - c_i^t + w_i^t) x_i^t \right]$$

$$- \sum_t \sum_j \left[\sum_n w_n^t y_{nj}^t + \sum_p w_p^t y_{pj}^t \right]$$

$$+ \sum_t \left[\sum_j \left[\sum_i S_{ij}^t y_{ij}^t + \sum_h S_{hj}^t y_{hj}^t + \sum_n S_{nj}^t y_{nj}^t + \sum_p S_{pj}^t y_{pj}^t \right] \right.$$

$$+ \sum_n \left[\sum_h S_{hn}^t yi_{hn}^t + \sum_i S_{in}^t yi_{in}^t \right.$$

$$\left. + \sum_p \left[\sum_h S_{hp}^t yi_{hp}^t + \sum_i S_{ip}^t yi_{ip}^t \right] \right]$$

$$- \sum_{i=1}^I \sum_h \sum_t r_{hi}^t \pi_{hi}^t - \sum_t \left[\sum_i F_i^t z_i^t + \sum_h F_h^t z_h^t + \sum_n F_n^t z_n^t \right.$$

$$\left. + \sum_p F_p^t z_p^t + \sum_p FR_p^t zr_p^t \right] - \sum_t \left[\sum_i (e_i^t \psi_i^t + ei_i^t \psi i_i^t) + \sum_p ei_p^t \psi_p^t \right] \quad (27.1)$$

S.t:

$$\pi_{hi}^t \leq u_h^t \cdot Q_h \cdot \sum_{\tau=1}^t z_i^\tau, \quad \forall t, i, h \tag{27.2}$$

$$\pi_{hi}^t \leq u_i^t \sum_{\tau=1}^T z_i^\tau, \quad \forall t, i, h \tag{27.3}$$

$$x_i^t \leq \sum_{\tau=1}^t \left[\sum_h \pi_{hi}^\tau + \psi_i^\tau \right], \quad \forall t, i \tag{27.4}$$

$$x_h^t \leq Q_h - \sum_i \sum_{\tau=1}^t \pi_{hi}^\tau, \quad \forall t, h \tag{27.5}$$

$$x_h^t = \sum_j y_{hj}^t + \sum_n yi_{hn}^t + \sum_p yi_{hp}^t, \quad \forall t, h \tag{27.6}$$

$$x_i^t = \sum_j y_{ij}^t + \sum_n yi_{in}^t + \sum_p yi_{ip}^t, \quad \forall t, i \tag{27.7}$$

$$\sum_j y_{nj}^t = \sum_h yi_{hn}^t + \sum_i yi_{in}^t, \quad \forall t, n \tag{27.8}$$

$$\sum_j y_{pj}^t = \sum_h yi_{hp}^t + \sum_i yi_{ip}^t, \quad \forall t, p \tag{27.9}$$

$$\sum_j y_{hj}^t + \sum_n yi_{hn}^t + \sum_p yi_{hp}^t \leq H_h \sum_{\tau=1}^t (1 - z_h^\tau), \quad \forall t, h \tag{27.10}$$

$$\sum_h yi_{hn}^t \leq H_n \sum_{\tau=1}^t (1 - z_n^\tau), \quad \forall t, n \tag{27.11}$$

$$\sum_j y_{nj}^t \leq H_n \sum_{\tau=1}^t \left(1 - z_n^\tau\right), \quad \forall t, n \tag{27.12}$$

$$\sum_j y_{ij}^t \leq \sum_{\tau=1}^t \psi i_i^\tau, \quad \forall t, i \tag{27.13}$$

$$\sum_j y_{pj}^t \leq \sum_{\tau=1}^t \psi_p^\tau, \quad \forall t, p \tag{27.14}$$

$$y_{ij}^t = D_{ij}^t v_{ij}^t, \quad \forall i, j, t \tag{27.15}$$

$$y_{hj}^t = D_{hj}^t v_{hj}^t, \quad \forall h, j, t \tag{27.16}$$

$$y_{nj}^t = D_{nj}^t v_{nj}^t, \quad \forall n, j, t \tag{27.17}$$

$$y_{pj}^t = D_{pj}^t v_{pj}^t, \quad \forall p, j, t \tag{27.18}$$

$$\sum_i v_{ij}^t + \sum_h v_{hj}^t + \sum_n v_{nj}^t + \sum_p v_{pj}^t = 1, \quad \forall t \tag{27.19}$$

$$\sum_{t}\left(\sum_{h}(c_h^t + w_h^t)x_h^t + \sum_{i}(c_i^t + w_i^t)x_i^t\right)$$

$$+\sum_{t}\sum_{j}\left(\sum_{n}w_n^t y_{nj}^t + \sum_{p}w_p^t y_{pj}^t\right)$$

$$+\sum_{t}\left(\sum_{j}\left(\sum_{i}S_{ij}^t y_{ij}^t + \sum_{h}S_{hj}^t y_{hj}^t + \sum_{n}S_{nj}^t y_{nj}^t + \sum_{p}S_{pj}^t y_{pj}^t\right)\right.$$

$$+\sum_{n}\left(\sum_{h}S_{hn}^t yi_{hn}^t + \sum_{i}S_{in}^t yi_{in}^t\right)$$

$$\left.+\sum_{p}\left(\sum_{h}S_{hp}^t yi_{hp}^t + \sum_{i}S_{ip}^t yi_{ip}^t\right)\right)$$

$$+\sum_{i=1}^{I-1}\sum_{h}\sum_{t}r_{hi}^t\pi_{hi}^t + \sum_{t}\sum_{p}FR_p^t zr_p^t$$

$$+\sum_{t}\sum_{i}(e_i^t\psi_i^t + ei_i^t\psi i_i^t) + \sum_{p}ei_p^t\psi_p^t \le B^t, \quad \forall t \qquad (27.20)$$

$$\sum_{t}z_h^t \le 1, \quad \forall h \qquad (27.21)$$

$$\sum_{t}z_i^t \le 1, \quad \forall i \qquad (27.22)$$

$$\sum_{t}z_p^t \le 1, \quad \forall p \qquad (27.23)$$

$$\sum_{t}z_n^t \le 1, \quad \forall n \qquad (27.24)$$

$$\sum_{\tau=1}^{t}z_p^t + zr_p^t \le 1, \quad \forall p,t \qquad (27.25)$$

$$x_i^t, x_h^t, y_{ij}^t, y_{hj}^t, y_{nj}^t, y_{pj}^t, yi_{in}^t, yi_{ip}^t, yi_{hn}^t, yi_{hp}^t, \pi_{hi}^t, \psi_i^t, \psi_p^t, \psi i_i^t \ge 0, \quad \forall i,h,t,j,n,p$$

$$z_n^t, z_h^t, zr_p^t, z_p^t, z_i^t, v_{pj}^t, v_{nj}^t, v_{ij}^t, v_{hj}^t \in \{0,1\}, \quad \forall i,h,t,j,n,p$$

Equation 27.1 is the objective function which indicates the income of the products produced in existing and new factories subtracted by the costs of inventory holding, transportation, production capacity transfer, creating and closing and renting the factories and warehouses, and creating new capacity in the whole facilities of the chain. Constraints (27.2) and (27.3) indicate the capacity which transfer from the existing factory to a new one in each period cannot exceed the allowance. Constraint (27.4) necessitates the model that the amount of production in each new factory doesn't exceed the aggregation of transferred capacity and created capacity in it. Constraint (27.5) shows that the amount of production in existing factory in each period doesn't exceed the initial capacity subtracted by the transferred capacity. Constraints (27.6) through (27.9) are the equilibrium of inputs and outputs for existing/new factories and warehouses. Constraints (27.10) through (27.14) indicate that the amount of new/existing warehouse inputs/outputs, or new/existing factory's warehouse inputs/outputs before closing, shouldn't be more than the warehouse or the factory's warehouse capacity. Constraints (27.15) through (27.18) necessitate the model to supply all customers' demands. Constraint (27.19) indicates that each customer should be allocated to only one factory or warehouse. Constraint (27.20) guarantees that the operational costs for all chain facilities in each period won't exceed the operational budget. Constraints (27.21) through (27.24) necessitate each factory or warehouse to be opened or closed only once. Constraint (27.25) indicates that in each period it is forbidden to rent and create a warehouse simultaneous. The last constraints show the sign and the kind of the model variables.

Two benefits could be stated for the proposed model:

- Linear form of the whole equations that makes it possible to be solved with solver software

- Being applicable for rolling horizon problems, that is, all data could be updated, and planning with new data is possible

27.3.2 Optimization by Genetic Algorithm

Genetic algorithms are a set of optimization algorithms that seek to improve performance by sampling areas of the parameters space that are more likely to lead to better solutions. The primary advantage of GAs lies in their capacity to move randomly from one feasible layout to another, without being drawn into local optima in which other algorithms are often trapped. Genetic algorithms employ a random, yet directed search for locating the globally optimal solution.

Typically, a set of GAs requires a representation scheme to encode feasible solutions to the optimization problem. Usually a solution is represented as a linear string called chromosome whose length varies with each application. Some measure of fitness is applied to the solutions in order to construct better solutions. There are three basic operators in the basic GA system: reproduction (or selection), crossover, and mutation. Reproduction is a process in which strings are duplicated according to their fitness magnitude. Crossover is a process in which the newly reproduced strings are randomly coupled, and whereby each couple of strings partially exchanges information. Mutation is the occasional random alteration of the value of one of the bits in the string.

In general, a GA contains a fixed-size population of potential solutions over the search space. These potential solutions of the search space are encoded as binary, integer, or floating-point strings and called chromosomes. The initial population can be created randomly or based on the problem specific knowledge. In each evolution step, a new population is created from the preceding one using the following procedures:

TABLE 27.1

Standard Genetic Algorithm

Genetic Algorithm
$t=0$
initialize - population $P(t)$
evaluate $P(t)$
while (*the termination criterion is not met*) do
begin
$t=t+1$
select $P(t)$ from $P(t-1)$ based on fitness
recombine $P(t)$
mutate $P(t)$
evaluate $P(t)$
End

- *Evaluation:* each chromosome of the old population is evaluated using a fitness function and given a value to denote its merit.
- *Selection:* chromosomes with better fitness are selected to generate next population. Commonly used selection schemes include proportional and tournament selection.
- *Recombination (crossover):* parts of two chromosomes selected based of fitness are swapped to generate trait-preserving offsprings. One-point, two-point and uniform crossovers are frequently seen in many GA applications.
- *Mutation:* parts of a chromosome are randomly changed to prevent the early maturity of a population. The applicable mutation operator depends on the data type of a gene. For example, with a binary gene, the gene value may be randomly flipped. For a real-coded gene, a random noise of different distribution types may be added or subtracted.

The above procedures are iterated for many generations until a satisfactory solution is found or a terminated criterion is met. A pseudo-code of a standard GA is shown in Table 27.1.

GAs have the following advantages over traditional search methods: (1) GAs directly work with a coding of the parameter set; (2) search is carried out from a population of points instead of a single one as in the case of the local search or simulated annealing algorithm; (3) pay-off information is used instead of derivatives or auxiliary knowledge; and (4) probabilistic transition rules are used instead of deterministic ones. GAs have been successfully applied to a diverse set of optimization problems. The flow chart of the proposed algorithm is presented in Figure 27.2.

27.4 Discussions

In this chapter a facility relocation model in supply chain is discussed. Regarding the specific conditions of supply chain and some assumptions in facility relocation, a mathematical model is proposed. The distinguished points of such a model are the ability of solving

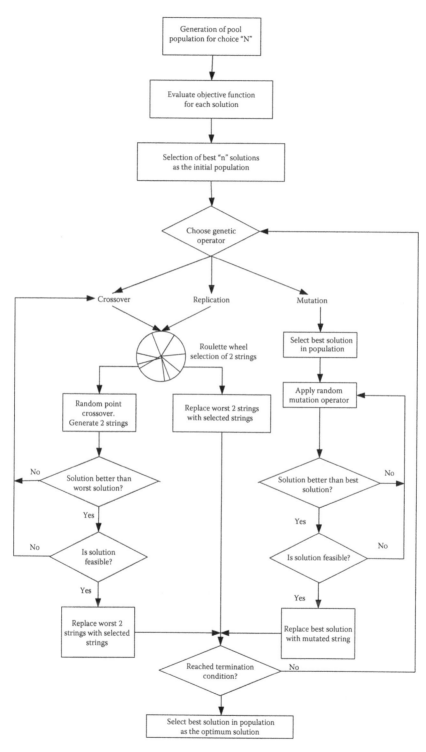

FIGURE 27.2
The flowchart of the proposed algorithm.

with software and also the possibility of correction in future decisions according to the changes on variable of the model. Moreover, because of the large dimensions of the model, genetic algorithm is proposed to decrease the solving time.

References

Chan W-T, Chua DKH, Kannan G. Construction resource scheduling with genetic algorithms. *ASCE Journal of Construction Engineering and Management* 1996;122(2):125–132. doi:10.1046/j.1365-232X.2002.00237

Colebrook M, Siciliaa J. Undesirable facility location problems on multi-criteria networks. *Computers & Operations Research* 2007;4(5):1491–1514. doi:10.1016/j.cor.2005.06.010

Batta R, Huang WV. On the synthesis of advertising and relocation decisions for facility. *Computers and Industrial Engineering* 1989;16(1):179–187. doi:10.1016/0360-8352(89)90020-X

Brimberg J, Wesolowsky GO. Facility location with closest rectangular distances. *Naval Research Logistics* 2000;47(1):77–84. doi: 10.1002/(SICI)1520-6750(200002)47:1<77:AID-NAV5>3.0.CO;2-#

Gero JS, Kazakov V. Learning and reusing information in space layout planning problems using genetic engineering. *Artificial Intelligence in Engineering* 1997;11(3):329–334. doi:10.1016/S0954-1810(96)00051-9

Heizer J, Render B. *Operations Management*. Prentice-Hall, USA, 1999.

Hinojosaa Y, Kalcsicsb J, Nickelc S, Puertod J, Veltene S. Dynamic supply chain design with inventory. *Journal of Computers & Operations Research* 2006. Available online 6 October. doi:10.1016/j.cor.2006.03.017

Huang WV, Batta R, Babu AJG. Relocation promotion problem with Euclidean distance. *European Journal of Operational Research* 1990;46(1):61–72. doi:10.1016/0377-2217(90)90298-P

Koopmans T-C, Beckmann M. Assignment problems and the location of economic activities. *Econometrica*, 1957;25(1):53–76. doi:10.2307j100139 Econometrica

Koumousis VK, Georgiou PG. Genetic algorithms in discrete optimization of steel truss roofs. *ASCE Journal of Computing in Civil Engineering* 1994;8(3):309–325. doi:10.1002/1099-1794(200009)9:4

Li H, Love PED. Improved genetic algorithms for time-cost optimization. *ASCE Construction Engineering and Management* 1997;123(3):233–237. portal.acm.org/citation.cfm? id=1326361.1326 455&coll=GUIDE&dl=&CFID=15151515

Li H, Love PED, Ogunlana S. Comparing genetic algorithms and non-linear optimisation for labor and equipment assignment. *Building Research and Information* 1998;26(8):322. doi:0961-3218

Lin BMT, Tseng SS. Generating the best K sequences in relocation problems. *European Journal of Operational Research* 1993;69(1):131–137. doi:10.1016/0377-2217(93)90098-8

Meller R-D, Gau K-Y. Facility layout objective functions and robust layouts. *International Journal of Production Research* 1996;34(10):2727–2742. doi:10.1080/00207549608905055

Moon G, Kim GP. Effects of relocation to AS/RS storage location policy with production quantity variation. *Computers and Industrial Engineering* 2001;40:1–13. doi:10.1016/S0360-8352(00)00005-X

Muther R. *Systematic Layout Planning*. VNR Company, 1969.

Nagendra S, Jestin D, Haftka RT, Watson LT. Improved genetic algorithm for the design of stiffened composite panels. *Computers and Structures* 1996;58(3):543–555. doi:10.1016/0045-7949(95)00160-I

Nozick LK, Turnquist MA. Integrating inventory impacts into a fixed-charge model for locating distribution centers. *Transportation Research Part E* 1998;34:173–186. doi:10.1016/S1366-5545(98)00010-6

Nozick LK. The fixed charge facility location problem with coverage restrictions. *Transportation Research Part E* 2001;37:281–296. doi:10.1016/S1366-5545(00)00018-1

Pierreval H, Paris JL. From "simulation optimization" to "simulation configuration" of systems. *Simulation Modelling Practice and Theory* 2003;11:5–19. doi:10.1016/S1569-190X(02)00096-5

Rad PF, James BM. The layout of temporary construction facilities. *Cost Engineering* 1983;25(2):19–27. doi:10.1111/0885-9507.00215

Rheault M, Drolet JR, Abdulnour G. 1996. Dynamic cellular manufacturing system. *Computers and Industrial Engineering* 1996;31(1/2):143–146. doi:10.1016/0360-8352(96)00098-8

Tommelein ID, Levitt RE, Confrey T. Sight Plan experiments: Alternate strategies for site layout design. *ASCE Journal of Computing in Civil Engineering* 1991;5(1):42–63. doi:10.1061/(ASCE)0887-2801(1991)5:1(42)

Tompkins J, White J. *Facilities Planning*. Wiley, New Jersey, United States, 1984.

Yapicioglua H, Smitha AE, Dozierb G. Solving the semi-desirable facility location problem using bi-objective particle swarm. *European Journal of Operational Research* 2007;177(2):733–749. doi:10.1016/j.ejor.2005.11.020

Yeh I-C. Construction-site layout using annealed neural network. *ASCE Journal of Computing in Civil Engineering* 1995;9(3):201–208. doi:10.1061/(ASCE)0887-3801(1995)9:3(201)

28

Reconfigurable Supply Chain: Immunity-Based Control Model

SUMMARY In this chapter, immunity-based control framework is designed for reconfigurable supply chain. The control framework implements a self-organizing distributed system for controlling replacements of facilities with schedule arising as an emergent behavior due to local interactions between facilities.

28.1 Introduction

Recently, artificial intelligence based methods have been applied for solving facility layout problems. This chapter describes a control framework that is developed by adopting the ideas of a biological novel, an artificial immune system (AIS), which is a multi-agent distributed system with distributed memory and specific mechanisms for learning behaviors. The self-organization and distribution features of AIS impart a high degree of robustness that inspired the implementation of the control framework based on its properties. Robustness is an inherent property of a system that has no central processing resource and where decisions are made as a consequence of local interactions. This implies that each facility could, in principle, suffer minor failures of its sensors and still be able to extract information by meeting other facilities. As the overall system is globally robust, the goal satisfaction algorithm is also distributed and a catastrophic failure of a facility will result in an unaccomplished goal that will be picked up by another facility automatically, without central intervention.

AIS exhibits the properties of human immune system for performing complicated candidate locations, for example, learning strategies, adaptive control, and memory managements. It has found applications in various fields, including artificial intelligence-based systems (Hunt and Cooke, 1995; Dasgupta, 1998; Tarakanov and Skormin, 2002), immunity-based computing system (Sokolova, 2003; Tieri et al, 2003), network-based intrusion detection systems (Kim and Bentley, 1999), fault tolerance systems (Bradley and Tyrrell, 2002; Canham and Tyrrell, 2002), autonomous agents (Jun et al, 1999; Meshref and VanLandingham, 2000), artificial immune system based intelligent multi-agent model (AISMAM) for mine detection (Sathyanath and Sahin, 2002) and immunology-derived distributed autonomous robotics architecture (IDARA) for heterogeneous groups of agents (Singh and Thayer, 2001). The proposed control framework addresses how individual facility with unique behaviors or talents can be exploited through communication and cooperation with each other in achieving goals. Facilities are able to determine various kinds of responses by perception of the changing environment so as to achieve goals efficiently. The highly distributed and adaptive properties of the human immune system are adopted to develop the control framework that has the ability to manage, coordinate and schedule a replacement of facilities in an automated supply chain. Hence, an intelligent multi-agent, fault tolerant and self-organizing transportation system that is robust and able to learn to achieve goals independently can be derived for warehousing operations.

28.2 Problem Description

Immunity is defined as resistance to infectious diseases. A collection of specialized organs, tissues and immune cells distributed throughout most of the human body form the complex functional human immune system. These components are inter-related and acting in a highly coordinated and specific manner when they recognize, eliminate and remember foreign macromolecules and cell (Elgert, 1996).

The main functionality of the immune system is to distinguish self, which is some normal pattern of activity or stable behavior of the system, from non self. This biological discrimination protects the human body by recognizing and defending against foreign antigens such as bacteria and virus. The ability of different foreign antigens to cause infection varies greatly and is related to how easily they can be controlled by the immune system (Eales, 1997).

This controllability of the immune system over foreign antigens comprises a group of defense mechanisms, which may fight against the antigens sequentially, gradually increasing the overall effectiveness of the immune response. The defense mechanisms are classified into innate immunity and acquired immunity (Sheehan, 1997; Cruse and Lewis, 1999).

Innate immunity is inborn and unchanging, is the first line of defense against infectious. In addition to provide early defense against infectious, innate immune responses enhance adaptive immune responses against the infectious agents. Adaptive immunity takes a longer time to develop. It is highly specific for antigens and remembers the antigens that a body has encountered previously. A general response that occurs after the first exposure of an individual to a foreign antigen is known as primary immune response, which is slower and less protective. On reoccurrence of the same antigens, a much faster and stronger secondary immune response is resulted. The ability of adaptive immunity to mount more rapid and effective responses to repeat encounters with the same antigen is achieved by the mechanism of immunological memory where immune cells proliferate and differentiate into memory cells during clonal expansion (Cadavid, 2003).

28.2.1 Artificial Immune Systems

The key characteristics of human immune system include specificity (Hofmeyr and Forrest, 1999), diversity (Fukuda, et al, 1998), memory (Smith et al., 1998; Kim and Bentley, 2003), discrimination (Beltran and Nino, 2002; Ayara et al., 2003) and self-organization (Watanabe et al., 1998), enabling the immune system to explore very high dimensional spaces efficiently that have great potential for solving engineering problems. Ongoing research related to multi-agent systems has established the emerging benefits of AIS, which is an engineering analogy of the human immune system.

The functionality that AIS delivers includes recognition, feature extraction, diversity, learning, memory, distributed detection, self-regulation, threshold mechanism, co-stimulation, dynamic protection and probabilistic detection (Dasgupta, 1999). In the human immune system, this is achieved via a chemical dynamic system held in homeostasis by the interaction of a number of cell types. Invasion by pathogens triggers a perturbation to the homeostasis, which results in the classical immune response. The result of the transient perturbation is to reinforce the chemical modes sustained in the underlying homeostatic system. In AIS, these modes act as a form of distributed memory ready to be triggered if another abnormality is encountered. Central to the system is a random combinatorial system that generates new families of the chemicals from a basic set of approximately 107 building blocks.

The success of a combination in fighting a pathogen results in the new combination being added to the distributed memory via a clonal amplification mechanism (Chowdhury, 1999). The distributed system has the capacity to learn by exploring the shape space of the receptors that coat typical pathogens or infected cells.

The shapes of these receptors parameterize the goals that the system tries to achieve. In AIS, the ability to adapt and learn is achieved by self-organization and self-improvement where agents have the autonomy in achieving goals (KrishnaKumar and Neidhoefer, 1999).

This combinatory process is the mechanism for exploring the associated probability function of success over the shape space. Success is measured by the capacity of the system to destroy pathogens without harming the host too much (Segel and Bar-Or, 1999). The system thus has to achieve an optimal balance between generating strong destructive agents using random search and the damage that these can do to the host.

These AIS-based distributed multi-agent studied have proven the applicability and feasibility of adopting the AIS novel to control the replacement of facility in relocation operations.

28.3 Immunity-Based Model

In the proposed problem of this chapter we encounter a supply chain that includes varied facilities. In the assumed chain, some factories exist that have warehouses too, that is, each factory is a combination of two facilities: warehouse facilities and manufacturing facilities. Each factory prepares its required materials and parts whether from local provider or imports them from foreign countries. The imported parts and materials by seaports, airports, or other places of delivering imports are being sent to the factories of chain. Each factory can send the products to customers (distributors) directly from its warehouse or first send the products to transshipment points (or transshipment terminals) for sending them to costumers (retailers). Also, each factory can send the products to other existing warehouses directly or by transshipment points, instead of keeping the products in its own warehouse, for delivering them to customers. A configuration of supply chain network is being shown in Figure 28.1.

During programming horizon some variations happen in some of the primary conditions of the problem. These changes motivate to the possibility of supply chain facilities location replacements, because of cost reduction, rival increase, and services improvement. In any of these locations, a factory or warehouse will be built, or a warehouse is rent, or a contractor will be in charge of it. Beside, some candidate points are considered for new factories and warehouses locations. Anyway, each of new locations is one of the implemental decisions. Maybe in those locations new capacities are added or the existed facilities capacity would transfer to them. But it is assumed if no decision on closing a factory or warehouse is got; no capacity would be transferred from it to other factories or warehouses.

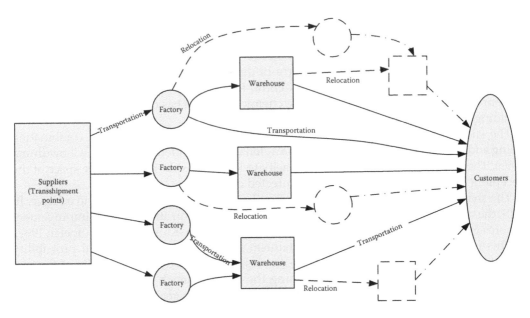

FIGURE 28.1
Supply chain network.

Furthermore, new facilities capacity can be created independently and there is no obligation to transfer that capacity from the existed facilities to them.

Deciding about from which transshipment terminal each customer could be serviced, is not the desired decisions of this problem; but by solving this problem the customers who are covered by each factory or warehouse will be identified. In each time period, each customer is being serviced by one factory or warehouse. The amount of customers' demands in each period, a little depends on which supply chain facilities their required goods are from; but this amount is not stochastic and it is deterministic. In this problem it is assumed that just one product is manufactured and distributed.

In this problem the distance between the existed supply chain facilities and candidate locations for building facilities and also their related transportation costs are cleared. Also variable costs of inventory manufacturing and holding in each period, in both existed facilities and new facilities are identified. Each existed facilities is revealed and in any time period only a limited segment of each facilities capacity could be added or declined. Operational budget is limited. It is not necessary that the whole facilities locations be transferred at the end of the programming horizon.

28.3.1 An Immunity-based Control Framework

The control of multiple facilities in a transportation system encompasses a number of major domains, including high-level planning, coordination, resource allocation, information consolidation, path planning, collision avoidance and locomotion. To effectively accomplish all these control criteria, facility's relocation with autonomous decision-making and individual behaviors are proposed to develop a fully distributed multi-agent system based on AIS.

In the immunity-based control framework, facilities are considered as immune cells and candidate locations are considered as antigens. While the major function of the human immune system is to protect the human body from the invasion of foreign antigens, the main operation of facilities in an automated supply chain is to handle and complete the assigned candidate locations. Each candidate location is specified with a complexity function to indicate the level of difficulty and skills required by facilities to handle such a candidate location. On the other hand, every facility contains a set of fundamental capabilities in the default stage. The basic actions a facility performs include exploring the surrounding environment and communicating with each other. These abilities are quantified by the sensory circle and communication circle parameters. Each facility is also capable of exchanging information with one another that are in close proximity defined by the communication circle.

Sensing and communicating are the two basic parameters common to all facilities. Additionally, facilities contain a set of capabilities that determines their intelligence in tackling candidate locations. The basic unit of capability is defined as the atomic ability. Facilities are able to manipulate the atomic abilities in order to create a new set of intelligence in coping with new problems.

28.3.2 The Control Framework

The prime objective of designing a fully distributed transportation system is to assign autonomy to individual facility in order to achieve a global goal.

The control framework proposed in this chapter adopts the biological theory of human immune system for manipulating facilities' internal behaviors. Facilities are able to provide different responses from its perception of the environment. A conceptual framework, depicted in Figure 28.2, is derived to encapsulate a facility candidate location exploration routine.

When carrying out control actions for multiple facilities, this immunity-based framework inherits the following characteristics and capabilities:

- Robustness: There is no full dependency between any facilities. Failure of a facility will not cripple the overall system.

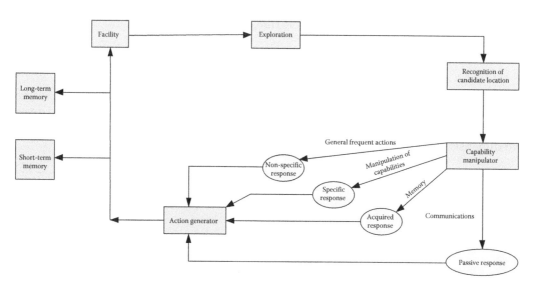

FIGURE 28.2
A conceptual framework for facility relocation control.

- Self-organization: The autonomous decision-making and communication capabilities with no central intervention allow facilities to determine responses to tackle candidate locations independently. This in turn develops an adaptive and distributed system under a fully decentralized control.

- Diversity: Upon manipulation of the atomic abilities for a specific response, re-combining the sequences of the sets of fundamental capabilities or re-allocating each atomic ability unit, facilities develop new sets of capabilities or knowledge for problems with higher complexity.

The first mission of facilities upon deployment in a supply chain is to explore their surrounding environment within their sensory circles. There is no centralized control or initial plan that dictates the facilities on the candidate locations they should first complete. Facilities use the measure of binding affinity to recognize and approach candidate locations. The binding affinity (β) is quantified by the distance between a facility and a specified candidate location, and the frequency of the candidate location occurrence. When a candidate location is recognized by a facility, it will then manipulate its capabilities to tackle the candidate location. This manipulation of capability allows facilities to perform appropriate responses and actions to move to the candidate location with maximum efficiency and in minimal time. Binding affinity is formally defined as follows:

F is the set of facilities indexed by j:

$$F = \{F_1, F_2, ..., F_j\}, \tag{28.1}$$

where $j = 1, 2, ..., n$

C is the set of candidate locations indexed by i:

$$C = \{C_1, C_2, ..., C_i\}, \tag{28.2}$$

where $i = 1, 2, ..., m$

binding affinity, β, is a function of d_{ij} and G_{ij}:

$$\beta_{ij} = f(d_{ij}, G_{ij}) \tag{28.3}$$

$$\beta_{ij} = w_1(d_{ij})^{-1} + w_2(G_{ij}) \tag{28.4}$$

where w_1 and w_2 are the weights for the parameters d_{ij} and G_{ij}, respectively. d_{ij} is the Euclidean distance measured between a candidate location and a facility, where $P(F_j)$ gives the existed location of the facility and $P(C_i)$ gives the candidate location of the facility in a two dimensional plane:

$$d_{ij} = P(F_j) - P(C_i) \tag{28.5}$$

Working with Cartesian coordinates, the distance between a facility's existed location and a candidate location is computed as follows:

$$d_{ij} = \sqrt{(x_i - x_j)^2 + (y_i - y_j)^2} \tag{28.6}$$

G_{ij} is the occurrence index of candidate location i in relation to the same class of candidate location, k, encountered by the facility j. $O_{i,k}$ is the number of occurrences of candidate location categorized as class k encountered by facility j and $O_{C(k)}$ is the total number of candidate location in class k located in the supply chain. Hence, G_{ij} is computed as follows:

$$G_{ij} = \frac{O_{i,k}}{O_{C(k)}} \tag{28.7}$$

Four types of responses, in relation to the human immune system, have been defined in the conceptual framework. They are non-specific, specific, acquired and passive responses. Non-specific response is equivalent to innate immunity of the immune system. Innate immunity is the first general defense that provides resistance to a variety of antigens, non-specific response of facilities deals with general candidate locations that occur frequently. Grouping and counting of facilities capacities are examples of non-specific responses in typical warehousing operation.

Facility recognizes candidate locations through matching of their capabilities with candidate location complexity. Specific response is carried out when candidate locations, such as goods reallocation, goods delivery and searching of goods, are detected by a facility. These kind of specific and distinct problems need to be solved by more advanced skills that are not as simple and direct as non-specific responses. Facilities are therefore required to manipulate their fundamental capabilities to cope with such non-general candidate locations. The capability manipulation includes re-combination of atomic ability or varying the sequences of a fundamental capability set. This matching of facility's capabilities with candidate location complexity mechanism is similar to antigen-specific acquired immunity where immune cells are activated to eliminate a particular kind of antigen. After a new set of skills has been generated through manipulation or re-combination of capabilities, the new knowledge for that specific distinct candidate location will go to facility's long-term memory for repeat occurrence.

One distinct feature of the human immune system is the significant difference between innate and acquired responses. Innate immune systems respond in the same way on re-occurrence with the same antigen where acquired immune systems respond better to each successive encounter with an antigen (Abbas and Lichtman, 2001). Skills acquired from capability manipulation during the first occurrence of a specific candidate location are stored in memory. On re-encountering the same specific candidate location, a facility will carry out a much faster and stronger acquired response. The mechanism in bringing out acquired response is the same as non-specific response. In addition, acquired responses allow facility to advance and improve their skills on re-encountering the same candidate locations.

In a distributed multi-agent system, cooperative work through communication between agents is an indispensable operation. Passive response is an action in responds to other facilities requests where the activated facility is assumed to have no suitable capability towards the requested job or teamwork is necessary for the requested job. This response is similar to the idea of vaccination of the human immune system. Vaccination is the process of intentionally eliciting acquired active immunity in an individual by administration of a vaccine (Elgert, 1996).

Following the concept of vaccination, responding facilities receive information entirely from the initiating facility in order to complete the requested job. The finalized response is established by an action generator to produce appropriate actions to tackle the candidate location encountered. The action generator coordinates and manages a chain of actions that is necessary to complete the candidate location. Knowledge acquired from capability manipulation will be put in long-term memory as an acquired response for the next occurrence of the same problem. On the other hand, knowledge transferred by other facilities during the execution of cooperative work and information regarding supply chain or candidate location will be put in short-term memory for temporary use.

Traditionally, facilities are inter-leaved by a central controller that coordinate and assign work orders for each facility throughout the whole operation. The central controller of these traditional facilities plans and commands all the activities involved in an operation. They established for each facility, with the best suitability that the facility has full capability to handle the candidate location, and to determine which candidate locations it should tackle in an operation. Unlike the traditional management of facilities, facilities in our AIS-based control have full autonomy in planning and determining their own duties. This autonomy starts with random exploration where no pre-defined targets being set to constrained the facility. A generic function of binding affinity is used to control the processes of candidate location recognition and tackling. After a candidate location is recognized by one or more facilities, the corresponding capability manipulator then activates a facility to execute appropriate responses to achieve the candidate location with its defaulted fundamental capabilities.

Different categories of candidate locations require capability of different levels. Besides the autonomous and distributed nature of our framework that outweighs the traditional centralized facilities locations, the other unique feature of our framework is the incorporation of the capability manipulator. It allows a facility to adepts and decides its appropriate actions in achieving its goals. Through the capability manipulator a facility manipulates its capability set in tackling candidate locations with different complexities. The descriptions of the capability sets in association with the four responses are given in Table 28.1.

For simple and straight-forward candidate locations, facilities are able to provide a fast and straight forward nonspecific response. For candidate locations of a higher complexity, facilities are able to re-arrange or re-combine the sequence of their fundamental capability in order to generate new responses that are specific to these complex candidate locations. The manipulated capability is then stored in facilities' memory as acquired response for future use. Hence, facilities under the AIS-based control can recognize candidate location independently, tackle candidate location with specificity, acquire new capability from fundamental capability, and memorize tackled candidate locations and provide a stronger and faster response for their next occurrence.

28.3.3 Strategic Behavioral Control for Facilities Relocation

The control framework derived in the previous section underpins how individual facility executes different responses towards various problems independently. In a multi-agent system, teamwork is an important and frequently occurred activity. Here, teamwork means when two or more agents work cooperatively to achieve a common goal. When a group

TABLE 28.1

Different Responses with Their Corresponding Capability Sets

Response	Capability Set
Non-specific	Fundamental capability pre-defined with atomic abilities in default stage
Specific	Candidate location specific capability manipulated from the fundamental capability in tackling complex candidate location
Acquired	New capability set manipulated by specific response
Passive	Transferred capability by other facilities through communication in cooperative work

of agents are working together, a crucial aspect is to have understanding and agreement among agents through communication. Hence, the behavior of facilities, which is characterized by unique behavior states, is studied to project their relocation strategies during operations. Through these behavioral states, a facility is able to determine its behavior in conjunction with the state information of other cooperating facilities obtained via communication; thereby an overall strategic plan is developed based on the mutual understanding between facilities. Facilities alter their behaviors by monitoring the dynamic environment. In different stages of an operation cycle, facilities change their behavior states to perform different activities. Figure 28.3 shows the behavior model of a facility given in the form of a state transition diagram that defines the change of behavior of a facility in response to help and request for help in operation.

Six different behaviors have been identified to characterize the strategies that are taken by a facility under different operating conditions that are represented in Table 28.2.

Immunologist Niels Jerne proposed the network theory of regulation in 1973 (Elgert, 1996), which suggested that an antibody produced to a foreign antigen elicits an anti-idiotypic antibody that acts to control further production of anti idiotypic antibody. The immune system is therefore kept in balance without the presence of antigens and the return to equilibrium is represented by an immune response. This balancing mechanism of antibody leads to an important concept of automatic control of antibody concentration that stimulates or suppresses an immune response.

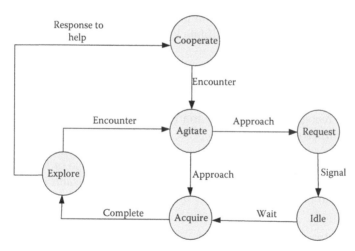

FIGURE 28.3
A behavioral state transition diagram of a facility.

TABLE 28.2

The Different Strategies Taken by a Facility Based on Its Behavioral States

Behavior	Strategy
Explore	Explore and search for candidate locations randomly in the surrounding environment
Agitate	Approach targeted candidate location when a candidate location has been found
Acquire	Tackle the candidate location
Request	Signal other facilities for help when teamwork is required
Cooperate	Response to participate in a teamwork operation
Idle	Wait for help from other facilities

According to this immunological concept, the group behavior of facilities is regulated by the facilities concentration in response to a particular candidate location. A concentration level is given to every candidate location in the default stage. The higher is the candidate location concentration level, the larger is the group of facilities needed to complete the job. Hence, a candidate location concentration level equal to unity indicates that a candidate location can be completed solely by a facility.

Initially, it is assumed that all facilities are in the explore state, searching for candidate location when they are deployed in a supply chain. Once a candidate location has been found by a particular facility, it changes its behavioral state to the agitate state and approaches the candidate location. The facility is stimulated by the candidate location and will trigger suitable response through capability manipulation. While the facility is approaching the targeted candidate location, the concentration level of that candidate location is being checked. If the candidate location concentration is greater than unity, the facility will send signal to request for help and change to the idle state until there are enough facilities to complete the candidate location. Facilities which are within the communication circles of the requesting facility will receive the request and only facilities in the explore state will respond to the request and change their states from explore to co-operate state. Facilities that are stimulated by and have triggered responses towards other candidate locations are not able to participate in another teamwork operation. On the other hand, if there are enough number of facilities responding to the same cooperative candidate location, other facilities that are approaching that candidate location but in a further location will leave this candidate location and change the behavioral state back to the explore state to look for new candidate locations.

28.3.4 The Algorithm

This section provides algorithmic descriptions of the key functions, namely exploration, candidate location recognition, capability manipulation and action generation of a facility, F_j, in the control framework.

Step 1 Exploration: Let SC_j be the magnitude of the sensory circle of F_j. Let δ_j be the candidate location detection index, which is a Boolean value that indicates if any candidate location is located within the sensory circle SC_j of a facility, F_j, such that:

$$\delta_j = \begin{cases} TRUE, & if \left| P(F_j) - P(C_\alpha) \right| < SC_j \\ FALSE, & if \left| P(F_j) - P(C_\alpha) \right| > SC_j \end{cases}, \quad C_\alpha \in C$$

$$IF(\delta_j = FALSE)$$

Identify the set of possible next location, S_j, for the facility F_j where S is a set of co-ordinates that is defined by the perimeter of a facility's current position $P(F_0) = (x_0, y_0)$. The size of the perimeter is given by the radius r, measured from $P(F_0)$, where

$$S_j = \left\{ ((x - x_j)^2 + (y - y_j)^2) = r^2 \right\}$$

$$RAN(nP_j) \in S_j$$

Note that a facility will move in a pseudo-random motion during exploration when no candidate locations are detected within the sensory circle (i.e., when $\delta_j = FALSE$). The next location of a facility is generated by the pseudo-random generator function RAN () that returns the next location nP_j, from the set S_j based on discrete uniform distribution.

Step 2 Recognition of a candidate location: When δ_j becomes TRUE meaning that at least one candidate location has been detected by F_j, the set of detected candidate locations is denoted by τ. F_j then determines a specific candidate location, C_i, to tackle using the binding affinity function, β_{ij} (Equation 28.4). In general, the candidate location with the highest affinity value will be chosen.

$$IF(\delta_j = TRUE)$$

$$\tau_j = \{C_\alpha\},$$

where $|P(F_j) - P(C\alpha)| < SC_j, C\alpha \in C$

$$C_i = C_\alpha(\max[\beta_{ij}])$$

$$D = |P(C_i) - P(Z)|, Z \in S_j$$

$$nP_j = \min|D|$$

Step 3 Capability manipulation: When F_j is in the right position to tackle C_i, the capability manipulator will then operate. The concentration level for tackling C_i is being examined initially to check whether it is a co-operative task or not.

```
IF (concentration=1)
        Match[F_j(fundamental capability), C_i(complexity)]
        IF(Match=TRUE)
                Non-specific response
        ELSE
                Match[F_j(acquired capability), C_i(complexity)]
                IF (Match=TRUE)
                        Acquired response
                ELSE
                        new Capablity = Manipulate (fundamental capability)
                        Specicic response
ELSE
new Capablity = Communicate(transferred capablity)
Passive response
```

Match is a function that investigates the capability of F_j with the complexity of C_i. Since the capability of a facility is represented by a chain of atomic abilities, *Match* returns true if the pattern of the complexity chain of C_i matches exactly with a segment of atomic abilities in the facility's capability chain. *Manipulate* is a function that rearranges the sequence of atomic abilities of a facility with a view to generate a new capability that is required to move to a new candidate location. *Communicate* allows a facility to transfer necessary capabilities to another facility in tackling cooperative tasks.

Step 4 Action generation: The action generator function gathers the information provided by the capability manipulator and feedbacks the appropriate actions, A, to the facility, F_j, for execution. $A_{ij} \in A$, where A_{ij} is one of non-specific response OR acquired response OR specific response OR passive response

```
IF(A_ij=Specific response)
    Long-term memory = Store(new Capability)
ELSEIF(A_ij=Passive response)
    Short-term memory = Store(new Capability)
```

Store is a function that saves the new capability either generated from specific response or passive response into a facility's memory.

28.4 Discussions

This chapter presents an immunity-based control framework to achieve a high level of performance from a multi-facility system implementing a flexible facility relocation system within a possible facility of the future. The control framework implements a self-organizing distributed system for controlling replacements of facilities with schedule arising as an emergent behavior due to local interactions between facilities. Facilities are regarded as independent agents with basic intelligence to achieve goals by exploring the environment. The strategic changes of facilities' behavioral states in response to the changing environment and identification of candidate locations allow effective group behavior and communication between facilities. A self-organized and fully autonomous system is achieved through the manipulations of individual facility capabilities and the ability to assert various responses. The ability to memorize encountered candidate locations and to communicate information with others permits facilities to propagate information in a highly distributed manner. As such, the replacement of AIS-based controlled facilities can therefore quickly adapt to and accommodate a dynamic environment by independent decision-making and interfacility communication.

References

Abbas AK, Lichtman AH. *Basic immunology: Functions and disorders of the immune system*. Philadelphia: W.B. Saunders Co., 2001.

Ayara M, Timmis J, de Lemos R, de Castro LN, Duncan R. Negative selection: How to generate detectors. In: Timmis J and Bentley PJ (eds.), *ICRAIS 2002*. University of Kent at Canterbury Printing Unit, 2002:89–98. ISBN 1902671325.

Beltran O, Nino F. A change detection software agent based on immune mixed selection. *Evolutionary Computation* 2002;1(12–17):693–698.

Bradley DW, Tyrrell AM. Immunotronics—novel finite state machine architecture with built-in self-test using self-nonself differentiation. *IEEE Trans. Evol. Comput* 2002:6(3):227–238.

Cadavid LF. *Overview of the Immune System*. Immunology Lecture 3, Spring, 2003.

Canham RO, Tyrrell AM. A multilayered immune system for hardware fault tolerance within an embryonic array. In: *1st International Conference on Artificial Immune Systems*, Canterbury, 2002:3–11.

Chowdhury D. Immune network: an example of complex adaptive systems. In: Dasgupta D (ed.), *Artificial Immune Systems and Their Applications*, Springer, Heidelberg, Berlin, pp. 89–104, 1999.

Cruse JM, Lewis RE. *Atlas of immunology*. Boca Raton, FL: CRC Press, 1999.

Dasgupta D. An artificial immune system as a multi-agent decision support system. In: *IEEE International Conference On Systems, Man, and Cybernetics*, San Diego, 1998;4:3816–3820.

Dasgupta D. *Artificial Immune Systems and Their Applications*. Springer, Germany, 1999.

Eales LJ. *Immunology For Life Scientists: A Basic Introduction: A Student Learning Approach*. Chichester: John Wiley & Sons, 1997.

Elgert KD. *Immunology: Understanding the Immune System*, New York: Wiley-Liss, 1996.

Fukuda T, Mori K, Tsukiyama M. Parallel search for multimodal function optimization with diversity and learning of immune algorithm. In: Dasgupta D (ed.), *Artificial Immune Systems and Their Applications*, Springer/Verlag, Heidelberg, Berlin, 1998:210–219.

Hofmeyr SA, Forrest S. Immunity by design: an artificial immune system. In: *Proceedings of the Genetic and Evolutionary Computation Conference*, Morgan Kaufmann, San Mateo, 1999:1289–1296.

Hunt JE, Cooke DE. An adaptive, distributed learning system based on the immune system. In: *Proceedings of the IEEE International Conference on Systems Man and Cybernetics*, Vancouver, BC, Canada, 1995:2494–2499.

Jun JH, Lee DW, Sim KB. Realization of cooperative strategies and swarm behavior in distributed autonomous robotic systems using artificial immune system. In: *IEEE International Conference on Systems, Man, and Cybernetics*, Tokyo, Japan, 1999;6(October 12–15):614–619.

Kim J, Bentley P. The human immune system and network intrusion detection. In: *Seventh European Conference on Intelligent Techniques and Soft Computing (EUFIT'99)*, Aachen, Germany, 1999.

Kim J, Bentley P. Immune memory in the dynamic clonal selection algorithm. In: *ICRAIS 2003 Session II*, London, UK, 2003:59–67.

KrishnaKumar K, Neidhoefer J. Immunized adaptive critic for an autonomous aircraft control application. In: Dasgupta D (ed.), *Artificial Immune Systems and Their Applications*, Springer, Heidelberg, Berlin, 1999:221–240.

Meshref H, VanLandingham H. Artificial immune systems: application to autonomous agents. In: *IEEE International Conference on Systems, Man, and Cybernetics*, Nashville, TN, USA, 2000;1:61–66.

Sathyanath S, Sahin F. AISIMAM—an artificial immune system based intelligent multi agent model and its application to a mine detection problem. In: *ICARIS 2002 Session I*, University of Kent at Canterbury, UK, 2002:22–31.

Segel LA, Bar-Or RL. Immunology viewed as the study of an autonomous decentralized system. In: Dasgupta D (ed.), *Artificial Immune Systems and Their Applications*, Springer, Heidelberg, Berlin, 1999:65–88.

Sheehan C. *Clinical immunology: Principles and laboratory diagnosis*, second edition, Lippincott, Philadelphia, 1997.

Singh S, Thayer S. Immunology directed methods for distributed robotics: A novel, immunity-based architecture for robust control & coordination. In: *Proceedings of SPIE: Mobile Robots XVI*, Boston, MA, USA, 2001;4573(Nov):1–12.

Smith DJ, Forrest S, Perelson AS. Immunological memory is associative. In: Dasgupta D (ed.), *Artificial Immune Systems and Their Applications*, Springer/Verlag, Heidelberg, Berlin, 1998:105–112.

Sokolova L. Index design by immunocomupting. In: *ICRAIS 2003*, USA, 2003:120–127.

Tarakanov A, Skormin VA. Pattern recognition by immunocomputing. In: *Proceedings Congress Evolutionary Computation, CEC'02*, Honolulu, HI, USA, 2002;1:938–943.

Tieri P, Valensin S, Franceschi C, Morandi C, Castellani GC. Memory and selectivity in evolving scale-free immune networks. In: *ICRAIS 2003*, USA, 2003:93–101.

Watanabe Y, Ishiguro A, Uchikawa Y. Decentralized behavior arbitration mechanism for autonomous mobile robot using immune network. In: Dasgupta D (ed.), *Artificial Immune Systems and Their Applications*, Springer/Verlag, Heidelberg, Berlin, 1998:187–209.

Index

A

Activity-based costing model, supply chain management, 111

Agent-based model
distributor-retailer layer, 274, 276
electronic supply chain management (e-SCM), 273–274
manufacturer layer, 273, 275
retailer-customer layer, 274, 277
supplier layer, 273, 274

Agro-food industry; *see also* Food supply chain
problem definition, 163–166
supply chain for, 161–163

AHP, *see* Analytic hierarchy process (AHP)

AI, *see* Artificial intelligence (AI)

Algorithm; *see also* Genetic algorithm (GA) model
analytic hierarchy process (AHP), 13–15, 93–94
clustering, 265–266
data mining, 21
immunity-based model, 366–368
k-means, 180, 184, 196–197
k-means results, 197–198
Vehicle Routing Problem with Time Windows (VRPTW), 217

American Association Reverse Logistics Executive Council, 189, 203, 243

Analytic hierarchy process (AHP)
algorithm, 13–15, 93–94
criteria-criteria pair-wise comparison matrix, 15, 65, 94
decision making approach, 60–61
identifying relationships and weights of criteria with, 62–65
multi-objective optimization using AHP, 200
network item-criteria matrix, 64, 65
parameter-criteria matrix, 14
priority setting, 61–62, 91–92
relative importance scale, 64, 93
strategic marketing model, 10, 13–15
supplier selection, 55–56
weighing objectives by, 91

Analytic network process (ANP), supplier selection, 56

Artificial immune system (AIS), 357, 358–359

Artificial immune system based intelligent multi-agent model (AISMAM), 357

Artificial intelligence (AI), 19, 343, 357

Asset utilization, 85

Association analysis, data mining, 178

B

Benders Decomposition, 135

Binding affinity, 362, 364

Bi-stage model, *see* Multiple Depots, Multiple Traveling Salesmen Problem (MDMTSP)

C

Capability indices, 51

Capacity constraints, 66

Characterization and discrimination, data mining, 177–178

Classification, data mining, 178

Cluster analysis, data mining, 178

Clustering; *see also* Return items' clustering model
algorithms, 265–266
definition of, 264
distance measure as component of, 266–269
types of, 264–265

Competition, global, and supply chains, 103–104

Competitors risk, 117

Comprehensive model (proposed) for reverse supply chain
analytical example, 253–256
assumptions, 246
constraints, 252–253
decision variables, 248
indices, 246
mathematical formulation, 248–251
for minimizing costs, 246–253, 257
objective function, 251–252
parameters, 247

Consistency index (CI), 14, 62, 63, 92, 94

Consistency ratio (CR), 14, 62, 63, 92, 94

Cost functions, designing, 112; *see also* Utility-based model

CRM, *see* Customer relationship management

Printed and bound by CPI Group (UK) Ltd, Croydon, CR0 4YY

01/11/2024

01782600-0012